> SPSS Base 15.0 User's Guide

For more information about SPSS® software products, please visit our Web site at *http://www.spss.com* or contact

SPSS Inc.
233 South Wacker Drive, 11th Floor
Chicago, IL 60606-6412
Tel: (312) 651-3000
Fax: (312) 651-3668

Preface

SPSS 15.0

SPSS 15.0 is a comprehensive system for analyzing data. SPSS can take data from almost any type of file and use them to generate tabulated reports, charts and plots of distributions and trends, descriptive statistics, and complex statistical analyses.

This manual, the *SPSS Base 15.0 User's Guide*, documents the graphical user interface of SPSS for Windows. Examples using the statistical procedures found in SPSS Base 15.0 are provided in the Help system, installed with the software. Algorithms used in the statistical procedures are provided in PDF form and are available from the Help menu.

In addition, beneath the menus and dialog boxes, SPSS uses a command language. Some extended features of the system can be accessed only via command syntax. (Those features are not available in the Student Version.) Detailed command syntax reference information is available in two forms: integrated into the overall Help system and as a separate document in PDF form in the *SPSS 15.0 Command Syntax Reference*, also available from the Help menu.

SPSS Options

The following options are available as add-on enhancements to the full (not Student Version) SPSS Base system:

SPSS Regression Models™ provides techniques for analyzing data that do not fit traditional linear statistical models. It includes procedures for probit analysis, logistic regression, weight estimation, two-stage least-squares regression, and general nonlinear regression.

SPSS Advanced Models™ focuses on techniques often used in sophisticated experimental and biomedical research. It includes procedures for general linear models (GLM), linear mixed models, generalized linear models (GZLM), generalized estimating equations (GEE), variance components analysis, loglinear analysis, actuarial life tables, Kaplan-Meier survival analysis, and basic and extended Cox regression.

SPSS Tables™ creates a variety of presentation-quality tabular reports, including complex stub-and-banner tables and displays of multiple-response data.

SPSS Trends™ performs comprehensive forecasting and time series analyses with multiple curve-fitting models, smoothing models, and methods for estimating autoregressive functions.

SPSS Categories® performs optimal scaling procedures, including correspondence analysis.

SPSS Conjoint™ provides a realistic way to measure how individual product attributes affect consumer and citizen preferences. With SPSS Conjoint, you can easily measure the trade-off effect of each product attribute in the context of a set of product attributes—as consumers do when making purchasing decisions.

SPSS Exact Tests™ calculates exact p values for statistical tests when small or very unevenly distributed samples could make the usual tests inaccurate.

SPSS Missing Value Analysis™ describes patterns of missing data, estimates means and other statistics, and imputes values for missing observations.

SPSS Maps™ turns your geographically distributed data into high-quality maps with symbols, colors, bar charts, pie charts, and combinations of themes to present not only what is happening but where it is happening.

SPSS Complex Samples™ allows survey, market, health, and public opinion researchers, as well as social scientists who use sample survey methodology, to incorporate their complex sample designs into data analysis.

SPSS Classification Trees™ creates a tree-based classification model. It classifies cases into groups or predicts values of a dependent (target) variable based on values of independent (predictor) variables. The procedure provides validation tools for exploratory and confirmatory classification analysis.

SPSS Data Preparation™ provides a quick visual snapshot of your data. It provides the ability to apply validation rules that identify invalid data values. You can create rules that flag out-of-range values, missing values, or blank values. You can also save variables that record individual rule violations and the total number of rule violations per case. A limited set of predefined rules that you can copy or modify is provided.

Amos™ (analysis of **mo**ment **s**tructures) uses structural equation modeling to confirm and explain conceptual models that involve attitudes, perceptions, and other factors that drive behavior.

The SPSS family of products also includes applications for data entry, text analysis, classification, neural networks, and predictive enterprise services.

Installation

To install the SPSS Base system, run the License Authorization Wizard using the authorization code that you received from SPSS Inc. For more information, see the installation instructions supplied with the SPSS Base system.

Compatibility

SPSS is designed to run on many computer systems. See the installation instructions that came with your system for specific information on minimum and recommended requirements.

Serial Numbers

Your serial number is your identification number with SPSS Inc. You will need this serial number when you contact SPSS Inc. for information regarding support, payment, or an upgraded system. The serial number was provided with your Base system.

Customer Service

If you have any questions concerning your shipment or account, contact your local office, listed on the SPSS Web site at *http://www.spss.com/worldwide*. Please have your serial number ready for identification.

Training Seminars

SPSS Inc. provides both public and onsite training seminars. All seminars feature hands-on workshops. Seminars will be offered in major cities on a regular basis. For more information on these seminars, contact your local office, listed on the SPSS Web site at *http://www.spss.com/worldwide*.

Technical Support

The services of SPSS Technical Support are available to maintenance customers. Customers may contact Technical Support for assistance in using SPSS or for installation help for one of the supported hardware environments. To reach Technical Support, see the SPSS Web site at *http://www.spss.com*, or contact your local office, listed on the SPSS Web site at *http://www.spss.com/worldwide*. Be prepared to identify yourself, your organization, and the serial number of your system.

Additional Publications

Additional copies of SPSS product manuals may be purchased directly from SPSS Inc. Visit the SPSS Web Store at *http://www.spss.com/estore*, or contact your local SPSS office, listed on the SPSS Web site at *http://www.spss.com/worldwide*. For telephone orders in the United States and Canada, call SPSS Inc. at 800-543-2185. For telephone orders outside of North America, contact your local office, listed on the SPSS Web site.

The *SPSS Statistical Procedures Companion*, by Marija Norušis, has been published by Prentice Hall. A new version of this book, updated for SPSS 15.0, is planned. The *SPSS Advanced Statistical Procedures Companion*, also based on SPSS 15.0, is forthcoming. The *SPSS Guide to Data Analysis* for SPSS 15.0 is also in development. Announcements of publications available exclusively through Prentice Hall will be available on the SPSS Web site at *http://www.spss.com/estore* (select your home country, and then click Books).

Tell Us Your Thoughts

Your comments are important. Please let us know about your experiences with SPSS products. We especially like to hear about new and interesting applications using the SPSS Base system. Please send e-mail to *suggest@spss.com* or write to SPSS Inc., Attn.: Director of Product Planning, 233 South Wacker Drive, 11th Floor, Chicago, IL 60606-6412.

About This Manual

This manual documents the graphical user interface for the procedures included in the SPSS Base system. Illustrations of dialog boxes are taken from SPSS for Windows. Dialog boxes in other operating systems are similar. Detailed information about the command syntax for features in the SPSS Base system is available in two forms:

integrated into the overall Help system and as a separate document in PDF form in the *SPSS 15.0 Command Syntax Reference*, available from the Help menu.

Contacting SPSS

If you would like to be on our mailing list, contact one of our offices, listed on our Web site at *http://www.spss.com/worldwide*.

Contents

1 Overview   *1*

What's New in SPSS 15.0? 2
Windows 4
Menus 7
Status Bar 7
Dialog Boxes 8
Variable Names and Variable Labels in Dialog Box Lists 8
Dialog Box Controls 9
Subdialog Boxes. 10
Selecting Variables. 10
Variable List Icons 10
Getting Information about Variables in Dialog Boxes. 11
Basic Steps in Data Analysis 11
Statistics Coach 12
Finding Out More about SPSS. 12

2 Getting Help   *13*

Using the Help Table of Contents. 15
Using the Help Index. 16
Using Help Search 17
Getting Help on Dialog Box Controls 18
Getting Help on Output Terms 19
Using Case Studies. 20
Copying Help Text from a Pop-Up Window 20

3   Data Files   21

Opening Data Files 21
File Information. 58
Saving Data Files 59
Protecting Original Data 86
Virtual Active File 86

4   Distributed Analysis Mode   91

Distributed Analysis versus Local Analysis 91

5   Data Editor   103

Data View. 104
Variable View 105
Entering Data 120
Editing Data 122
Go to Case 126
Case Selection Status in the Data Editor 126
Data Editor Display Options. 127
Data Editor Printing. 128

6   Working with Multiple Data Sources   131

Basic Handling of Multiple Data Sources 132

Copying and Pasting Information between Datasets 134

Renaming Datasets. 134

7   *Data Preparation*   *135*

Variable Properties. 135

Defining Variable Properties 136

Multiple Response Sets 144

Copying Data Properties. 147

Identifying Duplicate Cases 157

Visual Binning. 160

8   *Data Transformations*   *171*

Computing Variables. 171

Functions 174

Missing Values in Functions 175

Random Number Generators 175

Count Occurrences of Values within Cases. 176

Recoding Values. 179

Recode into Same Variables 179

Recode into Different Variables 181

Rank Cases. 184

Automatic Recode 188

Date and Time Wizard. 191

Time Series Data Transformations. 206

Scoring Data with Predictive Models 213

9 File Handling and File Transformations 219

Sort Cases 219
Transpose. 220
Merging Data Files 221
Add Cases 222
Add Variables. 226
Aggregate Data 229
Split File 234
Select Cases 235
Weight Cases 240
Restructuring Data 241

10 Working with Output 265

Viewer 265
Using Output in Other Applications 273
Pasting Objects into the Viewer 276
Export Output 278
Viewer Printing. 291
Saving Output. 298

11 Draft Viewer 301

To Create Draft Output 302
Controlling Draft Output Format. 303
Fonts in Draft Output. 308
To Print Draft Output. 308

To Save Draft Viewer Output. 309

12 Pivot Tables 311

Manipulating a Pivot Table 311

Working with Layers. 316

Bookmarks. 319

Showing and Hiding Cells 320

Editing Results 322

Changing the Appearance of Tables 322

Table Properties 325

To Change Pivot Table Properties 325

Table Properties: General. 325

Table Properties: Footnotes 326

Table Properties: Cell Formats 327

Table Properties: Borders. 329

Table Properties: Printing 331

Font 331

Data Cell Widths. 332

Cell Properties 334

Cell Properties: Value. 334

Cell Properties: Alignment 336

Cell Properties: Margins. 337

Cell Properties: Shading 337

Footnote Marker. 338

Selecting Rows and Columns in Pivot Tables. 339

To Select a Row or Column in a Pivot Table. 339

Modifying Pivot Table Results 340

Printing Pivot Tables 341

To Print Hidden Layers of a Pivot Table 341

Controlling Table Breaks for Wide and Long Tables 341

How to Create a Chart from a Pivot Table 342

13 *Working with Command Syntax*   *345*

Syntax Rules 345

Pasting Syntax from Dialog Boxes 347

Copying Syntax from the Output Log 348

Editing Syntax in a Journal File 350

To Run Command Syntax 351

Multiple Execute Commands 352

14 *Frequencies*   *353*

Frequencies Statistics 355

Frequencies Charts 357

Frequencies Format 358

15 *Descriptives*   *359*

Descriptives Options 361

DESCRIPTIVES Command Additional Features 362

16 Explore   363

Explore Statistics 366
Explore Plots 367
Explore Options 369
EXAMINE Command Additional Features 369

17 Crosstabs   371

Crosstabs Layers 373
Crosstabs Clustered Bar Charts 374
Crosstabs Statistics 374
Crosstabs Cell Display 377
Crosstabs Table Format 379

18 Summarize   381

Summarize Options. 383
Summarize Statistics 384

19 Means   387

Means Options. 390

20 OLAP Cubes   393

OLAP Cubes Statistics 396
OLAP Cubes Differences. 399
OLAP Cubes Title 400

21 T Tests   401

Independent-Samples T Test. 401
Paired-Samples T Test 405
One-Sample T Test 408
T-TEST Command Additional Features. 410

22 One-Way ANOVA   411

One-Way ANOVA Contrasts 414
One-Way ANOVA Post Hoc Tests 415
One-Way ANOVA Options. 417
ONEWAY Command Additional Features. 418

23 GLM Univariate Analysis   421

GLM Model. 425
GLM Contrasts 427
GLM Profile Plots 429
GLM Post Hoc Comparisons 430
GLM Save. 433

GLM Options. 435
UNIANOVA Command Additional Features 436

24 *Bivariate Correlations* 439

Bivariate Correlations Options 442
CORRELATIONS and NONPAR CORR Command Additional Features 443

25 *Partial Correlations* 445

Partial Correlations Options 448
PARTIAL CORR Command Additional Features 448

26 *Distances* 451

Distances Dissimilarity Measures. 453
Distances Similarity Measures 454
PROXIMITIES Command Additional Features 455

27 *Linear Regression* 457

Linear Regression Variable Selection Methods. 462
Linear Regression Set Rule. 463
Linear Regression Plots 463
Linear Regression: Saving New Variables. 465
Linear Regression Statistics 468

Linear Regression Options 470
REGRESSION Command Additional Features. 471

28 *Ordinal Regression* 473

Ordinal Regression Options. 475
Ordinal Regression Output 476
Ordinal Regression Location Model 478
Ordinal Regression Scale Model. 480
PLUM Command Additional Features 481

29 *Curve Estimation* 483

Curve Estimation Models 487
Curve Estimation Save 488

30 *Discriminant Analysis* 489

Discriminant Analysis Define Range 492
Discriminant Analysis Select Cases 492
Discriminant Analysis Statistics 493
Discriminant Analysis Stepwise Method. 494
Discriminant Analysis Classification 496
Discriminant Analysis Save. 497
DISCRIMINANT Command Additional Features. 498

31 Factor Analysis 499

Factor Analysis Select Cases 504
Factor Analysis Descriptives. 504
Factor Analysis Extraction 505
Factor Analysis Rotation. 507
Factor Analysis Scores. 508
Factor Analysis Options 509
FACTOR Command Additional Features. 510

32 Choosing a Procedure for Clustering 511

33 TwoStep Cluster Analysis 513

TwoStep Cluster Analysis Options. 517
TwoStep Cluster Analysis Plots. 519
TwoStep Cluster Analysis Output 521

34 Hierarchical Cluster Analysis 523

Hierarchical Cluster Analysis Method. 526
Hierarchical Cluster Analysis Statistics. 527
Hierarchical Cluster Analysis Plots 528
Hierarchical Cluster Analysis Save New Variables 529
CLUSTER Command Syntax Additional Features 529

35 K-Means Cluster Analysis   531

K-Means Cluster Analysis Efficiency 535
K-Means Cluster Analysis Iterate 536
K-Means Cluster Analysis Save 537
K-Means Cluster Analysis Options 537
QUICK CLUSTER Command Additional Features 538

36 Nonparametric Tests   539

Chi-Square Test 540
Binomial Test 544
Runs Test 547
One-Sample Kolmogorov-Smirnov Test 550
Two-Independent-Samples Tests 553
Two-Related-Samples Tests 557
Tests for Several Independent Samples 561
Tests for Several Related Samples 564

37 Multiple Response Analysis   569

Multiple Response Define Sets 570
Multiple Response Frequencies 571
Multiple Response Crosstabs 574
Multiple Response Crosstabs Define Ranges 576
Multiple Response Crosstabs Options 576
MULT RESPONSE Command Additional Features 577

38 *Reporting Results*   *579*

Report Summaries in Rows 579
Report Summaries in Columns 587
REPORT Command Additional Features 593

39 *Reliability Analysis*   *595*

Reliability Analysis Statistics 597
RELIABILITY Command Additional Features 599

40 *Multidimensional Scaling*   *601*

Multidimensional Scaling Shape of Data 603
Multidimensional Scaling Create Measure 604
Multidimensional Scaling Model 605
Multidimensional Scaling Options 606
ALSCAL Command Additional Features 606

41 *Ratio Statistics*   *609*

Ratio Statistics 611

42 *ROC Curves*   *613*

ROC Curve Options 616

43 Overview of the Chart Facility 617

Creating and Modifying a Chart 617
Chart Definition Options 624

44 Utilities 631

Variable Information 631
Data File Comments 632
Variable Sets 633
Define Variable Sets 633
Use Variable Sets 634
Reordering Target Variable Lists 636

45 Options 637

General Options 638
Viewer Options 640
Draft Viewer Options 641
Output Label Options 643
Chart Options 645
Interactive Chart Options 649
Pivot Table Options 650
Data Options 652
Currency Options 654
Script Options 655

46 Customizing Menus and Toolbars 657

Menu Editor 657
Customizing Toolbars 658
Show Toolbars 658
To Customize Toolbars 659

47 Production Facility 665

Using the Production Facility 667
Export Options 668
User Prompts 670
Production Macro Prompting 672
Production Options 673
Format Control for Production Jobs 674
Running Production Jobs from a Command Line 676
Publish to Web 678
SmartViewer Web Server Login 679

48 SPSS Scripting Facility 681

To Run a Script 681
Scripts Included with SPSS 682
Autoscripts 683
Creating and Editing Scripts 684
To Edit a Script 685
Script Window 686
Starter Scripts 688

Creating Autoscripts. 690

How Scripts Work. 693

Table of Object Classes and Naming Conventions 696

New Procedure (Scripting). 701

Adding a Description to a Script 703

Scripting Custom Dialog Boxes. 703

Debugging Scripts 708

Script Files and Syntax Files 710

49 *Output Management System*   *713*

Output Object Types 717

Command Identifiers and Table Subtypes 719

Labels. 720

OMS Options 721

Logging 726

Excluding Output Display from the Viewer. 727

Routing Output to SPSS Data Files 727

OXML Table Structure. 736

OMS Identifiers 741

Appendices

A   Database Access Administrator   745

B   Customizing HTML Documents   747

To Add Customized HTML Code to Exported Output Documents 747
Content and Format of the Text File for Customized HTML 747
To Use a Different File or Location for Custom HTML Code 748

Index   751

Overview

SPSS for Windows provides a powerful statistical-analysis and data-management system in a graphical environment, using descriptive menus and simple dialog boxes to do most of the work for you. Most tasks can be accomplished simply by pointing and clicking the mouse.

In addition to the simple point-and-click interface for statistical analysis, SPSS for Windows provides:

Data Editor. The Data Editor is a versatile spreadsheet-like system for defining, entering, editing, and displaying data.

Viewer. The Viewer makes it easy to browse your results, selectively show and hide output, change the display order results, and move presentation-quality tables and charts between SPSS and other applications.

Multidimensional pivot tables. Your results come alive with multidimensional pivot tables. Explore your tables by rearranging rows, columns, and layers. Uncover important findings that can get lost in standard reports. Compare groups easily by splitting your table so that only one group is displayed at a time.

High-resolution graphics. High-resolution, full-color pie charts, bar charts, histograms, scatterplots, 3-D graphics, and more are included as standard features in SPSS.

Database access. Retrieve information from databases by using the Database Wizard instead of complicated SQL queries.

Data transformations. Transformation features help get your data ready for analysis. You can easily subset data; combine categories; add, aggregate, merge, split, and transpose files; and more.

Electronic distribution. Send e-mail reports to other people with the click of a button, or export tables and charts in HTML format for Internet and intranet distribution.

Online Help. Detailed tutorials provide a comprehensive overview; context-sensitive Help topics in dialog boxes guide you through specific tasks; pop-up definitions in pivot table results explain statistical terms; the Statistics Coach helps you find the procedures that you need; Case Studies provide hands-on examples of how to use statistical procedures and interpret the results.

Command language. Although most tasks can be accomplished with simple point-and-click gestures, SPSS also provides a powerful command language that allows you to save and automate many common tasks. The command language also provides some functionality that is not found in the menus and dialog boxes.

Complete command syntax documentation is integrated into the overall Help system and is available as a separate PDF document, *SPSS Command Syntax Reference*, which is also available from the Help menu.

What's New in SPSS 15.0?

Data Management

Custom variable attributes. In addition to the standard variable attributes (for example, value labels, missing values, measurement level), you can create your own custom variable attributes. You can display and edit these attributes directly in Variable View of the Data Editor. Like standard variable attributes, these custom attributes are saved with SPSS-format data files. For more information, see "Custom Variable Attributes" in Chapter 5 on p. 116.

Variable sets. You can now use variable sets to control which variables are displayed in the Data Editor as well as in dialog box variable lists. (In previous releases, variable sets affected only dialog box variable lists.) Variable sets make it easier to work with data files that contain a large number of variables. For more information, see "Define Variable Sets" in Chapter 44 on p. 633.

Export to Database Wizard. Create new database tables, replace values for selected fields, and add new fields to an existing table, all without having to write a single line of SQL code yourself. For more information, see "Exporting to a Database" in Chapter 3 on p. 68.

Export to Dimensions. Export to Dimensions creates an SPSS data file and a Dimensions metadata file that you can use to read the data into Dimensions applications, such as mrInterview and mrTables. This is particularly useful when "roundtripping" data between SPSS and Dimensions applications. For example, you can read an mrInterview data source into SPSS, calculate some new variables, and then save the data in a form that can be read by mrTables without loss of any of the original metadata attributes. For more information, see "Exporting to Dimensions" in Chapter 3 on p. 84.

Save data in CSV format. Save data in CSV (comma-separated values) format. CSV is a common data format recognized by many applications. For more information, see "Saving Data Files in External Formats" in Chapter 3 on p. 59.

Reporting

Export results in PDF format. Export output in PDF format, including Viewer outline headings as bookmarks in the PDF file. For more information, see "Export Output" in Chapter 10 on p. 278.

Control chart enhancements. You can now define rules for control charts to help you quickly identify points that are out of control.

More chart types in Chart Builder. The Chart Builder has been expanded to include histograms, boxplots, scatterplot matrices, overlay scatterplots, population pyramids, error bar charts, high-low-close charts, difference area charts, range bar charts, dot plots, charts of separate variables, and paneled charts. You can also create charts that were not previously available in SPSS, such as charts with dual, independent *y* axes.

Chart Editor enhancements. The Chart Editor now offers more control over your charts. Major features include an updated Variables tab for changing chart types easily, automatic control of white space, additional distribution curves for histograms, a tool for quickly rescaling axes, and the ability to use custom equations to create reference lines.

Programmatic control of output documents. You can now create, open, activate, save, and close Viewer and Draft Viewer documents with command syntax using `OUTPUT NEW`, `OUTPUT NAME`, `OUTPUT ACTIVATE`, `OUTPUT OPEN`, `OUTPUT SAVE`, and `OUTPUT CLOSE`.

Statistical Enhancements

Ordinal Regression. This procedure, previously available as part of the Advanced Models add-on option, is now available in the SPSS Base system. For more information, see "Ordinal Regression " in Chapter 28 on p. 473.

PMML model files with transformations. You can now include transformations in PMML model files and merge information from model files using the TMS BEGIN-TMS END and TMS MERGE commands.

Generalized Linear Models. The Generalized Linear Models procedure expands the general linear model so that the dependent variable is linearly related to the factors and covariates via a specified link function. Moreover, the model allows for the dependent variable to have a non-normal distribution. This procedure is available in the Advanced Models option.

Generalized Estimating Equations. The Generalized Estimating Equations procedure extends the generalized linear model to allow for analysis of repeated measurements. This procedure is available in the Advanced Models option.

Complex Samples Ordinal Regression. The Complex Samples Ordinal Regression procedure performs regression analysis on a binary or ordinal dependent variable for samples drawn by complex sampling methods. Optionally, you can request analyses for a subpopulation. This procedure is available in the Complex Samples option.

Optimal Binning. The Optimal Binning procedure discretizes one or more scale variables by distributing the values of each variable into bins. Bin formation is optimal with respect to a categorical guide variable that "supervises" the binning process. Bins can then be used instead of the original data values for further analysis. This procedure is available in the Data Preparation option.

Programmability Extension

The Programmability Extension now allows you to write to the active dataset and create custom pivot tables and custom procedures. For more information go to *http://www.spss.com/devcentral*

Windows

There are a number of different types of windows in SPSS:

Data Editor. The Data Editor displays the contents of the data file. You can create new data files or modify existing data files with the Data Editor. If you have more than one data file open, there is a separate Data Editor window for each data file.

Viewer. All statistical results, tables, and charts are displayed in the Viewer. You can edit the output and save it for later use. A Viewer window opens automatically the first time you run a procedure that generates output.

Draft Viewer. In the Draft Viewer, you can display output as simple text (instead of interactive pivot tables).

Pivot Table Editor. Output that is displayed in pivot tables can be modified in many ways with the Pivot Table Editor. You can edit text, swap data in rows and columns, add color, create multidimensional tables, and selectively hide and show results.

Chart Editor. You can modify high-resolution charts and plots in chart windows. You can change the colors, select different type fonts or sizes, switch the horizontal and vertical axes, rotate 3-D scatterplots, and even change the chart type.

Text Output Editor. Text output that is not displayed in pivot tables can be modified with the Text Output Editor. You can edit the output and change font characteristics (type, style, color, size).

Syntax Editor. You can paste your dialog box choices into a syntax window, where your selections appear in the form of command syntax. You can then edit the command syntax to use special features of SPSS that are not available through dialog boxes. You can save these commands in a file for use in subsequent SPSS sessions.

Script Editor. Scripting and OLE automation allow you to customize and automate many tasks in SPSS. Use the Script Editor to create and modify basic scripts.

Figure 1-1
Data Editor and Viewer

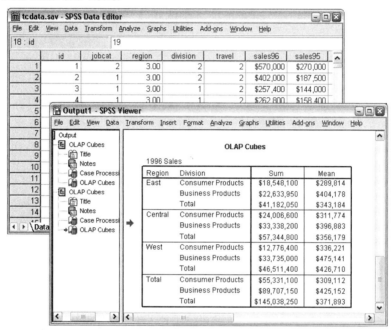

Designated Window versus Active Window

If you have more than one open Viewer window, output is routed to the **designated** Viewer window. If you have more than one open Syntax Editor window, command syntax is pasted into the designated Syntax Editor window. The designated windows are indicated by a plus sign in the icon in the title bar. You can change the designated windows at any time.

The designated window should not be confused with the **active** window, which is the currently selected window. If you have overlapping windows, the active window appears in the foreground. If you open a window, that window automatically becomes the active window and the designated window.

Changing the Designated Window

▶ Make the window that you want to designate the active window (click anywhere in the window).

▶ Click the Designate Window button on the toolbar (the plus sign icon).

or

▶ From the menus choose:
Utilities
  Designate Window

Note: For Data Editor windows, the active Data Editor window determines the dataset that is used in subsequent calculations or analyses. There is no "designated" Data Editor window. For more information, see "Basic Handling of Multiple Data Sources" in Chapter 6 on p. 132.

Menus

Many of the tasks that you want to perform with SPSS are available through menu selections. Each window in SPSS has its own menu bar with menu selections that are appropriate for that window type.

The Analyze and Graphs menus are available in all windows, making it easy to generate new output without having to switch windows.

Status Bar

The status bar at the bottom of each SPSS window provides the following information:

Command status. For each procedure or command that you run, a case counter indicates the number of cases processed so far. For statistical procedures that require iterative processing, the number of iterations is displayed.

Filter status. If you have selected a random sample or a subset of cases for analysis, the message Filter on indicates that some type of case filtering is currently in effect and not all cases in the data file are included in the analysis.

Weight status. The message Weight on indicates that a weight variable is being used to weight cases for analysis.

Split File status. The message Split File on indicates that the data file has been split into separate groups for analysis, based on the values of one or more grouping variables.

Showing and Hiding the Status Bar

▶ From the menus choose:
View
  Status Bar

Dialog Boxes

Most menu selections open dialog boxes. You use dialog boxes to select variables and options for analysis.

Dialog boxes for statistical procedures and charts typically have two basic components:

Source variable list. A list of variables in the active dataset. Only variable types that are allowed by the selected procedure are displayed in the source list. Use of short string and long string variables is restricted in many procedures.

Target variable list(s). One or more lists indicating the variables that you have chosen for the analysis, such as dependent and independent variable lists.

Variable Names and Variable Labels in Dialog Box Lists

You can display either variable names or variable labels in dialog box lists.

■ To control the display of variable names or labels, choose Options from the Edit menu in any window.

■ To define or modify variable labels, use Variable View in the Data Editor.

■ For data that are imported from database sources, field names are used as variable labels.

■ For long labels, position the mouse pointer over the label in the list to view the entire label.

■ If no variable label is defined, the variable name is displayed.

Figure 1-2
Variable labels displayed in a dialog box

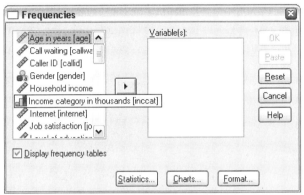

Dialog Box Controls

There are five standard controls in most dialog boxes:

OK. Runs the procedure. After you select your variables and choose any additional specifications, click OK to run the procedure and close the dialog box.

Paste. Generates command syntax from the dialog box selections and pastes the syntax into a syntax window. You can then customize the commands with additional features that are not available from dialog boxes.

Reset. Deselects any variables in the selected variable list(s) and resets all specifications in the dialog box and any subdialog boxes to the default state.

Cancel. Cancels any changes that were made in the dialog box settings since the last time it was opened and closes the dialog box. Within a session, dialog box settings are persistent. A dialog box retains your last set of specifications until you override them.

Help. Provides context-sensitive Help. This control takes you to a Help window that contains information about the current dialog box. You can also get help on individual dialog box controls by right-clicking the control.

Subdialog Boxes

Because most procedures provide a great deal of flexibility, not all of the possible choices can be contained in a single dialog box. The main dialog box usually contains the minimum information that is required to run a procedure. Additional specifications are made in subdialog boxes.

In the main dialog box, controls with an ellipsis (...) after the name indicate that a subdialog box will be displayed.

Selecting Variables

To select a single variable, simply highlight it on the source variable list and click the right arrow button next to the target variable list. If there is only one target variable list, you can double-click individual variables to move them from the source list to the target list.

You can also select multiple variables:

■ To select multiple variables that are grouped together on the variable list, click the first variable and then Shift-click the last variable in the group.

■ To select multiple variables that are not grouped together on the variable list, click the first variable, then Ctrl-click the next variable, and so on.

Variable List Icons

The icons that are displayed next to variables in dialog box lists provide information about the variable type and measurement level.

Measurement Level	Data Type			
	Numeric	String	Date	Time
Scale		n/a		
Ordinal				
Nominal				

Getting Information about Variables in Dialog Boxes

▶ Right-click a variable in the source or target variable list.

▶ Choose Variable Information.

Figure 1-3
Variable information

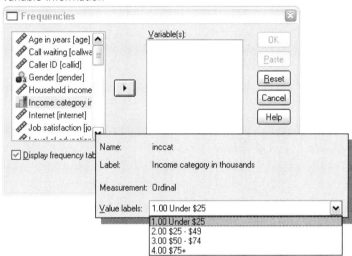

Basic Steps in Data Analysis

Analyzing data with SPSS is easy. All you have to do is:

Get your data into SPSS. You can open a previously saved SPSS data file, you can read a spreadsheet, database, or text data file, or you can enter your data directly in the Data Editor.

Select a procedure. Select a procedure from the menus to calculate statistics or to create a chart.

Select the variables for the analysis. The variables in the data file are displayed in a dialog box for the procedure.

Run the procedure and look at the results. Results are displayed in the Viewer.

Statistics Coach

If you are unfamiliar with SPSS or with the statistical procedures that are available in SPSS, the Statistics Coach can help you get started by prompting you with simple questions, nontechnical language, and visual examples that help you select the basic statistical and charting features that are best suited for your data.

To use the Statistics Coach, from the menus in any SPSS window choose:

Help
  Statistics Coach

The Statistics Coach covers only a selected subset of procedures in the SPSS Base system. It is designed to provide general assistance for many of the basic, commonly used statistical techniques.

Finding Out More about SPSS

For a comprehensive overview of SPSS basics, see the online tutorial. From any SPSS menu choose:

Help
  Tutorial

Getting Help

Help is provided in many different forms:

Help menu. The Help menu in most SPSS windows provides access to the main Help system, plus tutorials and technical reference material.

- **Topics.** Provides access to the Contents, Index, and Search tabs, which you can use to find specific Help topics.

- **Tutorial.** Illustrated, step-by-step instructions on how to use many of the basic features in SPSS. You don't have to view the whole tutorial from start to finish. You can choose the topics you want to view, skip around and view topics in any order, and use the index or table of contents to find specific topics.

- **Case Studies.** Hands-on examples of how to create various types of statistical analyses and how to interpret the results. The sample data files used in the examples are also provided so that you can work through the examples to see exactly how the results were produced. You can choose the specific procedure(s) that you want to learn about from the table of contents or search for relevant topics in the index.

- **Statistics Coach.** A wizard-like approach to guide you through the process of finding the procedure that you want to use. After you make a series of selections, the Statistics Coach opens the dialog box for the statistical, reporting, or charting procedure that meets your selected criteria. The Statistics Coach provides access to most statistical and reporting procedures in the Base system and many charting procedures.

- **Command Syntax Reference.** Detailed command syntax reference information is available in two forms: integrated into the overall Help system and as a separate document in PDF form in the *SPSS Command Syntax Reference*, available from the Help menu.

- **Statistical Algorithms.** The algorithms used for most statistical procedures are available in two forms: integrated into the overall Help system and as a separate document in PDF form in *SPSS Algorithms*, available on the manuals CD. For links to specific algorithms in the Help system, select Algorithms on the Help menu.

Context-sensitive Help. In many places in the user interface, you can get context-sensitive Help.

- **Dialog box Help buttons.** Most dialog boxes have a Help button that takes you directly to a Help topic for that dialog box. The Help topic provides general information and links to related topics.

- **Dialog box context menu Help.** Many dialog boxes provide context-sensitive Help for individual controls and features. Right-click on any control in a dialog box and select What's This? from the context menu to display a description of the control and directions for its use. (If What's This? does not appear on the context menu, then this form of Help is not available for that dialog box.)

- **Pivot table context menu Help.** Right-click on terms in an activated pivot table in the Viewer and select What's This? from the context menu to display definitions of the terms.

- **Case Studies.** Right-click on a pivot table and select Case Studies from the context menu to go directly to a detailed example for the procedure that produced that table. (If Case Studies does not appear on the context menu, then this form of Help is not available for that procedure.)

- **Command syntax.** In a command syntax window, position the cursor anywhere within a syntax block for a command and press F1 on the keyboard. A complete command syntax chart for that command will be displayed. Complete command syntax documentation is available from the links in the list of related topics and from the Help Contents tab.

- **Scripting and OLE automation.** In a script window (File menu, New or Open, Script), the Help menu provides access to information on the scripting language and SPSS OLE automation objects, methods, and properties. Context-sensitive Help in a script window is available with F1 or F2 (object browser).

Microsoft Internet Explorer Settings

Most Help features in this application use technology based on Microsoft Internet Explorer. Some versions of Internet Explorer (including the version provided with Microsoft Windows XP, Service Pack 2) will by default block what it considers to be "active content" in Internet Explorer windows on your local computer. This default setting may result in some blocked content in Help features. To see all Help content, you can change the default behavior of Internet Explorer.

▶ From the Internet Explorer menus choose:
  Tools
    Internet Options...

▶ Click the Advanced tab.

▶ Scroll down to the Security section.

▶ Select (check) Allow active content to run in files on My Computer.

Other Resources

If you can't find the information you want in the Help system, these other resources may have the answers you need:

- **SPSS for Windows Developer's Guide.** Provides information and examples for the developer's tools included with SPSS for Windows, including OLE automation, third-party API, input/output DLL, Production Facility, and Scripting Facility. The *Developer's Guide* is available in PDF form in the *SPSS\developer* directory on the installation CD.

- **Technical Support Web site.** Answers to many common problems can be found at *http://support.spss.com*. (The Technical Support Web site requires a login ID and password. Information on how to obtain an ID and password is provided at the URL listed above.)

Using the Help Table of Contents

▶ In any window, from the menus choose:
  Help
    Topics

▶ Click the Contents tab.

▶ Double-click items with a book icon to expand or collapse the contents.

▶ Click an item to go to that Help topic.

Figure 2-1
Help window with Contents tab displayed

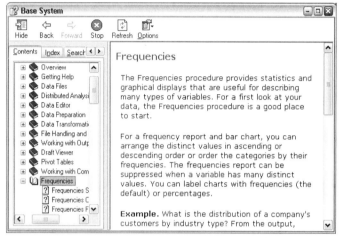

Using the Help Index

▶ In any window, from the menus choose:
Help
  Topics

▶ Click the Index tab.

▶ Enter a term to search for in the index.

▶ Double-click the topic that you want.

The Help index uses incremental search to find the text that you enter and selects the closest match in the index.

Figure 2-2

Index tab and incremental search

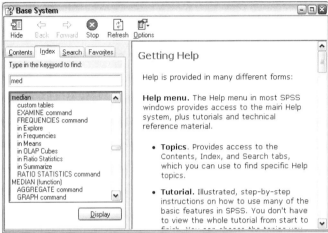

Using Help Search

The Search tab provides full-text search that includes topic titles, topic text, and index text. Topics are ranked based on how many times the search term(s) appear in the topic and/or its index.

Figure 2-3
Help Search tab

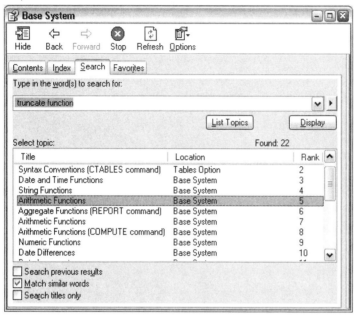

- Full-text search is most effective if you can refine your search criteria to limit the number of topics in the list. For example, if you know you want to find a function for truncating numeric values, "truncate function" will produce a better list of potential topics than simply "truncate."

- Topics with a command name in parentheses indicate that the Help topic is for command syntax. Topics without a command name in parentheses are generally Help topics for the dialog boxes and other graphical user interface topics.

Getting Help on Dialog Box Controls

▶ Right-click on the dialog box control that you want information about.

▶ Choose What's This? from the pop-up context menu.

A description of the control and how to use it is displayed in a pop-up window. General information about a dialog box is available from the Help button in the dialog box.

Figure 2-4
Dialog box control Help with right mouse button

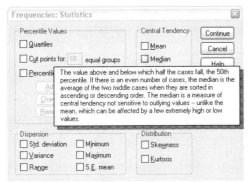

Getting Help on Output Terms

▶ Double-click the pivot table to activate it.

▶ Right-click on the term that you want explained.

▶ Choose What's This? from the context menu.

A definition of the term is displayed in a pop-up window.

Figure 2-5
Activated pivot table glossary Help with right mouse button

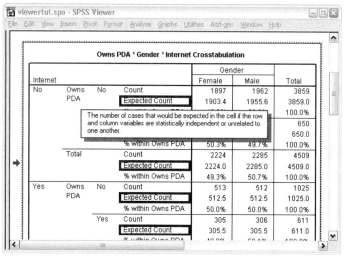

Using Case Studies

▶ Right-click on a pivot table in the Viewer window.

▶ Choose Case Studies from the pop-up context menu.

Copying Help Text from a Pop-Up Window

▶ Right-click anywhere in the pop-up window.

▶ Choose Copy from the context menu.

The entire text of the pop-up window is copied.

Data Files

Data files come in a wide variety of formats, and this software is designed to handle many of them, including:

- Spreadsheets created with Excel and Lotus
- Database tables from many database sources, including Oracle, SQLServer, Access, dBASE, and others
- Tab-delimited and other types of simple text files
- Data files in SPSS format created on other operating systems
- SYSTAT data files
- SAS data files
- Stata data files

Opening Data Files

In addition to files saved in SPSS format, you can open Excel, SAS, Stata, tab-delimited and other files without converting the files to an intermediate format or entering data definition information.

Opening a data file makes it the active dataset. If you already have one or more open data files, they remain open and available for subsequent use in the session. Clicking anywhere in the Data Editor window for an open data file will make it the active dataset.

In distributed analysis, mode using a remote server to process commands and run procedures, the available data files, folders, and drives are dependent on what is available on or from the remote server. The current server name is indicated at the top of the dialog box. You will not have access to data files on your local computer unless you specify the drive as a shared device and the folders containing your data files as shared folders. For more information, see "Distributed Analysis Mode" in Chapter 4 on p. 91.

To Open Data Files

▶ From the menus choose:
File
 Open
  Data...

▶ In the Open Data dialog box, select the file that you want to open.

▶ Click Open.

Optionally, you can:

- Read variable names from the first row for spreadsheet files.

- Specify a range of cells to read for spreadsheet files.

- Specify a sheet within an Excel file to read (Excel 5 or later).

For information on reading data from databases, see "Reading Database Files" on p. 25. For information on reading data from text data files, see "Text Wizard" on p. 43.

Data File Types

SPSS. Opens data files saved in SPSS format, including SPSS for Windows, Macintosh, UNIX, and also the DOS product SPSS/PC+.

SPSS/PC+. Opens SPSS/PC+ data files.

SYSTAT. Opens SYSTAT data files.

SPSS Portable. Opens data files saved in SPSS portable format. Saving a file in portable format takes considerably longer than saving the file in SPSS format.

Excel. Opens Excel files.

Lotus 1-2-3. Opens data files saved in 1-2-3 format for release 3.0, 2.0, or 1A of Lotus.

SYLK. Opens data files saved in SYLK (symbolic link) format, a format used by some spreadsheet applications.

dBASE. Opens dBASE-format files for either dBASE IV, dBASE III or III PLUS, or dBASE II. Each case is a record. Variable and value labels and missing-value specifications are lost when you save a file in this format.

SAS Long File Name. SAS versions 7–9 for Windows, long extension.

SAS Short File Name. SAS versions 7–9 for Windows, short extension.

SAS v6 for Windows. SAS version 6.08 for Windows and OS2.

SAS v6 for UNIX. SAS version 6 for UNIX (Sun, HP, IBM).

SAS Transport. SAS transport file.

Stata. Stata versions 4–8.

Text. ASCII text file.

Opening File Options

Read variable names. For spreadsheets, you can read variable names from the first row of the file or the first row of the defined range. The values are converted as necessary to create valid variable names, including converting spaces to underscores.

Worksheet. Excel 5 or later files can contain multiple worksheets. By default, the Data Editor reads the first worksheet. To read a different worksheet, select the worksheet from the drop-down list.

Range. For spreadsheet data files, you can also read a range of cells. Use the same method for specifying cell ranges as you would with the spreadsheet application.

Reading Excel 5 or Later Files

The following rules apply to reading Excel 5 or later files:

Data type and width. Each column is a variable. The data type and width for each variable are determined by the data type and width in the Excel file. If the column contains more than one data type (for example, date and numeric), the data type is set to string, and all values are read as valid string values.

Blank cells. For numeric variables, blank cells are converted to the system-missing value, indicated by a period. For string variables, a blank is a valid string value, and blank cells are treated as valid string values.

Variable names. If you read the first row of the Excel file (or the first row of the specified range) as variable names, values that don't conform to variable naming rules are converted to valid variable names, and the original names are used as variable labels. If you do not read variable names from the Excel file, default variable names are assigned.

Reading Older Excel Files and Other Spreadsheets

The following rules apply to reading Excel files prior to version 5 and other spreadsheet data:

Data type and width. The data type and width for each variable are determined by the column width and data type of the first data cell in the column. Values of other types are converted to the system-missing value. If the first data cell in the column is blank, the global default data type for the spreadsheet (usually numeric) is used.

Blank cells. For numeric variables, blank cells are converted to the system-missing value, indicated by a period. For string variables, a blank is a valid string value, and blank cells are treated as valid string values.

Variable names. If you do not read variable names from the spreadsheet, the column letters (*A*, *B*, *C*, ...) are used for variable names for Excel and Lotus files. For SYLK files and Excel files saved in R1C1 display format, the software uses the column number preceded by the letter *C* for variable names (*C1*, *C2*, *C3*, ...).

Reading dBASE Files

Database files are logically very similar to SPSS-format data files. The following general rules apply to dBASE files:

- Field names are converted to valid variable names.

- Colons used in dBASE field names are translated to underscores.

- Records marked for deletion but not actually purged are included. The software creates a new string variable, *D_R*, which contains an asterisk for cases marked for deletion.

Reading Stata Files

The following general rules apply to Stata data files:

- **Variable names.** Stata variable names are converted to SPSS variable names in case-sensitive form. Stata variable names that are identical except for case are converted to valid SPSS variable names by appending an underscore and a sequential letter (*_A*, *_B*, *_C*, ..., *_Z*, *_AA*, *_AB*, ..., etc.).

- **Variable labels.** Stata variable labels are converted to SPSS variable labels.

- **Value labels.** Stata value labels are converted to SPSS value labels, except for Stata value labels assigned to "extended" missing values.

- **Missing values.** Stata "extended" missing values are converted to system-missing values.

- **Date conversion.** Stata date format values are converted to SPSS DATE format (d-m-y) values. Stata "time-series" date format values (weeks, months, quarters, etc.) are converted to simple numeric (F) format, preserving the original, internal integer value, which is the number of weeks, months, quarters, etc., since the start of 1960.

Reading Database Files

You can read data from any database format for which you have a database driver. In local analysis mode, the necessary drivers must be installed on your local computer. In distributed analysis mode (available with SPSS Server), the drivers must be installed on the remote server. For more information, see "Distributed Analysis Mode" in Chapter 4 on p. 91.

To Read Database Files

▶ From the menus choose:
File
  Open Database
    New Query...

▶ Select the data source.

▶ If necessary (depending on the data source), select the database file and/or enter a login name, password, and other information.

▶ Select the table(s) and fields. (For OLE DB data sources, you can only select one table.)

▶ Specify any relationships between your tables.

▶ Optionally:

- Specify any selection criteria for your data.

- Add a prompt for user input to create a parameter query.

- Save your constructed query before running it.

To Edit Saved Database Queries

▶ From the menus choose:
   File
     Open Database
       Edit Query...

▶ Select the query file (**.spq*) that you want to edit.

▶ Follow the instructions for creating a new query.

To Read Database Files with Saved Queries

▶ From the menus choose:
   File
     Open Database
       Run Query...

▶ Select the query file (**.spq*) that you want to run.

▶ If necessary (depending on the database file), enter a login name and password.

▶ If the query has an embedded prompt, enter other information if necessary (for example, the quarter for which you want to retrieve sales figures).

Selecting a Data Source

Use the first screen of the Database Wizard to select the type of data source to read.

ODBC Data Sources

If you do not have any ODBC data sources configured, or if you want to add a new data source, click Add ODBC Data Source. In distributed analysis mode (available with SPSS Server), this button is not available. To add data sources in distributed analysis mode, see your system administrator.

An ODBC data source consists of two essential pieces of information: the driver that will be used to access the data and the location of the database you want to access. To specify data sources, you must have the appropriate drivers installed. For local analysis mode, you can install drivers from the CD-ROM for this product:

■ **SPSS Data Access Pack.** Installs drivers for a variety of database formats. This feature is available on the AutoPlay menu.

■ **Microsoft Data Access Pack.** Installs drivers for Microsoft products, including Microsoft Access. To install the Microsoft Data Access Pack, double-click Microsoft Data Access Pack in the Microsoft Data Access Pack folder on the CD-ROM.

Figure 3-1
Database Wizard dialog box

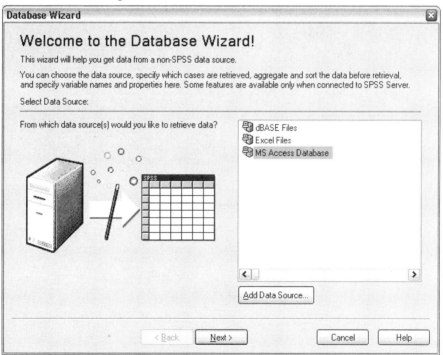

OLE DB Data Sources

To access OLE DB data sources, you must have the following items installed on the computer that is running SPSS:

■   .NET framework

■   Dimensions Data Model and OLE DB Access

Versions of these components that are compatible with this release of SPSS can be installed from the SPSS installation CD and are available on the AutoPlay menu.

■   Table joins are not available for OLE DB data sources. You can read only one table at a time.

■   You can add OLE DB data sources only in local analysis mode. To add OLE DB data sources in distributed analysis mode on a Windows server, consult your system administrator.

■   In distributed analysis mode (available with SPSS Server), OLE DB data sources are available only on Windows servers, and both .NET and the Dimensions Data Model and OLE DB Access must be installed on the server.

Figure 3-2
Database Wizard with access to OLE DB data sources

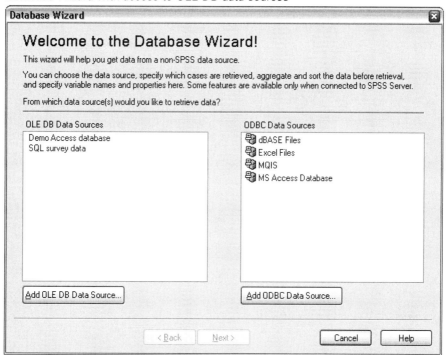

To add an OLE DB data source:

▶ Click Add OLE DB Data Source.

▶ In the Data Link Properties dialog box, click the Provider tab and select the OLE DB provider.

▶ Click Next or click the Connection tab.

▶ Select the database by entering the directory location and database name or by clicking the button to browse to a database. (A user name and password may also be required.)

▶ Click OK after entering all necessary information. (You can make sure the specified database is available by clicking the Test Connection button.)

▶ Enter a name for the database connection information. (This name will be displayed in the list of available OLE DB data sources.)

Figure 3-3
Save OLE DB Connection Information As dialog box

Save OLE DB Connection Information As	☒
Name	SQL survey data	OK
		Cancel

▶ Click OK.

This takes you back to the first screen of the Database Wizard, where you can select
the saved name from the list of OLE DB data sources and continue to the next step
of the wizard.

Deleting OLE DB Data Sources

To delete data source names from the list of OLE DB data sources, delete the UDL
file with the name of the data source in:

*[drive]:\Documents and Settings\[user login]\Local Settings\Application
Data\SPSS\UDL*

Database Login

If your ODBC database requires a password, the Database Wizard will prompt you
for your password before it can open the data source.

Figure 3-4
Login dialog box

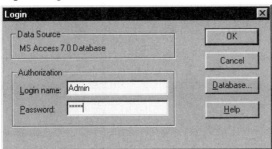

Selecting Data Fields

The Select Data dialog box controls which tables and fields are read. Database fields (columns) are read as variables.

If a table has any field(s) selected, all of its fields will be visible in the following Database Wizard windows, but only fields that are selected in this dialog box will be imported as variables. This enables you to create table joins and to specify criteria by using fields that you are not importing.

Figure 3-5
Database Wizard, selecting data

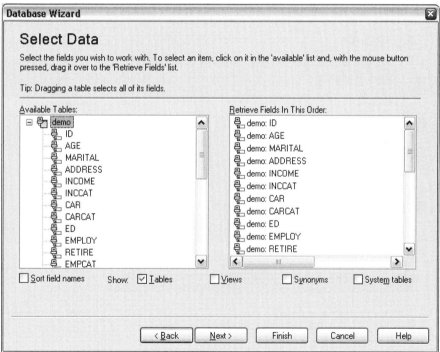

Displaying field names. To list the fields in a table, click the plus sign (+) to the left of a table name. To hide the fields, click the minus sign (−) to the left of a table name.

To add a field. Double-click any field in the Available Tables list, or drag it to the Retrieve Fields In This Order list. Fields can be reordered by dragging and dropping them within the fields list.

To remove a field. Double-click any field in the Retrieve Fields In This Order list, or drag it to the Available Tables list.

Sort field names. If this check box is selected, the Database Wizard will display your available fields in alphabetical order.

By default, the list of available tables displays only standard database tables. You can control the type of items that are displayed in the list:

- **Tables.** Standard database tables.
- **Views.** Views are virtual or dynamic "tables" defined by queries. These can include joins of multiple tables and/or fields derived from calculations based on the values of other fields.
- **Synonyms.** A synonym is an alias for a table or view, typically defined in a query.
- **System tables.** System tables define database properties. In some cases, standard database tables may be classified as system tables and will only be displayed if you select this option. Access to real system tables is often restricted to database administrators.

Note: For OLE DB data sources, you can select fields only from a single table. Multiple table joins are not supported for OLE DB data sources.

Creating a Relationship between Tables

The Specify Relationships dialog box allows you to define the relationships between the tables for ODBC data sources. If fields from more than one table are selected, you must define at least one join.

Figure 3-6
Database Wizard, specifying relationships

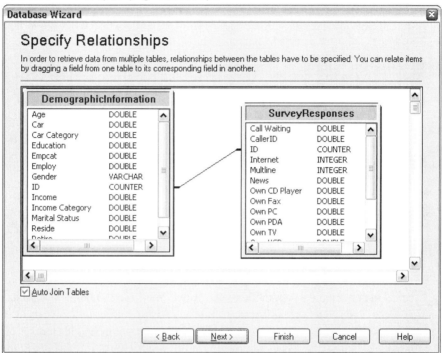

Establishing relationships. To create a relationship, drag a field from any table onto the field to which you want to join it. The Database Wizard will draw a **join line** between the two fields, indicating their relationship. These fields must be of the same data type.

Auto Join Tables. Attempts to automatically join tables based on primary/foreign keys or matching field names and data type.

Specifying join types. If outer joins are supported by your driver, you can specify inner joins, left outer joins, or right outer joins. To select the type of join, double-click the join line between the fields, and the wizard will display the Relationship Properties dialog box.

Relationship Properties

This dialog box allows you to specify which type of relationship joins your tables.

Figure 3-7
Relationship Properties dialog box

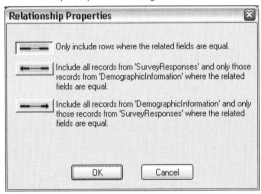

Inner joins. An inner join includes only rows where the related fields are equal. In this example, all rows with matching ID values in the two tables will be included.

Outer joins. In addition to one-to-one matching with inner joins, you can also use outer joins to merge tables with a one-to-many matching scheme. For example, you could match a table in which there are only a few records representing data values and associated descriptive labels with values in a table containing hundreds or thousands of records representing survey respondents. A left outer join includes all records from the table on the left and, from the table on the right, includes only those records in which the related fields are equal. In a right outer join, the join imports all records from the table on the right and, from the table on the left, imports only those records in which the related fields are equal.

Limiting Retrieved Cases

The Limit Retrieved Cases dialog box allows you to specify the criteria to select subsets of cases (rows). Limiting cases generally consists of filling the criteria grid with criteria. Criteria consist of two expressions and some relation between them. The expressions return a value of *true*, *false*, or *missing* for each case.

■ If the result is *true*, the case is selected.

■ If the result is *false* or *missing*, the case is not selected.

- Most criteria use one or more of the six relational operators (<, >, <=, >=, =, and <>).

- Expressions can include field names, constants, arithmetic operators, numeric and other functions, and logical variables. You can use fields that you do not plan to import as variables.

Figure 3-8
Database Wizard, limiting retrieved cases

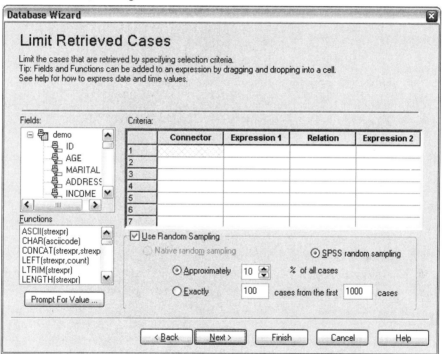

To build your criteria, you need at least two expressions and a relation to connect the expressions.

▶ To build an expression, choose one of the following methods:

- In an Expression cell, type field names, constants, arithmetic operators, numeric and other functions, or logical variables.

- Double-click the field in the Fields list.

- Drag the field from the Fields list to an Expression cell.

- Choose a field from the drop-down menu in any active Expression cell.

▶ To choose the relational operator (such as = or >), put your cursor in the Relation cell and either type the operator or choose it from the drop-down menu.

If the SQL contains WHERE clauses with expressions for case selection, dates and times in expressions need to be specified in a special manner (including the curly braces shown in the examples):

- Date literals should be specified using the general form {d 'yyyy-mm-dd'}.

- Time literals should be specified using the general form {t 'hh:mm:ss'}.

- Date/time literals (timestamps) should be specified using the general form {ts 'yyyy-mm-dd hh:mm:ss'}.

- The entire date and/or time value must be enclosed in single quotes. Years must be expressed in four-digit form, and dates and times must contain two digits for each portion of the value. For example January 1, 2005, 1:05 AM would be expressed as:

  {ts '2005-01-01 01:05:00'}

Functions. A selection of built-in arithmetic, logical, string, date, and time SQL functions is provided. You can drag a function from the list into the expression, or you can enter any valid SQL function. See your database documentation for valid SQL functions. A list of standard functions is available at:

http://msdn.microsoft.com/library/en-us/odbc/htm/odbcscalar_functions.asp

Use Random Sampling. This option selects a random sample of cases from the data source. For large data sources, you may want to limit the number of cases to a small, representative sample, which can significantly reduce the time that it takes to run procedures. Native random sampling, if available for the data source, is faster than SPSS random sampling, because SPSS random sampling must still read the entire data source to extract a random sample.

- **Approximately.** Generates a random sample of approximately the specified percentage of cases. Since this routine makes an independent pseudorandom decision for each case, the percentage of cases selected can only approximate the specified percentage. The more cases there are in the data file, the closer the percentage of cases selected is to the specified percentage.

■ **Exactly.** Selects a random sample of the specified number of cases from the specified total number of cases. If the total number of cases specified exceeds the total number of cases in the data file, the sample will contain proportionally fewer cases than the requested number.

Note: If you use random sampling, aggregation (available in distributed mode with SPSS Server) is not available.

Prompt For Value. You can embed a prompt in your query to create a **parameter query**. When users run the query, they will be asked to enter information (based on what is specified here). You might want to do this if you need to see different views of the same data. For example, you may want to run the same query to see sales figures for different fiscal quarters.

▶ Place your cursor in any Expression cell, and click Prompt For Value to create a prompt.

Creating a Parameter Query

Use the Prompt for Value dialog box to create a dialog box that solicits information from users each time someone runs your query. This feature is useful if you want to query the same data source by using different criteria.

Figure 3-9
Prompt for Value dialog box

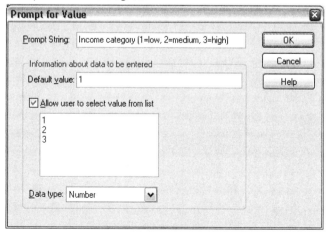

To build a prompt, enter a prompt string and a default value. The prompt string is displayed each time a user runs your query. The string should specify the kind of information to enter. If the user is not selecting from a list, the string should give hints about how the input should be formatted. An example is as follows: Enter a Quarter (Q1, Q2, Q3, ...).

Allow user to select value from list. If this check box is selected, you can limit the user to the values that you place here. Ensure that your values are separated by returns.

Data type. Choose the data type here (Number, String, or Date).

The final result looks like this:

Figure 3-10
User-defined prompt dialog box

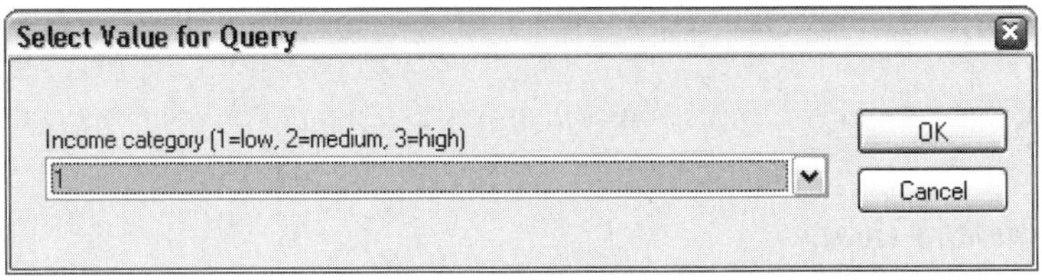

Aggregating Data

If you are in distributed mode, connected to a remote server (available with SPSS Server), you can aggregate the data before reading it into SPSS.

Figure 3-11
Database Wizard, aggregating data

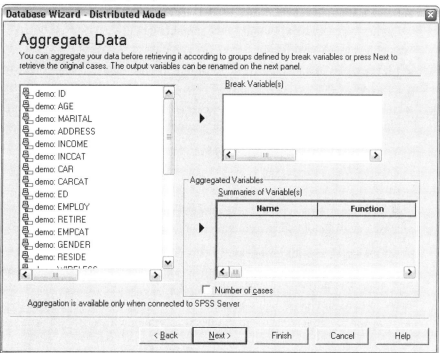

You can also aggregate data after reading it into SPSS, but preaggregating may save time for large data sources.

▶ To create aggregated data, select one or more break variables that define how cases are grouped.

▶ Select one or more aggregated variables.

▶ Select an aggregate function for each aggregate variable.

▶ Optionally, create a variable that contains the number of cases in each break group.

Note: If you use SPSS random sampling, aggregation is not available.

Defining Variables

Variable names and labels. The complete database field (column) name is used as the variable label. Unless you modify the variable name, the Database Wizard assigns variable names to each column from the database in one of two ways:

- If the name of the database field forms a valid, unique variable name, the name is used as the variable name.

- If the name of the database field does not form a valid, unique variable name, a new, unique name is automatically generated.

Click any cell to edit the variable name.

Converting strings to numeric values. Select the Recode to Numeric box for a string variable if you want to automatically convert it to a numeric variable. String values are converted to consecutive integer values based on alphabetical order of the original values. The original values are retained as value labels for the new variables.

Width for variable-width string fields. This option controls the width of variable-width string values. By default, the width is 255 bytes, and only the first 255 bytes (typically 255 characters in single-byte languages) will be read. The width can be up to 32,767 bytes. Although you probably don't want to truncate string values, you also don't want to specify an unnecessarily large value, which will cause SPSS processing to be inefficient.

Figure 3-12
Database Wizard, defining variables

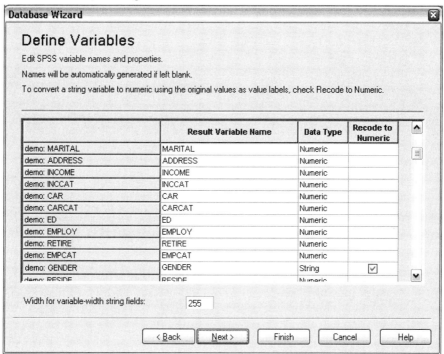

Sorting Cases

If you are in distributed mode, connected to a remote server (available with SPSS Server), you can sort the data before reading it into SPSS.

Figure 3-13
Database Wizard, sorting cases

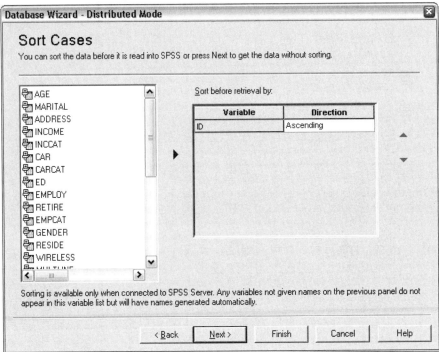

You can also sort data after reading it into SPSS, but presorting may save time for large data sources.

Results

The Results dialog box displays the SQL Select statement for your query.

■ You can edit the SQL Select statement before you run the query, but if you click the Back button to make changes in previous steps, the changes to the Select statement will be lost.

■ To save the query for future use, use the Save query to file section.

■ To paste complete GET DATA syntax into a syntax window, select Paste it into the syntax editor for further modification. Copying and pasting the Select statement from the Results window will not paste the necessary command syntax.

Note: The pasted syntax contains a blank space before the closing quote on each line of SQL that is generated by the wizard. These blanks are not superfluous. When the command is processed, all lines of the SQL statement are merged together in a very literal fashion. Without the space, there would be no space between the last character on one line and first character on the next line.

Figure 3-14
Database Wizard, results panel

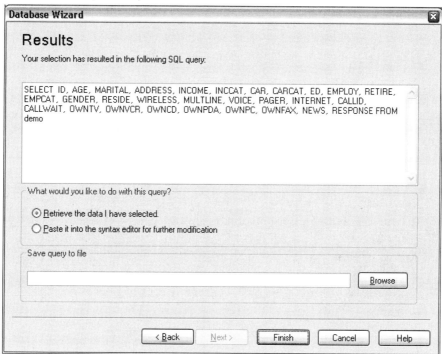

Text Wizard

The Text Wizard can read text data files formatted in a variety of ways:

- Tab-delimited files
- Space-delimited files
- Comma-delimited files
- Fixed-field format files

For delimited files, you can also specify other characters as delimiters between values, and you can specify multiple delimiters.

To Read Text Data Files

▶ From the menus choose:
File
   Read Text Data...

▶ Select the text file in the Open File dialog box.

▶ Follow the steps in the Text Wizard to define how to read the data file.

Text Wizard: Step 1

Figure 3-15
Text Wizard: Step 1

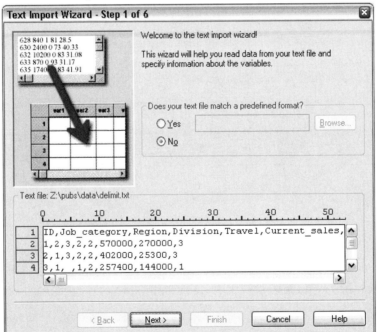

The text file is displayed in a preview window. You can apply a predefined format (previously saved from the Text Wizard) or follow the steps in the Text Wizard to specify how the data should be read.

Text Wizard: Step 2

Figure 3-16
Text Wizard: Step 2

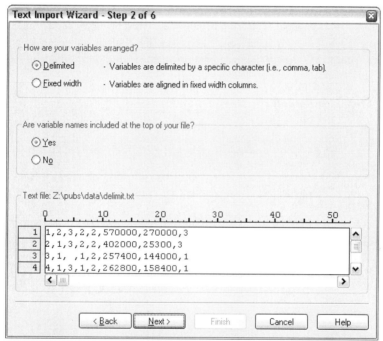

This step provides information about variables. A variable is similar to a field in a database. For example, each item in a questionnaire is a variable.

How are your variables arranged? To read your data properly, the Text Wizard needs to know how to determine where the data value for one variable ends and the data value for the next variable begins. The arrangement of variables defines the method used to differentiate one variable from the next.

- **Delimited.** Spaces, commas, tabs, or other characters are used to separate variables. The variables are recorded in the same order for each case but not necessarily in the same column locations.

- **Fixed width.** Each variable is recorded in the same column location on the same record (line) for each case in the data file. No delimiter is required between variables. In fact, in many text data files generated by computer programs, data values may appear to run together without even spaces separating them. The column location determines which variable is being read.

Are variable names included at the top of your file? If the first row of the data file contains descriptive labels for each variable, you can use these labels as variable names. Values that don't conform to variable naming rules are converted to valid variable names.

Text Wizard: Step 3 (Delimited Files)

Figure 3-17
Text Wizard: Step 3 (for delimited files)

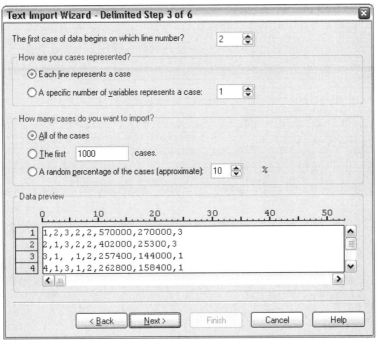

This step provides information about cases. A case is similar to a record in a database. For example, each respondent to a questionnaire is a case.

The first case of data begins on which line number? Indicates the first line of the data file that contains data values. If the top line(s) of the data file contain descriptive labels or other text that does not represent data values, this will *not* be line 1.

How are your cases represented? Controls how the Text Wizard determines where each case ends and the next one begins.

- **Each line represents a case.** Each line contains only one case. It is fairly common for each case to be contained on a single line (row), even though this can be a very long line for data files with a large number of variables. If not all lines contain the same number of data values, the number of variables for each case is determined by the line with the greatest number of data values. Cases with fewer data values are assigned missing values for the additional variables.

- **A specific number of variables represents a case.** The specified number of variables for each case tells the Text Wizard where to stop reading one case and start reading the next. Multiple cases can be contained on the same line, and cases can start in the middle of one line and be continued on the next line. The Text Wizard determines the end of each case based on the number of values read, regardless of the number of lines. Each case must contain data values (or missing values indicated by delimiters) for all variables, or the data file will be read incorrectly.

How many cases do you want to import? You can import all cases in the data file, the first *n* cases (*n* is a number you specify), or a random sample of a specified percentage. Since the random sampling routine makes an independent pseudo-random decision for each case, the percentage of cases selected can only approximate the specified percentage. The more cases there are in the data file, the closer the percentage of cases selected is to the specified percentage.

Text Wizard: Step 3 (Fixed-Width Files)

Figure 3-18
Text Wizard: Step 3 (for fixed-width files)

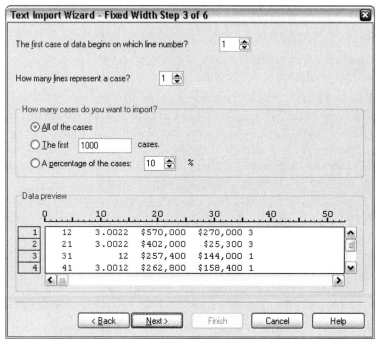

This step provides information about cases. A case is similar to a record in a database. For example, each respondent to questionnaire is a case.

The first case of data begins on which line number? Indicates the first line of the data file that contains data values. If the top line(s) of the data file contain descriptive labels or other text that does not represent data values, this will *not* be line 1.

How many lines represent a case? Controls how the Text Wizard determines where each case ends and the next one begins. Each variable is defined by its line number within the case and its column location. You need to specify the number of lines for each case to read the data correctly.

How many cases do you want to import? You can import all cases in the data file, the first *n* cases (*n* is a number you specify), or a random sample of a specified percentage. Since the random sampling routine makes an independent pseudo-random decision for each case, the percentage of cases selected can only approximate the specified percentage. The more cases there are in the data file, the closer the percentage of cases selected is to the specified percentage.

Text Wizard: Step 4 (Delimited Files)

Figure 3-19
Text Wizard: Step 4 (for delimited files)

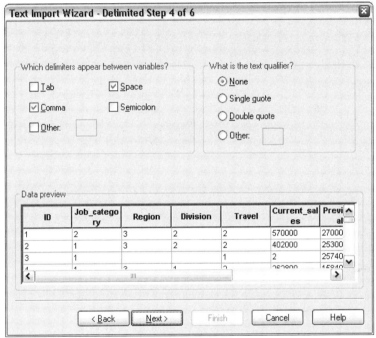

This step displays the Text Wizard's best guess on how to read the data file and allows you to modify how the Text Wizard will read variables from the data file.

Which delimiters appear between variables? Indicates the characters or symbols that separate data values. You can select any combination of spaces, commas, semicolons, tabs, or other characters. Multiple, consecutive delimiters without intervening data values are treated as missing values.

What is the text qualifier? Characters used to enclose values that contain delimiter characters. For example, if a comma is the delimiter, values that contain commas will be read incorrectly unless there is a text qualifier enclosing the value, preventing the commas in the value from being interpreted as delimiters between values. CSV-format data files exported from Excel use a double quotation mark (") as a text qualifier. The text qualifier appears at both the beginning and the end of the value, enclosing the entire value.

Text Wizard: Step 4 (Fixed-Width Files)

Figure 3-20
Text Wizard: Step 4 (for fixed-width files)

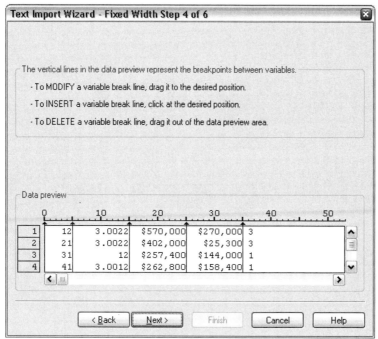

This step displays the Text Wizard's best guess on how to read the data file and allows you to modify how the Text Wizard will read variables from the data file. Vertical lines in the preview window indicate where the Text Wizard currently thinks each variable begins in the file.

Insert, move, and delete variable break lines as necessary to separate variables. If multiple lines are used for each case, select each line from the drop-down list and modify the variable break lines as necessary.

Notes:

For computer-generated data files that produce a continuous stream of data values with no intervening spaces or other distinguishing characteristics, it may be difficult to determine where each variable begins. Such data files usually rely on a data definition file or some other written description that specifies the line and column location for each variable.

The Text Wizard is designed for use with a monospaced (fixed-pitch) font to display the file contents. With nonmonospaced (proportional) fonts, the file content will not align properly. This font setting is controlled by the Text Output Font setting on the Viewer tab of the Options dialog box (Edit menu).

Text Wizard: Step 5

Figure 3-21
Text Wizard: Step 5

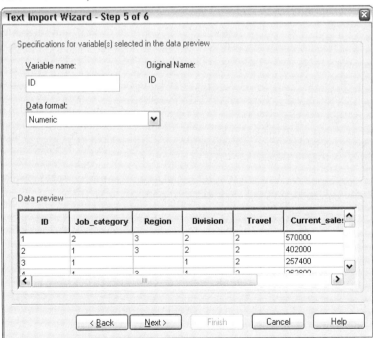

This steps controls the variable name and the data format that the Text Wizard will use to read each variable and which variables will be included in the final data file.

Variable name. You can overwrite the default variable names with your own variable names. If you read variable names from the data file, the Text Wizard will automatically modify variable names that don't conform to variable naming rules. Select a variable in the preview window and then enter a variable name.

Data format. Select a variable in the preview window and then select a format from the drop-down list. Shift-click to select multiple contiguous variables or Ctrl-click to select multiple noncontiguous variables.

Text Wizard Formatting Options

Formatting options for reading variables with the Text Wizard include:

Do not import. Omit the selected variable(s) from the imported data file.

Numeric. Valid values include numbers, a leading plus or minus sign, and a decimal indicator.

String. Valid values include virtually any keyboard characters and embedded blanks. For delimited files, you can specify the number of characters in the value, up to a maximum of 32,767. By default, the Text Wizard sets the number of characters to the longest string value encountered for the selected variable(s). For fixed-width files, the number of characters in string values is defined by the placement of variable break lines in step 4.

Date/Time. Valid values include dates of the general format *dd-mm-yyyy*, *mm/dd/yyyy*, *dd.mm.yyyy*, *yyyy/mm/dd*, *hh:mm:ss*, and a variety of other date and time formats. Months can be represented in digits, Roman numerals, or three-letter abbreviations, or they can be fully spelled out. Select a date format from the list.

Dollar. Valid values are numbers with an optional leading dollar sign and optional commas as thousands separators.

Comma. Valid values include numbers that use a period as a decimal indicator and commas as thousands separators.

Dot. Valid values include numbers that use a comma as a decimal indicator and periods as thousands separators.

Note: Values that contain invalid characters for the selected format will be treated as missing. Values that contain any of the specified delimiters will be treated as multiple values.

Text Wizard: Step 6

Figure 3-22
Text Wizard: Step 6

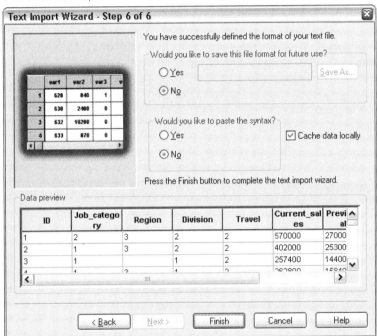

This is the final step of the Text Wizard. You can save your specifications in a file for use when importing similar text data files. You can also paste the syntax generated by the Text Wizard into a syntax window. You can then customize and/or save the syntax for use in other sessions or in production jobs.

Cache data locally. A data cache is a complete copy of the data file, stored in temporary disk space. Caching the data file can improve performance.

Reading Dimensions Data

You can read data from SPSS Dimensions products, including Quanvert, Quancept, and mrInterview.

To read Dimensions data sources, you must have the following items installed on the computer that is running SPSS:

■ .NET framework

■ Dimensions Data Model and OLE DB Access

Versions of these components that are compatible with this release of SPSS can be installed from the SPSS installation CD and are available on the AutoPlay menu. You can read Dimensions data sources only in local analysis mode. This feature is not available in distributed analysis mode using SPSS Server.

To read data from a Dimensions data source:

▶ In any open SPSS window, from the menus choose:
File
  Open Dimensions Data

▶ On the Connection tab of Data Link Properties, specify the metadata file, the case data type, and the case data file.

▶ Click OK.

▶ In the Dimensions Data Import dialog box, select the variables that you want to include and select any case selection criteria.

▶ Click OK to read the data.

Data Link Properties Connection Tab

To read a Dimensions data source, you need to specify:

Metadata Location. The metadata document file (*.mdd*) that contains questionnaire definition information.

Case Data Type. The format of the case data file. Available formats include:

■ Quancept Data File (DRS). Case data in a Quancept *.drs*, *.drz*, or *.dru* file.

■ Quanvert Database. Case data in a Quanvert database.

■ Dimensions Database (MS SQL Server). Case data in an SPSS MR relational database in SQL Server. This option can be used to read data that are collected by using mrInterview.

■ Dimensions XML Data File. Case data in an XML file.

Case Data Location. The file that contains the case data. The format of this file must be consistent with the selected case data type.

Figure 3-23
Data Link Properties: Connection tab

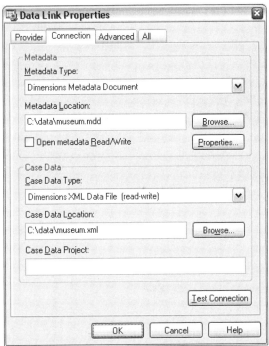

Note: The extent to which other settings on the Connection tab or any settings on the other Data Link Properties tabs may or may not affect reading Dimensions data into SPSS is not known, so we recommend that you do not change any of them.

Select Variables Tab

You can select a subset of variables to read. By default, all standard variables in the data source are displayed and selected.

■ **Show System variables.** Displays any "system" variables, including variables that indicate interview status (*in progress*, *completed*, *finish date*, etc.). You can then select any system variables that you want to include. By default, all system variables are excluded.

- **Show Codes variables.** Displays any variables that represent codes that are used for open-ended "Other" responses for categorical variables. You can then select any Codes variables that you want to include. By default, all Codes variables are excluded.

- **Show SourceFile variables.** Displays any variables that contain filenames of images of scanned responses. You can then select any SourceFile variables that you want to include. By default, all SourceFile variables are excluded.

Figure 3-24
Dimensions Data Import: Select Variables tab

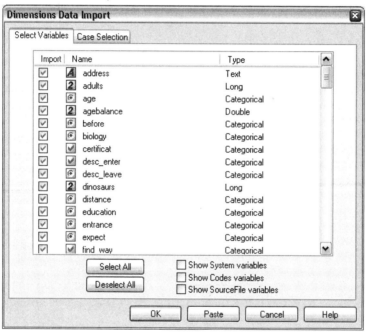

Case Selection Tab

For Dimensions data sources that contain system variables, you can select cases based on a number of system variable criteria. You do not need to include the corresponding system variables in the list of variables to read, but the necessary system variables must exist in the source data to apply the selection criteria. If the necessary system variables do not exist in the source data, the corresponding selection criteria are ignored.

Data collection status. You can select respondent data, test data, or both. You can also select cases based on any combination of the following interview status parameters:

■ Completed successfully

■ Active/in progress

■ Timed out

■ Stopped by script

■ Stopped by respondent

■ Interview system shutdown

■ Signal (terminated by a signal statement in the script)

Data collection finish date. You can select cases based on the data collection finish date.

■ **Start Date.** Cases for which data collection finished on or after the specified date are included.

■ **End Date.** Cases for which data collection finished before the specified date are included. This does *not* include cases for which data collection finished on the end date.

■ If you specify both a start date and end date, this defines a range of finish dates from the start date to (but not including) the end date.

Figure 3-25
Dimensions Data Import: Case Selection tab

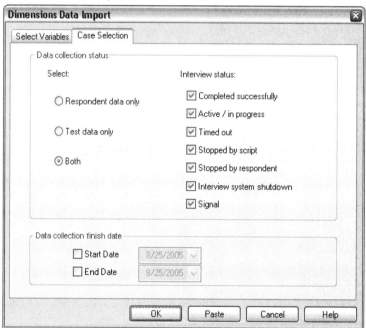

File Information

A data file contains much more than raw data. It also contains any variable definition information, including:

- Variable names
- Variable formats
- Descriptive variable and value labels

This information is stored in the dictionary portion of the data file. The Data Editor provides one way to view the variable definition information. You can also display complete dictionary information for the active dataset or any other data file.

To Display Data File Information

▶ From the menus in the Data Editor window choose:

File
  Display Data File Information

▶ For the currently open data file, choose Working File.

▶ For other data files, choose External File, and then select the data file.

The data file information is displayed in the Viewer.

Saving Data Files

In addition to saving data files in SPSS format, you can save data from SPSS in a wide variety of external formats, including:

■ Excel and other spreadsheet formats

■ Tab-delimited and CSV text files

■ SAS

■ Stata

■ Database tables

To Save Modified Data Files

▶ Make the Data Editor the active window (click anywhere in the window to make it active).

▶ From the menus choose:

File
  Save

The modified data file is saved, overwriting the previous version of the file.

Saving Data Files in External Formats

▶ Make the Data Editor the active window (click anywhere in the window to make it active).

▶ From the menus choose:
File
  Save As...

▶ Select a file type from the drop-down list.

▶ Enter a filename for the new data file.

To write variable names to the first row of a spreadsheet or tab-delimited data file:

▶ Click Write variable names to spreadsheet in the Save Data As dialog box.

To save value labels instead of data values in Excel 97 format:

▶ Click Save value labels where defined instead of data values in the Save Data As dialog box.

To save value labels to a SAS syntax file (active only when a SAS file type is selected):

▶ Click Save value labels into a .sas file in the Save Data As dialog box.

For information on exporting data to database tables, see "Exporting to a Database" on p. 68.

For information on exporting data for use in Dimensions applications, see "Exporting to Dimensions" on p. 84.

Saving Data: Data File Types

You can save data in the following formats:

SPSS (*.sav). SPSS format.

■ Data files saved in SPSS format cannot be read by versions of the software prior to version 7.5.

- When using data files with variable names longer than eight bytes in SPSS 10.x or 11.x, unique, eight-byte versions of variable names are used—but the original variable names are preserved for use in release 12.0 or later. In releases prior to SPSS 10.0, the original long variable names are lost if you save the data file.

- When using data files with string variables longer than 255 bytes in versions of SPSS prior to release 13.0, those string variables are broken up into multiple 255-byte string variables.

SPSS 7.0 (*.sav). SPSS 7.0 for Windows format. Data files saved in SPSS 7.0 format can be read by SPSS 7.0 and earlier versions of SPSS for Windows but do not include defined multiple response sets or Data Entry for Windows information.

SPSS/PC+ (*.sys). SPSS/PC+ format. If the data file contains more than 500 variables, only the first 500 will be saved. For variables with more than one defined user-missing value, additional user-missing values will be recoded into the first defined user-missing value.

SPSS Portable (*.por). SPSS portable format that can be read by other versions of SPSS and versions on other operating systems (for example, Macintosh or UNIX). Variable names are limited to eight bytes and are automatically converted to unique eight-byte names if necessary.

Tab-delimited (*.dat). Text files with values separated by tabs. (*Note*: Tab characters embedded in string values are preserved as tab characters in the tab-delimited file. No distinction is made between tab characters embedded in values and tab characters that separate values.)

Comma-delimited (*.csv). Text files with values separated by commas or semicolons. If the current SPSS decimal indicator is a period, values are separated by commas. If the current SPSS decimal indicator is a comma, values are separated by semicolons.

Fixed ASCII (*.dat). Text file in fixed format, using the default write formats for all variables. There are no tabs or spaces between variable fields.

Excel 97 and later (*.xls). Microsoft Excel 97/2000/XP spreadsheet file. The maximum number of variables is 256, and the maximum number of rows is 65,536.

Excel 2.1 (*.xls). Microsoft Excel 2.1 spreadsheet file. The maximum number of variables is 256, and the maximum number of rows is 16,384.

1-2-3 Release 3.0 (*.wk3). Lotus 1-2-3 spreadsheet file, release 3.0. The maximum number of variables that you can save is 256.

1-2-3 Release 2.0 (*.wk1). Lotus 1-2-3 spreadsheet file, release 2.0. The maximum number of variables that you can save is 256.

1-2-3 Release 1.0 (*.wks). Lotus 1-2-3 spreadsheet file, release 1A. The maximum number of variables that you can save is 256.

SYLK (*.slk). Symbolic link format for Microsoft Excel and Multiplan spreadsheet files. The maximum number of variables that you can save is 256.

dBASE IV (*.dbf). dBASE IV format.

dBASE III (*.dbf). dBASE III format.

dBASE II (*.dbf). dBASE II format.

SAS v7+ Windows short extension (*.sd7). SAS versions 7–8 for Windows short filename format.

SAS v7+ Windows long extension (*.sas7bdat). SAS versions 7–8 for Windows long filename format.

SAS v7+ for UNIX (*.ssd01). SAS v8 for UNIX.

SAS v6 for Windows (*.sd2). SAS v6 file format for Windows/OS2.

SAS v6 for UNIX (*.ssd01). SAS v6 file format for UNIX (Sun, HP, IBM).

SAS v6 for Alpha/OSF (*.ssd04). SAS v6 file format for Alpha/OSF (DEC UNIX).

SAS Transport (*.xpt). SAS transport file.

Stata Version 8 Intercooled (*.dta).

Stata Version 8 SE (*.dta).

Stata Version 7 Intercooled (*.dta).

Stata Version 7 SE (*.dta).

Stata Version 6 (*.dta).

Stata Versions 4–5 (*.dta).

Saving File Options

For spreadsheet, tab-delimited files, and comma-delimited files, you can write variable names to the first row of the file.

Saving Data Files in Excel Format

You can save your data in one of three Microsoft Excel file formats. The choice of format depends on the version of Excel that will be used to open the data. The Excel application cannot open an Excel file from a newer version of the application. For example, Excel 5.0 cannot open an Excel 2000 document. However, Excel 2000 can easily read an Excel 5.0 document.

There are a few limitations to the Excel file format that don't exist in SPSS. These limitations include:

- Variable information, such as missing values and variable labels, is not included in exported Excel files.

- When exporting to Excel 97 and later, an option is provided to include value labels instead of values.

- Because all Excel files are limited to 256 columns of data, only the first 256 variables are included in the exported file.

- Excel 4.0 and Excel 5.0/95 files are limited to 16,384 records, or rows of data. Excel 97–2000 files allow 65,536 records. If your data exceed these limits, a warning message is displayed and the data are truncated to the maximum size allowed by Excel.

Variable Types

The following table shows the variable type matching between the original data in SPSS and the exported data in Excel.

SPSS Variable Type	Excel Data Format
Numeric	0.00; #,##0.00; ...
Comma	0.00; #,##0.00; ...
Dollar	$#,##0_); ...
Date	d-mmm-yyyy
Time	hh:mm:ss
String	General

Saving Data Files in SAS Format

Special handling is given to various aspects of your data when saved as a SAS file. These cases include:

- Certain characters that are allowed in SPSS variable names are not valid in SAS, such as @, #, and $. These illegal characters are replaced with an underscore when the data are exported.

- SPSS variable names that contain multibyte characters (for example, Japanese or Chinese characters) are converted to variables names of the general form *Vnnn*, where *nnn* is an integer value.

- SPSS variable labels containing more than 40 characters are truncated when exported to a SAS v6 file.

- Where they exist, SPSS variable labels are mapped to the SAS variable labels. If no variable label exists in the SPSS data, the variable name is mapped to the SAS variable label.

- SAS allows only one value for system-missing, whereas SPSS allows numerous system-missing values. As a result, all system-missing values in SPSS are mapped to a single system-missing value in the SAS file.

Save Value Labels

You have the option of saving the values and value labels associated with your data file to a SAS syntax file. For example, when the value labels for the *cars.sav* data file are exported, the generated syntax file contains:

```
libname library 'd:\spss\' ;

proc format library = library ;
   value ORIGIN /* Country of Origin */
      1 = 'American'
      2 = 'European'
      3 = 'Japanese' ;
   value CYLINDER /* Number of Cylinders */
      3 = '3 Cylinders'
      4 = '4 Cylinders'
```

```
        5 = '5 Cylinders'

        6 = '6 Cylinders'

        8 = '8 Cylinders' ;

    value FILTER__  /* cylrec = 1 | cylrec = 2 (FILTER) */

        0 = 'Not Selected'

        1 = 'Selected' ;

proc datasets library = library ;

modify cars;

    format    ORIGIN ORIGIN.;

    format  CYLINDER CYLINDER.;

    format  FILTER__  FILTER__.;

quit;
```

This feature is not supported for the SAS transport file.

Variable Types

The following table shows the variable type matching between the original data in SPSS and the exported data in SAS.

SPSS Variable Type	SAS Variable Type	SAS Data Format
Numeric	Numeric	12
Comma	Numeric	12
Dot	Numeric	12
Scientific Notation	Numeric	12
Date	Numeric	(Date) for example, MMDDYY10, ...
Date (Time)	Numeric	Time18
Dollar	Numeric	12
Custom Currency	Numeric	12
String	Character	$8

Saving Data Files in Stata Format

■ Data can be written in Stata 5–8 format and in both Intercooled and SE format (versions 7 and 8 only).

■ Data files that are saved in Stata 5 format can be read by Stata 4.

■ The first 80 bytes of variable labels are saved as Stata variable labels.

■ For numeric variables, the first 80 bytes of value labels are saved as Stata value labels. For string variables, value labels are dropped.

■ For versions 7 and 8, the first 32 bytes of variable names in case-sensitive form are saved as Stata variable names. For earlier versions, the first eight bytes of variable names are saved as Stata variable names. Any characters other than letters, numbers, and underscores are converted to underscores.

■ SPSS variable names that contain multibyte characters (for example, Japanese or Chinese characters) are converted to variable names of the general form *Vnnn*, where *nnn* is an integer value.

■ For versions 5–6 and Intercooled versions 7–8, the first 80 bytes of string values are saved. For Stata SE 7–8, the first 244 bytes of string values are saved.

■ For versions 5–6 and Intercooled versions 7–8, only the first 2,047 variables are saved. For Stata SE 7–8, only the first 32,767 variables are saved.

SPSS Variable Type	Stata Variable Type	Stata Data Format
Numeric	Numeric	g
Comma	Numeric	g
Dot	Numeric	g
Scientific Notation	Numeric	g
Date*, Datetime	Numeric	D_m_Y
Time, DTime	Numeic	g (number of seconds)
Wkday	Numeric	g (1–7)
Month	Numeric	g (1–12)
Dollar	Numeric	g
Custom Currency	Numeric	g
String	String	s

*Date, Adate, Edate, SDate, Jdate, Qyr, Moyr, Wkyr

Saving Subsets of Variables

Figure 3-26
Save Data As Variables dialog box

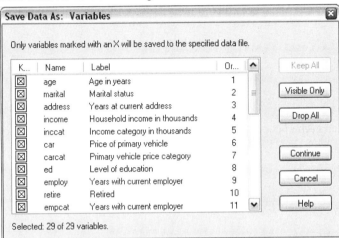

The Save Data As Variables dialog box allows you to select the variables that you want saved in the new data file. By default, all variables will be saved. Deselect the variables that you don't want to save, or click Drop All and then select the variables that you want to save.

Visible Only. Selects only variables in variable sets currently in use. For more information, see "Use Variable Sets" in Chapter 44 on p. 634.

To Save a Subset of Variables

▶ Make the Data Editor the active window (click anywhere in the window to make it active).

▶ From the menus choose:
File
    Save As...

▶ Click Variables.

▶ Select the variables that you want to save.

Exporting to a Database

You can use the Export to Database Wizard to:

- Replace values in existing database table fields (columns) or add new fields to a table.

- Append new records (rows) to a database table.

- Completely replace a database table or create a new table.

To export data to a database:

▶ From the menus in the Data Editor window for the dataset that contains the data you want to export, choose:

File
  Export to Database

▶ Select the database source.

▶ Follow the instructions in the export wizard to export the data.

Creating Database Fields from SPSS Variables

When creating new fields (adding fields to an existing database table, creating a new table, replacing a table), you can specify field names, data type, and width (where applicable).

Field name. The default field names are the same as the SPSS variable names. You can change the field names to any names allowed by the database format. For example, many databases allow characters in field names that aren't allowed in SPSS variable names, including spaces. Therefore, an SPSS variable name like *CallWaiting* could be changed to the field name *Call Waiting*.

Type. The export wizard makes initial data type assignments based on the standard ODBC data types or data types allowed by the selected database format that most closely matches the defined SPSS data format—but databases can make type distinctions that have no direct equivalent in SPSS, and vice versa. For example, most numeric values in SPSS are stored as double-precision floating-point values, whereas database numeric data types include float (double), integer, real, etc. In addition, many databases don't have equivalents to SPSS time formats. You can change the data type to any type available in the drop-down list.

As a general rule, the basic data type (string or numeric) for the SPSS variable should match the basic data type of the database field. If there is a data type mismatch that cannot be resolved by the database, an error results and no data are exported to the database. For example, if you export an SPSS string variable to a database field with a numeric data type, an error will result if any values of the string variable contain non-numeric characters.

Width. You can change the defined width for string (char, varchar) field types. Numeric field widths are defined by the data type.

By default, SPSS variable formats are mapped to database field types based on the following general scheme. Actual database field types may vary, depending on the database.

SPSS Variable Format	Database Field Type
Numeric	Float or Double
Comma	Float or Double
Dot	Float or Double
Scientific Notation	Float or Double
Date	Date or Datetime or Timestamp
Datetime	Datetime or Timestamp
Time, DTime	Float or Double (number of seconds)
Wkday	Integer (1–7)
Month	Integer (1–12)
Dollar	Float or Double
Custom Currency	Float or Double
String	Char or Varchar

User-Missing Values

There are two options for the treatment of user-missing values when data from SPSS variables are exported to database fields:

Export as valid values. User-missing values are treated as regular, valid, nonmissing values.

Export numeric user-missing as nulls and export string user-missing values as blank spaces. Numeric user-missing values are treated the same as system-missing values. String user-missing values are converted to blank spaces (strings cannot be system-missing).

Selecting a Data Source

In the first panel of the Export to Database Wizard, you select the data source to which you want to export data.

Figure 3-27
Export to Database Wizard, selecting a data source

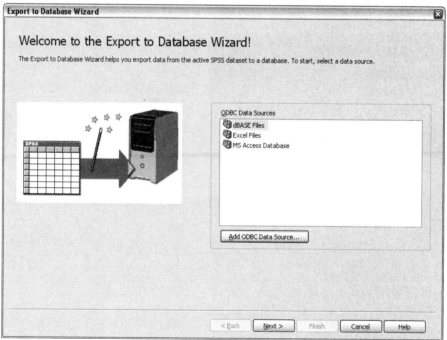

You can export data to any database source for which you have the appropriate ODBC driver. (*Note*: Exporting data to OLE DB data sources is not supported.)

If you do not have any ODBC data sources configured, or if you want to add a new data source, click Add ODBC Data Source. In distributed analysis mode (available with SPSS Server), this button is not available. To add data sources in distributed analysis mode, see your system administrator.

An ODBC data source consists of two essential pieces of information: the driver that will be used to access the data and the location of the database you want to access. To specify data sources, you must have the appropriate drivers installed. For local analysis mode, you can install drivers from the CD-ROM for this product:

■ **SPSS Data Access Pack.** Installs drivers for a variety of database formats. This feature is available on the AutoPlay menu.

■ **Microsoft Data Access Pack.** Installs drivers for Microsoft products, including Microsoft Access. To install the Microsoft Data Access Pack, double-click Microsoft Data Access Pack in the Microsoft Data Access Pack folder on the CD-ROM.

Some data sources may require a login ID and password before you can proceed to the next step.

Choosing How to Export the Data

After you select the data source, you indicate the manner in which you want to export the data.

Figure 3-28
Export to Database Wizard, choosing how to export

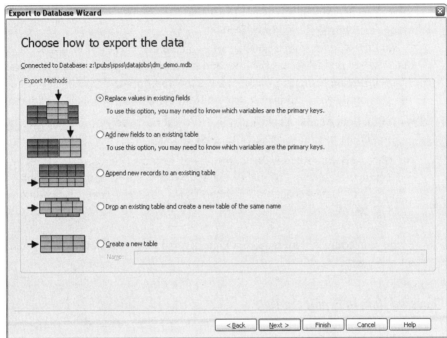

The following choices are available for exporting data to a database:

- **Replace values in existing fields.** Replaces values of selected fields in an existing table with values from the selected variables in the active dataset. For more information, see "Replacing Values in Existing Fields" on p. 77.

- **Add new fields to an existing table.** Creates new fields in an existing table that contain the values of selected variables in the active dataset. For more information, see "Adding New Fields" on p. 78. This option is not available for Excel files.

- **Append new records to an existing table.** Adds new records (rows) to an existing table containing the values from cases in the active dataset. For more information, see "Appending New Records (Cases)" on p. 79.

- **Drop an existing table and create a new table of the same name.** Deletes the specified table and creates a new table of the same name that contains selected variables from the active dataset. All information from the original table, including definitions of field properties (for example, primary keys, data types) is lost. For more information, see "Creating a New Table or Replacing a Table" on p. 81.

■ **Create a new table.** Creates a new table in the database containing data from selected variables in the active dataset. The name can be any value that is allowed as a table name by the data source. The name cannot duplicate the name of an existing table or view in the database. For more information, see "Creating a New Table or Replacing a Table" on p. 81.

Selecting a Table

When modifying or replacing a table in the database, you need to select the table to modify or replace. This panel in the Export to Database Wizard displays a list of tables and views in the selected database.

Figure 3-29
Export to Database Wizard, selecting a table or view

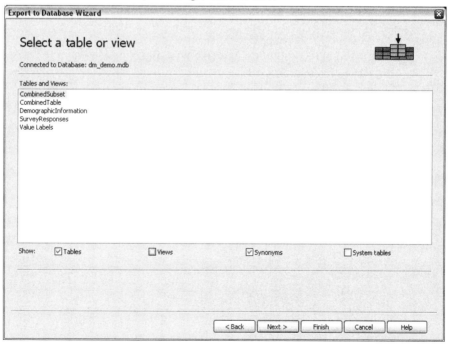

By default, the list displays only standard database tables. You can control the type of items that are displayed in the list:

- **Tables.** Standard database tables.

- **Views.** Views are virtual or dynamic "tables" defined by queries. These can include joins of multiple tables and/or fields derived from calculations based on the values of other fields. You can append records or replace values of existing fields in views, but the fields that you can modify may be restricted, depending on how the view is structured. For example, you cannot modify a derived field, add fields to a view, or replace a view.

- **Synonyms.** A synonym is an alias for a table or view, typically defined in a query.

- **System tables.** System tables define database properties. In some cases, standard database tables may be classified as system tables and will be displayed only if you select this option. Access to real system tables is often restricted to database administrators.

Selecting Cases to Export

Case selection in the Export to Database Wizard is limited either to all cases or to cases selected using a previously defined filter condition. If no case filtering is in effect, this panel will not appear, and all cases in the active dataset will be exported.

Figure 3-30
Export to Database Wizard, selecting cases to export

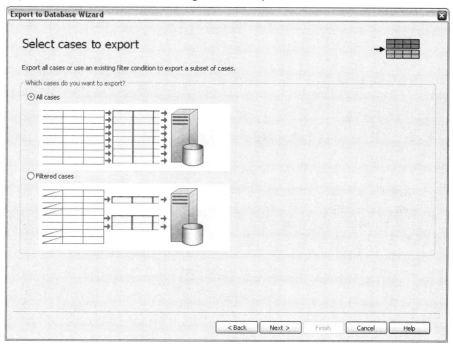

For information on defining a filter condition for case selection, see "Select Cases" on p. 235.

Matching Cases to Records

When adding fields (columns) to an existing table or replacing the values of existing fields, you need to make sure that each case (row) in the active dataset is correctly matched to the corresponding record in the database.

- In the database, the field or set of fields that uniquely identifies each record is often designated as the **primary key**.

- You need to identify which SPSS variable(s) correspond to the primary key field(s) or other fields that uniquely identify each record.

- The fields don't have to be the primary key in the database, but the field value or combination of field values must be unique for each case.

To match SPSS variables with fields in the database that uniquely identify each record:

▶ Drag and drop the SPSS variable(s) onto the corresponding database fields.

or

▶ Select a variable from the list of SPSS variables, select the corresponding field in the database table, and click Connect.

To delete a connection line:

▶ Select the connection line and press the Delete key.

Figure 3-31
Export to Database Wizard, matching cases to records

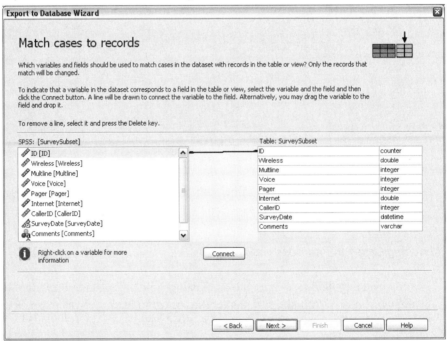

Note: The SPSS variable names and database field names may not be identical (since database field names may contain characters not allowed in SPSS variable names), but if the active dataset was created from the database table you are modifying, either the variable names or the variable labels will usually be at least similar to the database field names.

Replacing Values in Existing Fields

To replace values of existing fields in a database:

▶ In the Choose how to export the data panel of the Export to Database Wizard, select Replace values in existing fields.

▶ In the Select a table or view panel, select the database table.

▶ In the Match cases to records panel, match the SPSS variables that uniquely identify each case to the corresponding database field names.

▶ For each field for which you want to replace values, drag and drop the SPSS variable that contains the new values into the Source of values column, next to the corresponding database field name.

Figure 3-32
Export to Database Wizard, replacing values of existing fields

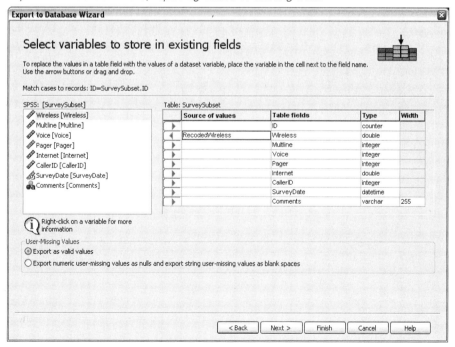

- As a general rule, the basic data type (string or numeric) for the SPSS variable should match the basic data type of the database field. If there is a data type mismatch that cannot be resolved by the database, an error results and no data is exported to the database. For example, if you export an SPSS string variable to a database field with a numeric data type (for example, double, real, integer), an error will result if any values of the string variable contain non-numeric characters. The letter *a* in the icon next to an SPSS variable denotes a string variable.

- You cannot modify the field name, type, or width. The original database field attributes are preserved; only the values are replaced.

Adding New Fields

To add new fields to an existing database table:

▶ In the Choose how to export the data panel of the Export to Database Wizard, select Add new fields to an existing table.

▶ In the Select a table or view panel, select the database table.

▶ In the Match cases to records panel, match the SPSS variables that uniquely identify each case to the corresponding database field names.

▶ Drag and drop the SPSS variables that you want to add as new fields to the Source of values column.

Figure 3-33
Export to Database Wizard, adding new fields to an existing table

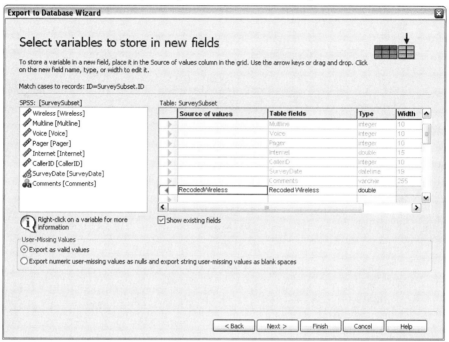

For information on field names and data types, see the section on creating database fields from SPSS variables in "Exporting to a Database" on p. 68.

Show existing fields. Select this option to display a list of existing fields. You cannot use this panel in the Export to Database Wizard to replace existing fields, but it may be helpful to know what fields are already present in the table. If you want to replace the values of existing fields, see "Replacing Values in Existing Fields" on p. 77.

Appending New Records (Cases)

To append new records (cases) to a database table:

▶ In the Choose how to export the data panel of the Export to Database Wizard, select Append new records to an existing table.

▶ In the Select a table or view panel, select the database table.

▶ Match variables in the active dataset to table fields by dragging and dropping SPSS variables to the Source of values column.

Figure 3-34
Export to Database Wizard, adding records (cases) to a table

The Export to Database Wizard will automatically select all variables that match existing fields, based on information about the original database table stored in the active dataset (if available) and/or variable names that are the same as field names. This initial automatic matching is intended only as a guide and does not prevent you from changing the way in which SPSS variables are matched with database fields.

When adding new records to an existing table, the following basic rules/limitations apply:

■ All cases (or all selected cases) in the active dataset are added to the table. If any of these cases duplicate existing records in the database, an error may result if a duplicate key value is encountered. For information on exporting only selected cases, see "Selecting Cases to Export" on p. 74.

■ You can use the values of new variables created in the session as the values for existing fields, but you cannot add new fields or change the names of existing fields. To add new fields to a database table, see "Adding New Fields" on p. 78.

■ Any excluded database fields or fields not matched to an SPSS variable will have no values for the added records in the database table. (If the Source of values cell is empty, there is no SPSS variable matched to the field.)

Creating a New Table or Replacing a Table

To create a new database table or replace an existing database table:

▶ In the Choose how to export the data panel of the export wizard, select Drop an existing table and create a new table of the same name or select Create a new table and enter a name for the new table.

▶ If you are replacing an existing table, in the Select a table or view panel, select the database table.

▶ Drag and drop SPSS variables into the Variables to save column.

▶ Optionally, you can designate variables/fields that define the primary key, change field names, and change the data type.

Figure 3-35
Export to Database Wizard, selecting variables for a new table

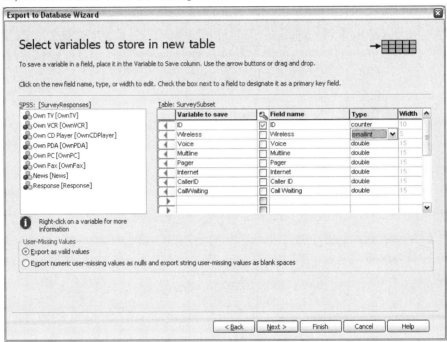

Primary key. To designate variables as the primary key in the database table, select the box in the column identified with the key icon.

- All values of the primary key must be unique or an error will result.

- If you select a single variable as the primary key, every record (case) must have a unique value for that variable.

- If you select multiple variables as the primary key, this defines a composite primary key, and the *combination* of values for the selected variables must be unique for each case.

For information on field names and data types, see the section on creating database fields from SPSS variables in "Exporting to a Database" on p. 68.

Completing the Database Export Wizard

The last panel of the Export to Database Wizard provides a summary that indicates what data will be exported and how it will be exported. It also gives you the option of either exporting the data or pasting the underlying command syntax to a syntax window.

Figure 3-36
Export to Database Wizard, Finish panel

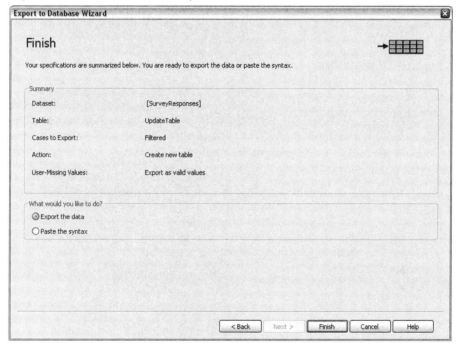

Summary Information

- **Dataset.** The SPSS session name for the dataset that will be used to export data. This information is primarily useful if you have multiple open data sources. Data sources opened using the graphical user interface (for example, the Database Wizard) are automatically assigned names such as *DataSet1*, *DataSet2*, etc. A data source opened using command syntax will have a dataset name only if one is explicitly assigned.

- **Table.** The name of the table to be modified or created.

- **Cases to Export.** Either all cases are exported or cases selected by a previously defined filter condition are exported. For more information, see "Selecting Cases to Export" on p. 74.

- **Action.** Indicates how the database will be modified (for example, create a new table, add fields or records to an existing table).

- **User-Missing Values.** User-missing values can be exported as valid values or treated the same as system-missing for numeric variables and converted to blank spaces for string variables. This setting is controlled in the panel in which you select the variables to export.

Exporting to Dimensions

The Export to Dimensions dialog box creates an SPSS data file and a Dimensions metadata file that you can use to read the data into Dimensions applications such as mrInterview and mrTables. This is particularly useful when "roundtripping" data between SPSS and Dimensions applications. For example, you can read an mrInterview data source into SPSS, calculate some new variables, and then save the data in a form that can be read by mrTables, without loss of any of the original metadata attributes.

To export data for use in Dimensions applications:

▶ From the menus in the Data Editor window that contains the data you want to export, choose:
File
  Export to Dimensions

▶ Click Data File to specify the name and location of the SPSS data file.

▶ Click Metadata File to specify the name and location of the Dimensions metadata file.

Figure 3-37
Export to Dimensions dialog box

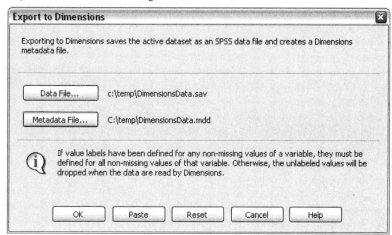

For new variables and datasets not created from Dimensions data sources, SPSS variable attributes are mapped to Dimensions metadata attributes in the metadata file according to the methods described in the SPSS SAV DSC documentation in the Dimensions Development Library.

If the active dataset was created from a Dimensions data source:

■ The new metadata file is created by merging the original metadata attributes with metadata attributes for any new variables, plus any changes to original variables that might effect their metadata attributes (for example, addition of, or changes to, value labels).

■ For original variables read from the Dimensions data source, any metadata attributes not recognized by SPSS are preserved in their original state. For example, SPSS converts grid variables to regular SPSS variables, but the metadata that defines these grid variables is preserved when you save the new metadata file.

■ If any Dimensions variables were automatically renamed to conform to SPSS variable naming rules, the metadata file maps the converted names back to the original Dimensions variable names.

The presence or absence of value labels can affect the metadata attributes of variables and consequently the way those variables are read by Dimensions applications. If value labels have been defined for any nonmissing values of a variable, they should be defined for all nonmissing values of that variable; otherwise, the unlabeled values will be dropped when the data file is read by Dimensions.

Protecting Original Data

To prevent the accidental modification or deletion of your original data, you can mark the file as read-only.

▶ From the Data Editor menus choose:

File
  Mark File Read Only

If you make subsequent modifications to the data and then try to save the data file, you can save the data only with a different filename, so the original data are not affected.

You can change the file permissions back to read-write by selecting Mark File Read Write from the File menu.

Virtual Active File

The virtual active file enables you to work with large data files without requiring equally large (or larger) amounts of temporary disk space. For most analysis and charting procedures, the original data source is reread each time you run a different procedure. Procedures that modify the data require a certain amount of temporary disk space to keep track of the changes, and some actions always require enough disk space for at least one entire copy of the data file.

Figure 3-38
Temporary disk space requirements

Action	GET FILE = 'v1-5.sav'. FREQUENCIES...	COMPUTE v6 = ... RECODE v4... REGRESSION... /SAVE ZPRED.	SORT CASES BY... or CACHE
Virtual Active File	<table><tr><td>v1</td><td>v2</td><td>v3</td><td>v4</td><td>v5</td></tr><tr><td>11</td><td>12</td><td>13</td><td>14</td><td>15</td></tr><tr><td>21</td><td>22</td><td>23</td><td>24</td><td>25</td></tr><tr><td>31</td><td>32</td><td>33</td><td>34</td><td>35</td></tr><tr><td>41</td><td>42</td><td>43</td><td>44</td><td>45</td></tr><tr><td>51</td><td>52</td><td>53</td><td>54</td><td>55</td></tr><tr><td>61</td><td>62</td><td>63</td><td>64</td><td>65</td></tr></table>	<table><tr><td>v1</td><td>v2</td><td>v3</td><td>v4</td><td>v5</td><td>v6</td><td>zpre</td></tr><tr><td>11</td><td>12</td><td>13</td><td>14</td><td>15</td><td>16</td><td>1</td></tr><tr><td>21</td><td>22</td><td>23</td><td>24</td><td>25</td><td>26</td><td>2</td></tr><tr><td>31</td><td>32</td><td>33</td><td>34</td><td>35</td><td>36</td><td>3</td></tr><tr><td>41</td><td>42</td><td>43</td><td>44</td><td>45</td><td>46</td><td>4</td></tr><tr><td>51</td><td>52</td><td>53</td><td>54</td><td>55</td><td>56</td><td>5</td></tr><tr><td>61</td><td>62</td><td>63</td><td>64</td><td>65</td><td>66</td><td>6</td></tr></table>	<table><tr><td>v1</td><td>v2</td><td>v3</td><td>v4</td><td>v5</td><td>v6</td><td>zpre</td></tr><tr><td>11</td><td>12</td><td>13</td><td>14</td><td>15</td><td>16</td><td>1</td></tr><tr><td>21</td><td>22</td><td>23</td><td>24</td><td>25</td><td>26</td><td>2</td></tr><tr><td>31</td><td>32</td><td>33</td><td>34</td><td>35</td><td>36</td><td>3</td></tr><tr><td>41</td><td>42</td><td>43</td><td>44</td><td>45</td><td>46</td><td>4</td></tr><tr><td>51</td><td>52</td><td>53</td><td>54</td><td>55</td><td>56</td><td>5</td></tr><tr><td>61</td><td>62</td><td>63</td><td>64</td><td>65</td><td>66</td><td>6</td></tr></table>
Data Stored in Temporary Disk Space	None	<table><tr><td>v4</td><td>v6</td><td>zpre</td></tr><tr><td>14</td><td>16</td><td>1</td></tr><tr><td>24</td><td>26</td><td>2</td></tr><tr><td>34</td><td>36</td><td>3</td></tr><tr><td>44</td><td>46</td><td>4</td></tr><tr><td>54</td><td>56</td><td>5</td></tr><tr><td>64</td><td>66</td><td>6</td></tr></table>	<table><tr><td>v1</td><td>v2</td><td>v3</td><td>v4</td><td>v5</td><td>v6</td><td>zpre</td></tr><tr><td>11</td><td>12</td><td>13</td><td>14</td><td>15</td><td>16</td><td>1</td></tr><tr><td>21</td><td>22</td><td>23</td><td>24</td><td>25</td><td>26</td><td>2</td></tr><tr><td>31</td><td>32</td><td>33</td><td>34</td><td>35</td><td>36</td><td>3</td></tr><tr><td>41</td><td>42</td><td>43</td><td>44</td><td>45</td><td>46</td><td>4</td></tr><tr><td>51</td><td>52</td><td>53</td><td>54</td><td>55</td><td>56</td><td>5</td></tr><tr><td>61</td><td>62</td><td>63</td><td>64</td><td>65</td><td>66</td><td>6</td></tr></table>

Actions that don't require any temporary disk space include:

■ Reading SPSS data files

■ Merging two or more SPSS data files

■ Reading database tables with the Database Wizard

■ Merging an SPSS data file with a database table

■ Running procedures that read data (for example, Frequencies, Crosstabs, Explore)

Actions that create one or more columns of data in temporary disk space include:

■ Computing new variables

■ Recoding existing variables

■ Running procedures that create or modify variables (for example, saving predicted values in Linear Regression)

Actions that create an entire copy of the data file in temporary disk space include:

■ Reading Excel files

■ Running procedures that sort data (for example, Sort Cases, Split File)

- Reading data with GET TRANSLATE or DATA LIST commands
- Using the Cache Data facility or the CACHE command
- Launching other applications from SPSS that read the data file (for example, AnswerTree, DecisionTime)

Note: The GET DATA command provides functionality comparable to DATA LIST without creating an entire copy of the data file in temporary disk space. The SPLIT FILE command in command syntax does not sort the data file and therefore does not create a copy of the data file. This command, however, requires sorted data for proper operation, and the dialog box interface for this procedure will automatically sort the data file, resulting in a complete copy of the data file. (Command syntax is not available with the Student Version.)

Actions that create an entire copy of the data file by default:

- Reading databases with the Database Wizard
- Reading text files with the Text Wizard

The Text Wizard provides an optional setting to automatically cache the data. By default, this option is selected. You can turn it off by deselecting Cache data locally. For the Database Wizard you can paste the generated command syntax and delete the CACHE command.

Creating a Data Cache

Although the virtual active file can vastly reduce the amount of temporary disk space required, the absence of a temporary copy of the "active" file means that the original data source has to be reread for each procedure. For large data files read from an external source, creating a temporary copy of the data may improve performance. For example, for data tables read from a database source, the SQL query that reads the information from the database must be reexecuted for any command or procedure that needs to read the data. Since virtually all statistical analysis procedures and charting procedures need to read the data, the SQL query is reexecuted for each procedure you run, which can result in a significant increase in processing time if you run a large number of procedures.

If you have sufficient disk space on the computer performing the analysis (either your local computer or a remote server), you can eliminate multiple SQL queries and improve processing time by creating a data cache of the active file. The data cache is a temporary copy of the complete data.

Note: By default, the Database Wizard automatically creates a data cache, but if you use the GET DATA command in command syntax to read a database, a data cache is not automatically created. (Command syntax is not available with the Student Version.)

To Create a Data Cache

▶ From the menus choose:
File
　Cache Data...

▶ Click OK or Cache Now.

OK creates a data cache the next time the program reads the data (for example, the next time you run a statistical procedure), which is usually what you want because it doesn't require an extra data pass. Cache Now creates a data cache immediately, which shouldn't be necessary under most circumstances. Cache Now is useful primarily for two reasons:

■ A data source is "locked" and can't be updated by anyone until you end your session, open a different data source, or cache the data.

■ For large data sources, scrolling through the contents of the Data View tab in the Data Editor will be much faster if you cache the data.

To Cache Data Automatically

You can use the SET command to automatically create a data cache after a specified number of changes in the active data file. By default, the active data file is automatically cached after 20 changes in the active data file.

▶ From the menus choose:
File
　New
　　Syntax

▶ In the syntax window, type SET CACHE n (where *n* represents the number of changes in the active data file before the data file is cached).

▶ From the menus in the syntax window choose:
Run
  All

Note: The cache setting is not persistent across sessions. Each time you start a new session, the value is reset to the default of 20.

Distributed Analysis Mode

Distributed analysis mode allows you to use a computer other than your local (or desktop) computer for memory-intensive work. Because remote servers that are used for distributed analysis are typically more powerful and faster than your local computer, distributed analysis mode can significantly reduce computer processing time. Distributed analysis with a remote server can be useful if your work involves:

- Large data files, particularly data read from database sources.

- Memory-intensive tasks. Any task that takes a long time in local analysis mode may be a good candidate for distributed analysis.

Distributed analysis affects only data-related tasks, such as reading data, transforming data, computing new variables, and calculating statistics. Distributed analysis has no effect on tasks related to editing output, such as manipulating pivot tables or modifying charts.

Note: Distributed analysis is available only if you have both a local version and access to a licensed server version of the software that is installed on a remote server.

Distributed Analysis versus Local Analysis

Following are some guidelines for choosing distributed analysis or local analysis mode:

Database access. Jobs that perform database queries may run faster in distributed mode if the server has superior access to the database or if the server is running on the same machine as the database engine. Additionally, if the necessary database-access software is available only on the server, or if your network administrator does not permit you to download large data tables, you can access the database only in distributed mode.

Ratio of computation to output. Commands that have the most to gain from running in distributed mode are commands that perform a lot of computation and produce small output results (for example, few and small pivot tables, brief text results, or few and simple charts). The degree of improvement largely depends on the computing power of the remote server.

Small jobs. Jobs that run quickly in local mode will almost always run slower in distributed mode because of inherent client/server overhead.

Charts. Case-oriented charts, such as scatterplots, regression residual plots, and sequence charts, require raw data on your local computer. For large data files or database tables, this process can result in slower performance in distributed mode because the data have to be sent from the remote server to your local computer. Other charts are based on summarized or aggregated data and should perform adequately because the aggregation is performed on the server.

Interactive graphics. If raw data are saved with interactive graphics (an optional setting), large amounts of data may be transferred from the remote server to your local computer, significantly increasing the time that it takes to save your results.

Pivot tables. Large pivot tables may take longer to create in distributed mode. This situation is particularly true for the OLAP Cubes procedure and for tables that contain individual case data, such as the tables that are available in the Summarize procedure.

Text output. The more text that is produced, the slower the output will be in distributed mode because the text is produced on the remote server and copied to your local computer for display. Text results have low overhead, however, and tend to transmit quickly.

Server Login

The Server Login dialog box allows you to select the computer that processes commands and runs procedures. You can select your local computer or a remote server.

Figure 4-1
Server Login dialog box

You can add, modify, or delete remote servers in the list. Remote servers usually require a user ID and password, and a domain name may also be necessary. Contact your system administrator for information about available servers, a user ID and password, domain names, and other connection information.

You can select a default server and save the user ID, domain name, and password that are associated with any server. You are automatically connected to the default server when you start a new session.

Adding and Editing Server Login Settings

Use the Server Login Settings dialog box to add or edit connection information for remote servers for use in distributed analysis mode.

Figure 4-2
Server Login Settings dialog box

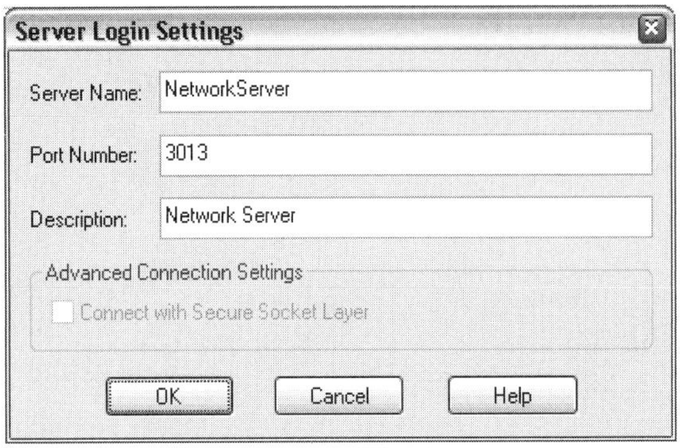

Contact your system administrator for a list of available servers, port numbers for the servers, and additional connection information. Do not use the Secure Socket Layer unless instructed to do so by your administrator.

Server Name. A server "name" can be an alphanumeric name that is assigned to a computer (for example, NetworkServer) or a unique IP address that is assigned to a computer (for example, 202.123.456.78).

Port Number. The port number is the port that the server software uses for communications.

Description. You can enter an optional description to display in the servers list.

Connect with Secure Socket Layer. Secure Socket Layer (SSL) encrypts requests for distributed analysis when they are sent to the remote SPSS server. Before you use SSL, check with your administrator. For this option to be enabled, SSL must be configured on your desktop computer and the server.

To Select, Switch, or Add Servers

▶ From the menus choose:
File
  Switch Server...

To select a default server:

▶ In the server list, select the box next to the server that you want to use.

▶ Enter the user ID, domain name, and password that were provided by your administrator.

Note: You are automatically connected to the default server when you start a new session.

To switch to another server:

▶ Select the server from the list.

▶ Enter your user ID, domain name, and password (if necessary).

Note: When you switch servers during a session, all open windows are closed. You will be prompted to save changes before the windows are closed.

To add a server:

▶ Get the server connection information from your administrator.

▶ Click Add to open the Server Login Settings dialog box.

▶ Enter the connection information and optional settings, and then click OK.

To edit a server:

▶ Get the revised connection information from your administrator.

▶ Click Edit to open the Server Login Settings dialog box.

▶ Enter the changes and click OK.

Opening Data Files from a Remote Server

Figure 4-3
Open Remote File dialog box

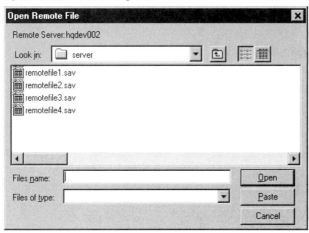

In distributed analysis mode, the Open Remote File dialog box replaces the standard Open File dialog box.

■ The contents of the list of available files, folders, and drives depends on what is available on or from the remote server. The current server name is indicated at the top of the dialog box.

■ In distributed analysis mode, you will not have access to files on your local computer unless you specify the drive as a shared device or specify the folders containing your data files as shared folders.

■ If the server is running a different operating system (for example, you are running Windows and the server is running UNIX), you probably won't have access to local data files in distributed analysis mode even if they are in shared folders.

To Open Data Files from a Remote Server

▶ If you aren't already connected to the remote server, log in to the remote server.

▶ Depending on the type of data file that you want to open, from the menus choose:
File
  Open
    Data...

or

File
  Open Database

or

File
  Read Text Data...

Saving Data Files from a Remote Server

Figure 4-4
Save Remote File dialog box

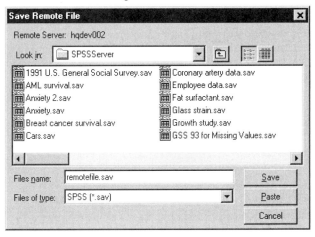

In distributed analysis mode, the Save Remote File dialog box replaces the standard Save File dialog box.

The contents of the list of available folders and drives depends on what is available on or from the remote server. The current server name is indicated at the top of the dialog box. You will not have access to folders on your local computer unless you specify the drive as a shared device and specify the folders as shared folders. If the server is running a different operating system (for example, you are running Windows and the server is running UNIX), you probably will not have access to local data files in distributed analysis mode even if they are in shared folders. If you want to save data files in a local folder, permissions for shared folders must include the ability to write to the folder.

To Save Data Files from a Remote Server

▶ Make the Data Editor the active window.

▶ From the menus choose:
File
   Save (or Save As...)

Data File Access in Local and Distributed Analysis Mode

The view of data files, folders (directories), and drives for both your local computer and the network is based on the computer that you are currently using to process commands and run procedures—which is not necessarily the computer in front of you.

Local analysis mode. When you use your local computer as your "server," the view of data files, folders, and drives in the file access dialog box (for opening data files) is similar to what you see in other applications or in Windows Explorer. You can see all of the data files and folders on your computer and any files and folders on mounted network drives.

Distributed analysis mode. When you use another computer as a "remote server" to run commands and procedures, the view of data files, folders, and drives represents the view from the remote server computer. Although you may see familiar folder names (such as *Program Files*) and drives (such as *C*), these items are *not* the folders and drives on your computer; they are the folders and drives on the remote server.

Figure 4-5
Local and remote views

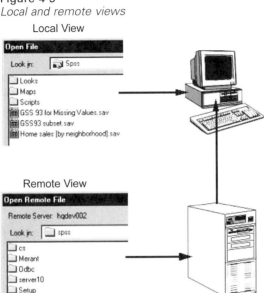

Local View

Remote View

In distributed analysis mode, you will not have access to data files on your local computer unless you specify the drive as a shared device or specify the folders containing your data files as shared folders. If the server is running a different operating system (for example, you are running Windows and the server is running UNIX), you probably won't have access to local data files in distributed analysis mode even if they are in shared folders.

Distributed analysis mode is not the same as accessing data files that reside on another computer on your network. You can access data files on other network devices in local analysis mode or in distributed analysis mode. In local mode, you access other devices from your local computer. In distributed mode, you access other network devices from the remote server.

If you're not sure if you're using local analysis mode or distributed analysis mode, look at the title bar in the dialog box for accessing data files. If the title of the dialog box contains the word *Remote* (as in Open Remote File), or if the text Remote Server: [server name] appears at the top of the dialog box, you're using distributed analysis mode.

Note: This situation affects only dialog boxes for accessing data files (for example, Open Data, Save Data, Open Database, and Apply Data Dictionary). For all other file types (for example, Viewer files, syntax files, and script files), the local view is always used.

To Set Sharing Permissions for a Drive or Folder

▶ In My Computer, click the folder (directory) or drive that you want to share.

▶ From the File menu, choose Properties.

▶ Click the Sharing tab, and then click Shared As.

For more information about sharing drives and folders, see the Help for your operating system.

Availability of Procedures in Distributed Analysis Mode

In distributed analysis mode, procedures are available for use only if they are installed on both your local version and the version on the remote server.

If you have optional components installed locally that are not available on the remote server and you switch from your local computer to a remote server, the affected procedures will be removed from the menus and the corresponding command syntax will result in errors. Switching back to local mode will restore all affected procedures.

Using UNC Path Specifications

With the Windows NT server version of SPSS, relative path specifications for data files are relative to the current server in distributed analysis mode, not relative to your local computer. A path specification (such as *c:\mydocs\mydata.sav*) does not point to a directory and file on your *C* drive; it points to a directory and file on the remote server's hard drive. If the directory and/or file do not exist on the remote server, an error results in command syntax, as in the following example:

```
GET FILE='c:\mydocs\mydata.sav'.
```

If you are using the Windows NT server version of SPSS, you can use universal naming convention (UNC) specifications when accessing data files with command syntax. The general form of a UNC specification is:

```
\\servername\sharename\path\filename
```

- *Servername* is the name of the computer that contains the data file.
- *Sharename* is the folder (directory) on that computer that is designated as a shared folder.
- *Path* is any additional folder (subdirectory) path below the shared folder.
- *Filename* is the name of the data file.

An example is as follows:

```
GET FILE='\\hqdev001\public\july\sales.sav'.
```

If the computer does not have a name assigned to it, you can use its IP address, as in:

```
GET FILE='\\204.125.125.53\public\july\sales.sav'.
```

Even with UNC path specifications, you can access data files only from devices and folders that are designated as shared. When you use distributed analysis mode, this situation includes data files on your local computer.

UNIX servers. On UNIX platforms, there is no equivalent to the UNC path, and all directory paths must be absolute paths that start at the root of the server; relative paths are not allowed. For example, if the data file is located in *bin/spss/data* and the current directory is also *bin/spss/data*, `GET FILE='sales.sav'` is not valid; you must specify the entire path, as in:

```
GET FILE='/bin/spss/data/sales.sav'.
```

Data Editor

The Data Editor provides a convenient, spreadsheet-like method for creating and editing data files. The Data Editor window opens automatically when you start a session.

The Data Editor provides two views of your data:

- **Data View.** This view displays the actual data values or defined value labels.
- **Variable View.** This view displays variable definition information, including defined variable and value labels, data type (for example, string, date, or numeric), measurement level (nominal, ordinal, or scale), and user-defined missing values.

In both views, you can add, change, and delete information that is contained in the data file.

Data View

Figure 5-1
Data View

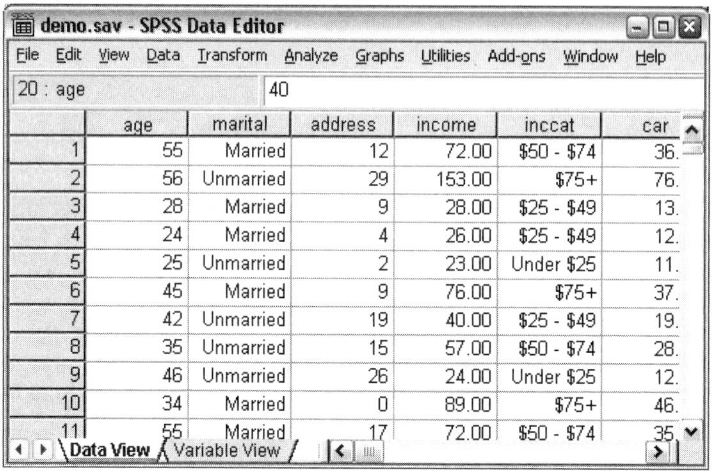

Many of the features of Data View are similar to the features that are found in spreadsheet applications. There are, however, several important distinctions:

- Rows are cases. Each row represents a case or an observation. For example, each individual respondent to a questionnaire is a case.

- Columns are variables. Each column represents a variable or characteristic that is being measured. For example, each item on a questionnaire is a variable.

- Cells contain values. Each cell contains a single value of a variable for a case. The cell is where the case and the variable intersect. Cells contain only data values. Unlike spreadsheet programs, cells in the Data Editor cannot contain formulas.

- The data file is rectangular. The dimensions of the data file are determined by the number of cases and variables. You can enter data in any cell. If you enter data in a cell outside the boundaries of the defined data file, the data rectangle is extended to include any rows and/or columns between that cell and the file boundaries. There are no "empty" cells within the boundaries of the data file. For numeric variables, blank cells are converted to the system-missing value. For string variables, a blank is considered a valid value.

Variable View

Figure 5-2
Variable View

	Name	Type	Width	Decimals	Label	Value
1	age	Numeric	4	0	Age in years	None
2	marital	Numeric	4	0	Marital status	{0, Unmar
3	address	Numeric	4	0	Years at curre	None
4	income	Numeric	8	2	Household inc	None
5	inccat	Numeric	8	2	Income catego	{1.00, Un(
6	car	Numeric	8	2	Price of primar	None
7	carcat	Numeric	8	2	Primary vehicl	{1.00, Ecc
8	ed	Numeric	4	0	Level of educat	{1, Did no
9	employ	Numeric	4	0	Years with curr	None
10	retire	Numeric	4	0	Retired	{0, No}...
11	empcat	Numeric	4	0	Years with curr	{1, Less t
12	jobsat	Numeric	4	0	Job satisfactio	{1, Highly
13	gender	String	2	0	Gender	{f, Female
14	reside	Numeric	4	0	Number of peo	None
15	wireless	Numeric	4	0	Wireless servi	{0, No}...
16	multline	Numeric	4	0	Multiple lines	{0, No}...

Variable View contains descriptions of the attributes of each variable in the data file. In Variable View:

- Rows are variables.
- Columns are variable attributes.

You can add or delete variables and modify attributes of variables, including the following attributes:

- Variable name
- Data type
- Number of digits or characters
- Number of decimal places
- Descriptive variable and value labels
- User-defined missing values

■ Column width

■ Measurement level

All of these attributes are saved when you save the data file.

In addition to defining variable properties in Variable View, there are two other methods for defining variable properties:

■ The Copy Data Properties Wizard provides the ability to use an external SPSS data file or another dataset that is available in the current session as a template for defining file and variable properties in the active dataset. You can also use variables in the active dataset as templates for other variables in the active dataset. Copy Data Properties is available on the Data menu in the Data Editor window.

■ Define Variable Properties (also available on the Data menu in the Data Editor window) scans your data and lists all unique data values for any selected variables, identifies unlabeled values, and provides an auto-label feature. This method is particularly useful for categorical variables that use numeric codes to represent categories—for example, 0 = *Male*, 1 = *Female*.

To Display or Define Variable Attributes

▶ Make the Data Editor the active window.

▶ Double-click a variable name at the top of the column in Data View, or click the Variable View tab.

▶ To define new variables, enter a variable name in any blank row.

▶ Select the attribute(s) that you want to define or modify.

Variable Names

The following rules apply to variable names:

■ Each variable name must be unique; duplication is not allowed.

■ Variable names can be up to 64 bytes long, and the first character must be a letter or one of the characters @, #, or $. Subsequent characters can be any combination of letters, numbers, a period (.), and nonpunctuation characters. Sixty-four bytes typically means 64 characters in single-byte languages (e.g., English, French,

German, Spanish, Italian, Hebrew, Russian, Greek, Arabic, and Thai) and 32 characters in double-byte languages (e.g., Japanese, Chinese, and Korean).

(*Note*: Letters include any nonpunctuation characters used in writing ordinary words in the languages supported in the character set of the platform on which SPSS is running.)

■ Variable names cannot contain spaces.

■ A # character in the first position of a variable name defines a scratch variable. You can only create scratch variables with command syntax. You cannot specify a # as the first character of a variable in dialog boxes that create new variables.

■ A $ sign in the first position indicates that the variable is a system variable. The $ sign is not allowed as the initial character of a user-defined variable.

■ The period, underscore, and the characters $, #, and @ can be used within variable names. For example, *A._$@#1* is a valid variable name.

■ Variable names ending with a period should be avoided, since the period may be interpreted as a command terminator. You can only create variables that end with a period in command syntax. You cannot create variables that end with a period in dialog boxes that create new variables.

■ Variable names ending in underscores should be avoided, since such names may conflict with names of variables automatically created by commands and procedures.

■ Reserved keywords cannot be used as variable names. Reserved keywords are: ALL, AND, BY, EQ, GE, GT, LE, LT, NE, NOT, OR, TO, and WITH.

■ Variable names can be defined with any mixture of uppercase and lowercase characters, and case is preserved for display purposes.

■ When long variable names need to wrap onto multiple lines in output, SPSS attempts to break the lines at underscores, periods, and where content changes from lower case to upper case.

Variable Measurement Level

You can specify the level of measurement as scale (numeric data on an interval or ratio scale), ordinal, or nominal. Nominal and ordinal data can be either string (alphanumeric) or numeric.

■ **Nominal.** A variable can be treated as nominal when its values represent categories with no intrinsic ranking (for example, the department of the company in which an employee works). Examples of nominal variables include region, zip code, and religious affiliation.

■ **Ordinal.** A variable can be treated as ordinal when its values represent categories with some intrinsic ranking (for example, levels of service satisfaction from highly dissatisfied to highly satisfied). Examples of ordinal variables include attitude scores representing degree of satisfaction or confidence and preference rating scores.

■ **Scale.** A variable can be treated as scale when its values represent ordered categories with a meaningful metric, so that distance comparisons between values are appropriate. Examples of scale variables include age in years and income in thousands of dollars.

Note: For ordinal string variables, the alphabetic order of string values is assumed to reflect the true order of the categories. For example, for a string variable with the values of *low, medium, high*, the order of the categories is interpreted as *high, low, medium*—which is not the correct order. In general, it is more reliable to use numeric codes to represent ordinal data.

For new variables created during a session, data read from external file formats, and SPSS data files that were created prior to version 8.0, default assignment of measurement level is based on the following rules:

■ Numeric variables with fewer than 24 unique values and string variables are set to nominal.

■ Numeric variables with 24 or more unique values are set to scale.

You can change the scale/nominal cutoff value for numeric variables in the Options dialog box. For more information, see "Interactive Chart Options" in Chapter 45 on p. 649.

The Define Variable Properties dialog box, available from the Data menu, can help you assign the correct measurement level. For more information, see "Assigning the Measurement Level" in Chapter 7 on p. 140.

Variable Type

Variable Type specifies the data type for each variable. By default, all new variables are assumed to be numeric. You can use Variable Type to change the data type. The contents of the Variable Type dialog box depend on the selected data type. For some data types, there are text boxes for width and number of decimals; for other data types, you can simply select a format from a scrollable list of examples.

Figure 5-3
Variable Type dialog box

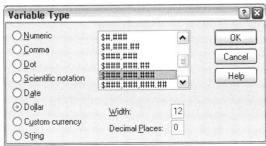

The available data types are as follows:

Numeric. A variable whose values are numbers. Values are displayed in standard numeric format. The Data Editor accepts numeric values in standard format or in scientific notation.

Comma. A numeric variable whose values are displayed with commas delimiting every three places and displayed with the period as a decimal delimiter. The Data Editor accepts numeric values for comma variables with or without commas or in scientific notation. Values cannot contain commas to the right of the decimal indicator.

Dot. A numeric variable whose values are displayed with periods delimiting every three places and with the comma as a decimal delimiter. The Data Editor accepts numeric values for dot variables with or without periods or in scientific notation. Values cannot contain periods to the right of the decimal indicator.

Scientific notation. A numeric variable whose values are displayed with an embedded E and a signed power-of-10 exponent. The Data Editor accepts numeric values for such variables with or without an exponent. The exponent can be preceded by E or D with an optional sign or by the sign alone—for example, 123, 1.23E2, 1.23D2, 1.23E+2, and 1.23+2.

Date. A numeric variable whose values are displayed in one of several calendar-date or clock-time formats. Select a format from the list. You can enter dates with slashes, hyphens, periods, commas, or blank spaces as delimiters. The century range for two-digit year values is determined by your Options settings (from the Edit menu, choose Options, and then click the Data tab).

Dollar. A numeric variable displayed with a leading dollar sign ($), commas delimiting every three places, and a period as the decimal delimiter. You can enter data values with or without the leading dollar sign.

Custom currency. A numeric variable whose values are displayed in one of the custom currency formats that you have defined in the Currency tab of the Options dialog box. Defined custom currency characters cannot be used in data entry but are displayed in the Data Editor.

String. A variable whose values are not numeric and therefore are not used in calculations. The values can contain any characters up to the defined length. Uppercase and lowercase letters are considered distinct. This type is also known as an alphanumeric variable.

To Define Variable Type

▶ Click the button in the *Type* cell for the variable that you want to define.

▶ Select the data type in the Variable Type dialog box.

▶ Click OK.

Input versus Display Formats

Depending on the format, the display of values in Data View may differ from the actual value as entered and stored internally. Following are some general guidelines:

■ For numeric, comma, and dot formats, you can enter values with any number of decimal positions (up to 16), and the entire value is stored internally. The Data View displays only the defined number of decimal places and rounds values with more decimals. However, the complete value is used in all computations.

■ For string variables, all values are right-padded to the maximum width. For a string variable with a maximum width of three, a value of *No* is stored internally as 'No ' and is not equivalent to ' No'.

■ For date formats, you can use slashes, dashes, spaces, commas, or periods as delimiters between day, month, and year values, and you can enter numbers, three-letter abbreviations, or complete names for month values. Dates of the general format *dd-mmm-yy* are displayed with dashes as delimiters and three-letter abbreviations for the month. Dates of the general format *dd/mm/yy* and *mm/dd/yy* are displayed with slashes for delimiters and numbers for the month. Internally, dates are stored as the number of seconds from October 14, 1582. The century range for dates with two-digit years is determined by your Options settings (from the Edit menu, choose Options, and then click the Data tab).

■ For time formats, you can use colons, periods, or spaces as delimiters between hours, minutes, and seconds. Times are displayed with colons as delimiters. Internally, times are stored as a number of seconds that represents a time interval. For example, 10:00:00 is stored internally as 36000, which is 60 (seconds per minute) x 60 (minutes per hour) x 10 (hours).

Variable Labels

You can assign descriptive variable labels up to 256 characters (128 characters in double-byte languages). Variable labels can contain spaces and reserved characters that are not allowed in variable names.

To Specify Variable Labels

▶ Make the Data Editor the active window.

▶ Double-click a variable name at the top of the column in Data View, or click the Variable View tab.

▶ In the *Label* cell for the variable, enter the descriptive variable label.

Value Labels

You can assign descriptive value labels for each value of a variable. This process is particularly useful if your data file uses numeric codes to represent non-numeric categories (for example, codes of 1 and 2 for *male* and *female*).

- Value labels can be up to 120 bytes.
- Value labels are not available for long string variables (string variables longer than eight characters).

Figure 5-4
Value Labels dialog box

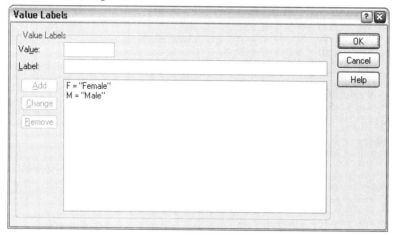

To Specify Value Labels

▶ Click the button in the *Values* cell for the variable that you want to define.

▶ For each value, enter the value and a label.

▶ Click Add to enter the value label.

▶ Click OK.

Inserting Line Breaks in Labels

Variable labels and value labels automatically wrap to multiple lines in pivot tables and charts if the cell or area isn't wide enough to display the entire label on one line, and you can edit results to insert manual line breaks if you want the label to wrap at a different point. You can also create variable labels and value labels that will *always* wrap at specified points and be displayed on multiple lines.

▶ For variable labels, select the Label cell for the variable in Variable View in the Data Editor.

▶ For value labels, select the Values cell for the variable in Variable View in the Data Editor, click the button in the cell, and in the Value Labels dialog box, select the label that you want to modify.

▶ At the place in the label where you want the label to wrap, type \n.

The \n is not displayed in pivot tables or charts; it is interpreted as a line break character.

Missing Values

Missing Values defines specified data values as **user-missing**. For example, you might want to distinguish between data that are missing because a respondent refused to answer and data that are missing because the question didn't apply to that respondent. Data values that are specified as user-missing are flagged for special treatment and are excluded from most calculations.

Figure 5-5
Missing Values dialog box

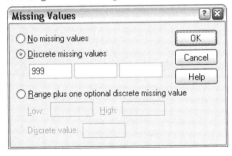

■ You can enter up to three discrete (individual) missing values, a range of missing values, or a range plus one discrete value.

■ Ranges can be specified only for numeric variables.

■ You cannot define missing values for long string variables (string variables that are longer than eight characters).

Missing values for string variables. All string values, including null or blank values, are considered to be valid unless you explicitly define them as missing. To define null or blank values as missing for a string variable, enter a single space in one of the fields under the Discrete missing values selection.

To Define Missing Values

▶ Click the button in the *Missing* cell for the variable that you want to define.

▶ Enter the values or range of values that represent missing data.

Column Width

You can specify a number of characters for the column width. Column widths can also be changed in Data View by clicking and dragging the column borders.

Column formats affect only the display of values in the Data Editor. Changing the column width does not change the defined width of a variable. If the defined and actual width of a value are wider than the column, asterisks (*) are displayed in Data View.

Variable Alignment

Alignment controls the display of data values and/or value labels in Data View. The default alignment is right for numeric variables and left for string variables. This setting affects only the display in Data View.

Applying Variable Definition Attributes to Multiple Variables

After you have defined variable definition attributes for a variable, you can copy one or more attributes and apply them to one or more variables.

Basic copy and paste operations are used to apply variable definition attributes. You can:

■ Copy a single attribute (for example, value labels) and paste it to the same attribute cell(s) for one or more variables.

■ Copy all attributes from one variable and paste them to one or more other variables.

■ Create multiple new variables with all the attributes of a copied variable.

Applying Variable Definition Attributes to Other Variables

To Apply Individual Attributes from a Defined Variable

▶ In Variable View, select the attribute cell that you want to apply to other variables.

▶ From the menus choose:
  Edit
    Copy

▶ Select the attribute cell(s) to which you want to apply the attribute. (You can select multiple target variables.)

▶ From the menus choose:
  Edit
    Paste

If you paste the attribute to blank rows, new variables are created with default attributes for all attributes except the selected attribute.

To Apply All Attributes from a Defined Variable

▶ In Variable View, select the row number for the variable with the attributes that you want to use. (The entire row is highlighted.)

▶ From the menus choose:
  Edit
    Copy

▶ Select the row number(s) for the variable(s) to which you want to apply the attributes. (You can select multiple target variables.)

▶ From the menus choose:
  Edit
    Paste

Generating Multiple New Variables with the Same Attributes

▶ In Variable View, click the row number for the variable that has the attributes that you want to use for the new variable. (The entire row is highlighted.)

▶ From the menus choose:
  Edit
    Copy

▶ Click the empty row number beneath the last defined variable in the data file.

▶ From the menus choose:
  Edit
    Paste Variables...

▶ In the Paste Variables dialog box, enter the number of variables that you want to create.

▶ Enter a prefix and starting number for the new variables.

▶ Click OK.

The new variable names will consist of the specified prefix plus a sequential number starting with the specified number.

Custom Variable Attributes

In addition to the standard variable attributes (for example, value labels, missing values, measurement level), you can create your own custom variable attributes. Like standard variable attributes, these custom attributes are saved with SPSS-format data files.

Creating Custom Variable Attributes

To create new custom attributes:

▶ In Variable View of the Data Editor, from the menus choose:
  Data
    New Custom Attribute...

▶ Drag and drop the variables to which you want to assign the new attribute to the Selected Variables list.

▶ Enter a name for the attribute. Attribute names must follow the same rules as variable names. For more information, see "Variable Names" on p. 106.

▶ Enter an optional value for the attribute. If you select multiple variables, the value is assigned to all selected variables. You can leave this blank and then enter values for each variable in Variable View.

Figure 5-6
New Custom Attribute dialog box

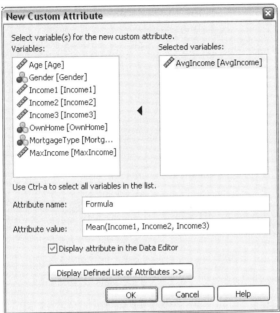

Display attribute in the Data Editor. Displays the attribute in Variable View of the Data Editor. For information on controlling the display of custom attributes, see "Displaying and Editing Custom Variable Attributes" below.

Display Defined List of Attributes. Displays a list of custom attributes already defined for the dataset. Attribute names that begin with a dollar sign ($) are reserved attributes that cannot be modified.

Displaying and Editing Custom Variable Attributes

Custom variable attributes can be displayed and edited in the Data Editor in Variable View.

Figure 5-7

Custom variable attributes displayed in Variable View

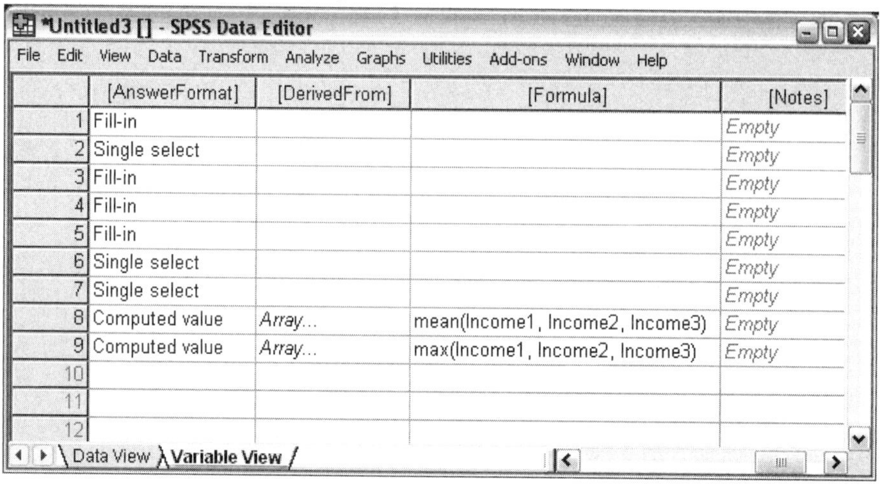

		[AnswerFormat]	[DerivedFrom]	[Formula]	[Notes]
	1	Fill-in			Empty
	2	Single select			Empty
	3	Fill-in			Empty
	4	Fill-in			Empty
	5	Fill-in			Empty
	6	Single select			Empty
	7	Single select			Empty
	8	Computed value	Array...	mean(Income1, Income2, Income3)	Empty
	9	Computed value	Array...	max(Income1, Income2, Income3)	Empty
	10				
	11				
	12				

- Custom variable attribute names are enclosed in square brackets.

- Attribute names that begin with a dollar sign are reserved and cannot be modified.

- A blank cell indicates that the attribute does not exist for that variable; the text *Empty* displayed in a cell indicates that the attribute exists for that variable but no value has been assigned to the attribute for that variable. Once you enter text in the cell, the attribute exists for that variable with the value you enter.

- The text *Array...* displayed in a cell indicates that this is an attribute array: an attribute that contains multiple values. Click the button in the cell to display the list of values.

To Display and Edit Custom Variable Attributes

▶ From the menus in Variable View of the Data Editor, choose:
View
  Display Custom Attributes...

▶ Select (check) the custom variable attributes you want to display.

Figure 5-8
Display Custom Attributes dialog box

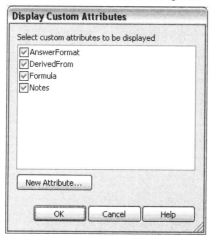

New Attribute. This opens the New Custom Attribute dialog box, which you can use to create additional custom variable attributes.

Variable Attribute Arrays

The text *Array...*—displayed in a cell for a custom variable attribute in Variable View or in the Custom Variable Properties dialog box in Define Variable Properties—indicates that this is an **attribute array**, an attribute that contains multiple values. For example, you could have an attribute array that identifies all of the source variables used to compute a derived variable. Click the button in the cell to display and edit the list of values.

Figure 5-9
Custom Attribute Array dialog box

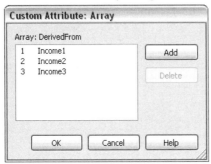

Entering Data

In Data View, you can enter data directly in the Data Editor. You can enter data in any order. You can enter data by case or by variable, for selected areas or for individual cells.

■   The active cell is highlighted.

■   The variable name and row number of the active cell are displayed in the top left corner of the Data Editor.

■   When you select a cell and enter a data value, the value is displayed in the cell editor at the top of the Data Editor.

■   Data values are not recorded until you press Enter or select another cell.

■   To enter anything other than simple numeric data, you must define the variable type first.

If you enter a value in an empty column, the Data Editor automatically creates a new variable and assigns a variable name.

Figure 5-10
Active dataset in Data View

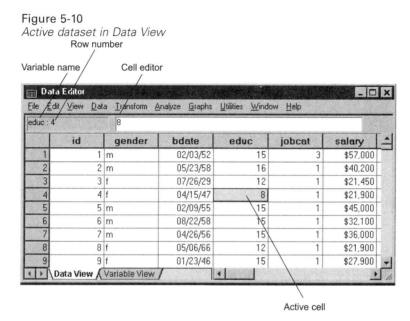

Active cell

To Enter Numeric Data

▶ Select a cell in Data View.

▶ Enter the data value. (The value is displayed in the cell editor at the top of the Data Editor.)

▶ To record the value, press Enter or select another cell.

To Enter Non-Numeric Data

▶ Double-click a variable name at the top of the column in Data View or click the Variable View tab.

▶ Click the button in the *Type* cell for the variable.

▶ Select the data type in the Variable Type dialog box.

▶ Click OK.

▶ Double-click the row number or click the Data View tab.

▶ Enter the data in the column for the newly defined variable.

To Use Value Labels for Data Entry

▶ If value labels aren't currently displayed in Data View, from the menus choose:
View
  Value Labels

▶ Click the cell in which you want to enter the value.

▶ Choose a value label from the drop-down list.

The value is entered, and the value label is displayed in the cell.

Note: This process works only if you have defined value labels for the variable.

Data Value Restrictions in the Data Editor

The defined variable type and width determine the type of value that can be entered in the cell in Data View.

■ If you type a character that is not allowed by the defined variable type, the Data Editor beeps and does not enter the character.

■ For string variables, characters beyond the defined width are not allowed.

■ For numeric variables, integer values that exceed the defined width can be entered, but the Data Editor displays either scientific notation or asterisks in the cell to indicate that the value is wider than the defined width. To display the value in the cell, change the defined width of the variable. (*Note*: Changing the column width does not affect the variable width.)

Editing Data

With the Data Editor, you can modify data values in Data View in many ways. You can:

■ Change data values.

■ Cut, copy, and paste data values.

■ Add and delete cases.

■ Add and delete variables.

■ Change the order of variables.

Replacing or Modifying Data Values

To Delete the Old Value and Enter a New Value

▶ In Data View, double-click the cell. (The cell value is displayed in the cell editor.)

▶ Edit the value directly in the cell or in the cell editor.

▶ Press Enter (or move to another cell) to record the new value.

Cutting, Copying, and Pasting Data Values

You can cut, copy, and paste individual cell values or groups of values in the Data Editor. You can:

■ Move or copy a single cell value to another cell.

■ Move or copy a single cell value to a group of cells.

■ Move or copy the values for a single case (row) to multiple cases.

■ Move or copy the values for a single variable (column) to multiple variables.

■ Move or copy a group of cell values to another group of cells.

Data Conversion for Pasted Values in the Data Editor

If the defined variable types of the source and target cells are not the same, the Data Editor attempts to convert the value. If no conversion is possible, the system-missing value is inserted in the target cell.

Converting numeric or date into string. Numeric (for example, numeric, dollar, dot, or comma) and date formats are converted to strings if they are pasted into a string variable cell. The string value is the numeric value as displayed in the cell. For example, for a dollar format variable, the displayed dollar sign becomes part of the string value. Values that exceed the defined string variable width are truncated.

Converting string into numeric or date. String values that contain acceptable characters for the numeric or date format of the target cell are converted to the equivalent numeric or date value. For example, a string value of 25/12/91 is converted to a valid date if the format type of the target cell is one of the day-month-year formats, but the value is converted to system-missing if the format type of the target cell is one of the month-day-year formats.

Converting date into numeric. Date and time values are converted to a number of seconds if the target cell is one of the numeric formats (for example, numeric, dollar, dot, or comma). Because dates are stored internally as the number of seconds since October 14, 1582, converting dates to numeric values can yield some extremely large numbers. For example, the date 10/29/91 is converted to a numeric value of 12,908,073,600.

Converting numeric into date or time. Numeric values are converted to dates or times if the value represents a number of seconds that can produce a valid date or time. For dates, numeric values that are less than 86,400 are converted to the system-missing value.

Inserting New Cases

Entering data in a cell in a blank row automatically creates a new case. The Data Editor inserts the system-missing value for all other variables for that case. If there are any blank rows between the new case and the existing cases, the blank rows become new cases with the system-missing value for all variables. You can also insert new cases between existing cases.

To Insert New Cases between Existing Cases

▶ In Data View, select any cell in the case (row) below the position where you want to insert the new case.

▶ From the menus choose:
Edit
  Insert Cases

A new row is inserted for the case, and all variables receive the system-missing value.

Inserting New Variables

Entering data in an empty column in Data View or in an empty row in Variable View automatically creates a new variable with a default variable name (the prefix *var* and a sequential number) and a default data format type (numeric). The Data Editor inserts the system-missing value for all cases for the new variable. If there are any empty columns in Data View or empty rows in Variable View between the new variable and the existing variables, these rows or columns also become new variables with the system-missing value for all cases. You can also insert new variables between existing variables.

To Insert New Variables between Existing Variables

▶ Select any cell in the variable to the right of (Data View) or below (Variable View) the position where you want to insert the new variable.

▶ From the menus choose:
Edit
  Insert Variable

A new variable is inserted with the system-missing value for all cases.

To Move Variables

▶ To select the variable, click the variable name in Data View or the row number for the variable in Variable View.

▶ Drag and drop the variable to the new location.

▶ If you want to place the variable between two existing variables: In Data View, drop the variable on the variable column to the right of where you want to place the variable, or in Variable View, drop the variable on the variable row below where you want to place the variable.

To Change Data Type

You can change the data type for a variable at any time by using the Variable Type dialog box in Variable View. The Data Editor will attempt to convert existing values to the new type. If no conversion is possible, the system-missing value is assigned. The conversion rules are the same as the rules for pasting data values to a variable with a different format type. If the change in data format may result in the loss of missing-value specifications or value labels, the Data Editor displays an alert box and asks whether you want to proceed with the change or cancel it.

Go to Case

The Go to Case dialog box allows you to go to the specified case (row) number in the Data Editor.

Figure 5-11
Go to Case dialog box

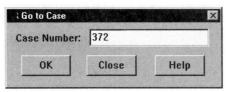

To Find a Case in the Data Editor

▶ Make the Data Editor the active window.

▶ From the menus choose:
  Data
   Go to Case...

▶ In the Go to Case dialog box, enter the Data Editor row number for the case.

Case Selection Status in the Data Editor

If you have selected a subset of cases but have not discarded unselected cases, unselected cases are marked in the Data Editor with a diagonal line (slash) through the row number.

Figure 5-12
Filtered cases in the Data Editor

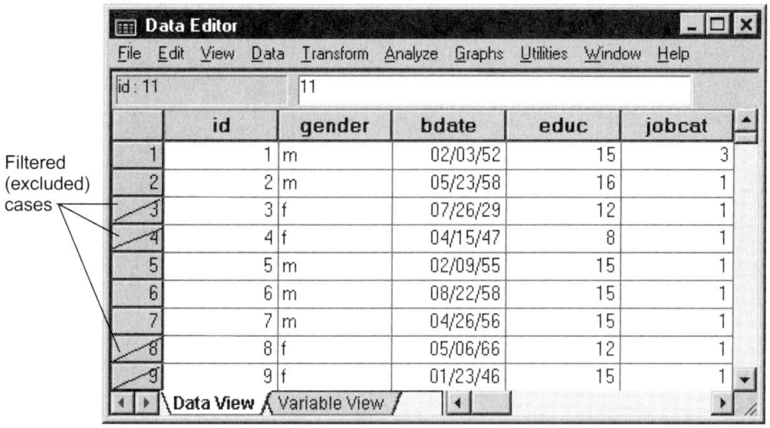

Filtered
(excluded)
cases

Data Editor Display Options

The View menu provides several display options for the Data Editor:

Fonts. This option controls the font characteristics of the data display.

Grid Lines. This option toggles the display of grid lines.

Value Labels. This option toggles between the display of actual data values and user-defined descriptive value labels. This option is available only in Data View.

Using Multiple Views

In Data View, you can create multiple views (panes) by using the splitters that are located above the horizontal scroll bar and to the right of the vertical scroll bar.

Figure 5-13
Data View pane splitters

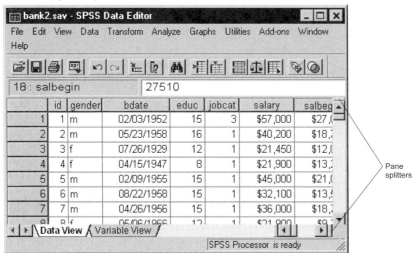

Pane
splitters

You can also use the Window menu to insert and remove pane splitters. To insert
splitters:

► In Data View, from the menus choose:
Window
  Split

Splitters are inserted above and to the left of the selected cell.

■ If the top left cell is selected, splitters are inserted to divide the current view
approximately in half, both horizontally and vertically.

■ If any cell other than the top cell in the first column is selected, a horizontal pane
splitter is inserted above the selected cell.

■ If any cell other than the first cell in the top row is selected, a vertical pane splitter
is inserted to the left of the selected cell.

Data Editor Printing

A data file is printed as it appears on the screen.

■ The information in the currently displayed view is printed. In Data View, the data
are printed. In Variable View, data definition information is printed.

- Grid lines are printed if they are currently displayed in the selected view.

- Value labels are printed in Data View if they are currently displayed. Otherwise, the actual data values are printed.

Use the View menu in the Data Editor window to display or hide grid lines and toggle between the display of data values and value labels.

To Print Data Editor Contents

▶ Make the Data Editor the active window.

▶ Click the tab for the view that you want to print.

▶ From the menus choose:
  File
    Print...

Working with Multiple Data Sources

Starting with SPSS 14.0, SPSS can have multiple data sources open at the same time, making it easier to:

- Switch back and forth between data sources.

- Compare the contents of different data sources.

- Copy and paste data between data sources.

- Create multiple subsets of cases and/or variables for analysis.

- Merge multiple data sources from various data formats (for example, spreadsheet, database, text data) without saving each data source in SPSS format first.

Basic Handling of Multiple Data Sources

Figure 6-1

Two data sources open at same time

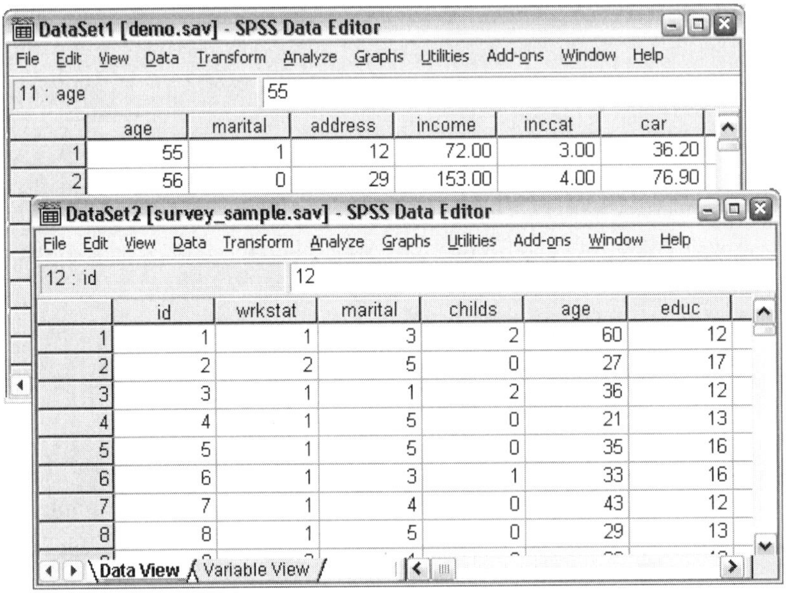

Each data source that you open is displayed in a new Data Editor window.

■ Any previously open data sources remain open and available for further use.

■ When you first open a data source, it automatically becomes the **active dataset**.

■ You can change the active dataset simply by clicking anywhere in the Data Editor window of the data source that you want to use or by selecting the Data Editor window for that data source from the Window menu.

■ Only the variables in the active dataset are available for analysis.

Figure 6-2

Variable list containing variables in the active dataset

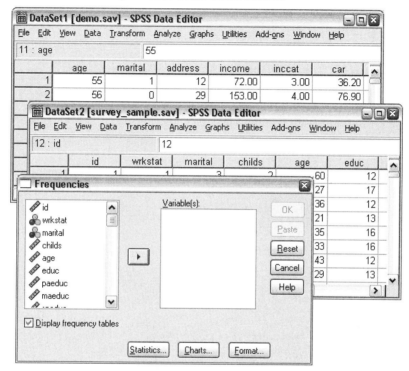

- You cannot change the active dataset when any dialog box that accesses the data is open (including all dialog boxes that display variable lists).

- At least one Data Editor window must be open during a session. When you close the last open Data Editor window, SPSS automatically shuts down, prompting you to save changes first.

Note: If you use command syntax to open data sources (for example, GET FILE, GET DATA), you need to name each dataset explicitly in order to have more than one data source open at the same time.

Copying and Pasting Information between Datasets

You can copy both data and variable definition attributes from one dataset to another dataset in basically the same way that you copy and paste information within a single data file.

- Copying and pasting selected data cells in Data View pastes only the data values, with no variable definition attributes.

- Copying and pasting an entire variable in Data View by selecting the variable name at the top of the column pastes all of the data and all of the variable definition attributes for that variable.

- Copying and pasting variable definition attributes or entire variables in Variable View pastes the selected attributes (or the entire variable definition) but does not paste any data values.

Renaming Datasets

When you open a data source through the menus and dialog boxes, each data source is automatically assigned a dataset name of *DataSetn*, where *n* is a sequential integer value, and when you open a data source using command syntax, no dataset name is assigned unless you explicitly specify one with DATASET NAME. To provide more descriptive dataset names:

▶ From the menus in the Data Editor window for the dataset whose name you want to change choose:

File
  Rename Dataset...

▶ Enter a new dataset name that conforms to SPSS variable naming rules. For more information, see "Variable Names" in Chapter 5 on p. 106.

Data Preparation

Once you've opened a data file or entered data in the Data Editor, you can start creating reports, charts, and analyses without any additional preliminary work. However, there are some additional data preparation features that you may find useful, including the ability to:

■ Assign variable properties that describe the data and determine how certain values should be treated.

■ Identify cases that may contain duplicate information and exclude those cases from analyses or delete them from the data file.

■ Create new variables with a few distinct categories that represent ranges of values from variables with a large number of possible values.

Variable Properties

Data entered in the Data Editor in Data View or read into SPSS from an external file format (such as an Excel spreadsheet or a text data file) lack certain variable properties that you may find very useful, including:

■ Definition of descriptive value labels for numeric codes (for example, 0 = *Male* and 1 = *Female*).

■ Identification of missing values codes (for example, 99 = *Not applicable*).

■ Assignment of measurement level (nominal, ordinal, or scale).

All of these variable properties (and others) can be assigned in Variable View in the Data Editor. There are also several utilities that can assist you in this process:

■ **Define Variable Properties** can help you define descriptive value labels and missing values. This is particularly useful for categorical data with numeric codes used for category values.

■ **Copy Data Properties** provides the ability to use an existing SPSS-format data file as a template for file and variable properties in the current data file. This is particularly useful if you frequently use external-format data files that contain similar content (such as monthly reports in Excel format).

Defining Variable Properties

Define Variable Properties is designed to assist you in the process of creating descriptive value labels for categorical (nominal, ordinal) variables. Define Variable Properties:

■ Scans the actual data values and lists all unique data values for each selected variable.

■ Identifies unlabeled values and provides an "auto-label" feature.

■ Provides the ability to copy defined value labels from another variable to the selected variable or from the selected variable to multiple additional variables.

Note: To use Define Variable Properties without first scanning cases, enter 0 for the number of cases to scan.

To Define Variable Properties

▶ From the menus choose:
Data
   Define Variable Properties...

Figure 7-1
Initial dialog box for selecting variables to define

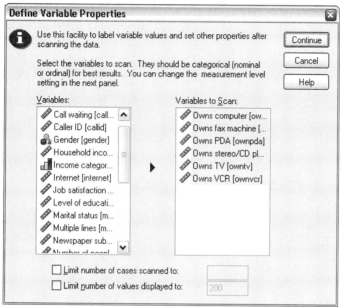

▶ Select the numeric or short string variables for which you want to create value labels or define or change other variable properties, such as missing values or descriptive variable labels.

Note: Long string variables (string variables with a defined width of more than eight characters) are not displayed in the variable list. Long string variables cannot have defined value labels or missing values categories.

▶ Specify the number of cases to scan to generate the list of unique values. This is particularly useful for data files with a large number of cases for which a scan of the complete data file might take a significant amount of time.

▶ Specify an upper limit for the number of unique values to display. This is primarily useful to prevent listing hundreds, thousands, or even millions of values for scale (continuous interval, ratio) variables.

▶ Click Continue to open the main Define Variable Properties dialog box.

▶ Select a variable for which you want to create value labels or define or change other variable properties.

▶ Enter the label text for any unlabeled values that are displayed in the Value Label grid.

▶ If there are values for which you want to create value labels but those values are not displayed, you can enter values in the *Value* column below the last scanned value.

▶ Repeat this process for each listed variable for which you want to create value labels.

▶ Click OK to apply the value labels and other variable properties.

Defining Value Labels and Other Variable Properties

Figure 7-2
Define Variable Properties, main dialog box

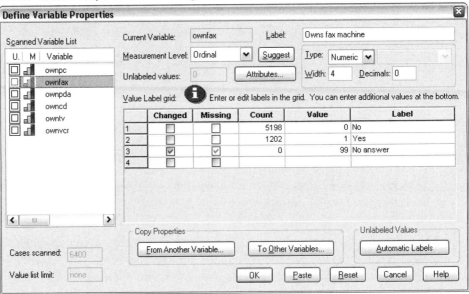

The Define Variable Properties main dialog box provides the following information for the scanned variables:

Scanned Variable List. For each scanned variable, a check mark in the *Unlabeled* (*U.*) column indicates that the variable contains values without assigned value labels.

To sort the variable list to display all variables with unlabeled values at the top of the list:

▶ Click the *Unlabeled* column heading under Scanned Variable List.

You can also sort by variable name or measurement level by clicking the corresponding column heading under Scanned Variable List.

Value Label Grid

- **Label.** Displays any value labels that have already been defined. You can add or change labels in this column.

- **Value.** Unique values for each selected variable. This list of unique values is based on the number of scanned cases. For example, if you scanned only the first 100 cases in the data file, then the list reflects only the unique values present in those cases. If the data file has already been sorted by the variable for which you want to assign value labels, the list may display far fewer unique values than are actually present in the data.

- **Count.** The number of times each value occurs in the scanned cases.

- **Missing.** Values defined as representing missing data. You can change the missing values designation of the category by clicking the check box. A check indicates that the category is defined as a user-missing category. If a variable already has a range of values defined as user-missing (for example, 90-99), you cannot add or delete missing values categories for that variable with Define Variable Properties. You can use Variable View in the Data Editor to modify the missing values categories for variables with missing values ranges. For more information, see "Missing Values" in Chapter 5 on p. 113.

- **Changed.** Indicates that you have added or changed a value label.

Note: If you specified 0 for the number of cases to scan in the initial dialog box, the Value Label grid will initially be blank, except for any preexisting value labels and/or defined missing values categories for the selected variable. In addition, the Suggest button for the measurement level will be disabled.

Measurement Level. Value labels are primarily useful for categorical (nominal and ordinal) variables, and some procedures treat categorical and scale variables differently; so it is sometimes important to assign the correct measurement level. However, by default, all new numeric variables are assigned the scale measurement level. Thus, many variables that are in fact categorical may initially be displayed as scale.

If you are unsure of what measurement level to assign to a variable, click Suggest.

Copy Properties. You can copy value labels and other variable properties from another variable to the currently selected variable or from the currently selected variable to one or more other variables.

Unlabeled Values. To create labels for unlabeled values automatically, click Automatic Labels.

Variable Label and Display Format

You can change the descriptive variable label and the display format.

- You cannot change the variable's fundamental type (string or numeric).

- For string variables, you can change only the variable label, not the display format.

- For numeric variables, you can change the numeric type (such as numeric, date, dollar, or custom currency), width (maximum number of digits, including any decimal and/or grouping indicators), and number of decimal positions.

- For numeric date format, you can select a specific date format (such as dd-mm-yyyy, mm/dd/yy, and yyyyddd)

- For numeric custom format, you can select one of five custom currency formats (CCA through CCE). For more information, see "Currency Options" in Chapter 45 on p. 654.

- An asterisk is displayed in the *Value* column if the specified width is less than the width of the scanned values or the displayed values for preexisting defined value labels or missing values categories.

- A period (.) is displayed if the scanned values or the displayed values for preexisting defined value labels or missing values categories are invalid for the selected display format type. For example, an internal numeric value of less than 86,400 is invalid for a date format variable.

Assigning the Measurement Level

When you click Suggest for the measurement level in the Define Variable Properties main dialog box, the current variable is evaluated based on the scanned cases and defined value labels, and a measurement level is suggested in the Suggest Measurement Level dialog box that opens. The Explanation area provides a brief description of the criteria used to provide the suggested measurement level.

Figure 7-3
Suggest Measurement Level dialog box

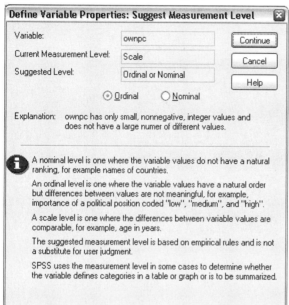

Note: Values defined as representing missing values are not included in the evaluation for measurement level. For example, the explanation for the suggested measurement level may indicate that the suggestion is in part based on the fact that the variable contains no negative values, whereas it may in fact contain negative values—but those values are already defined as missing values.

▶ Click Continue to accept the suggested level of measurement or Cancel to leave the measurement level unchanged.

Custom Variable Attributes

The Attributes button in Define Variable Properties opens the Custom Variable Attributes dialog box. In addition to the standard variable attributes, such as value labels, missing values, and measurement level, you can create your own custom variable attributes. Like standard variable attributes, these custom attributes are saved with SPSS-format data files.

Figure 7-4
Custom Variable Attributes

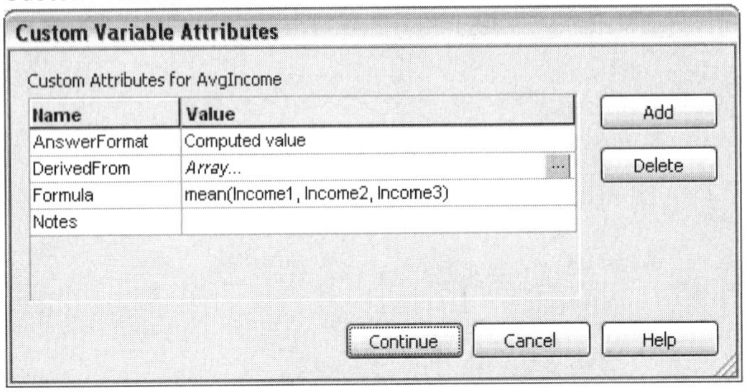

Name. Attribute names must follow the same rules as variable names. For more information, see "Variable Names" in Chapter 5 on p. 106.

Value. The value assigned to the attribute for the selected variable.

■ Attribute names that begin with a dollar sign are reserved and cannot be modified.

■ The text *Array...*, displayed in a Value cell, indicates that this is an **attribute array**, an attribute that contains multiple values. Click the button in the cell to display the list of values.

Copying Variable Properties

The Apply Labels and Level dialog box is displayed when you click From Another Variable or To Other Variables in the Define Variable Properties main dialog box. It displays all of the scanned variables that match the current variable's type (numeric or string). For string variables, the defined width must also match.

Figure 7-5
Apply Labels and Level dialog box

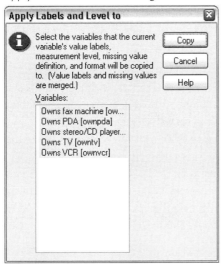

▶ Select a single variable from which to copy value labels and other variable properties (except variable label).

or

▶ Select one or more variables to which to copy value labels and other variable properties.

▶ Click Copy to copy the value labels and the measurement level.

■ Existing value labels and missing value categories for target variable(s) are not replaced.

■ Value labels and missing value categories for values not already defined for the target variable(s) are added to the set of value labels and missing value categories for the target variable(s).

■ The measurement level for the target variable(s) is always replaced.

■ If either the source or target variable has a defined range of missing values, missing values definitions are not copied.

Multiple Response Sets

Custom Tables and the Chart Builder support a special kind of "variable" called a **multiple response set**. Multiple response sets aren't really "variables" in the normal sense. You can't see them in the Data Editor, and other procedures don't recognize them. Multiple response sets use multiple variables to record responses to questions where the respondent can give more than one answer. Multiple response sets are treated like categorical variables, and most of the things you can do with categorical variables, you can also do with multiple response sets.

Multiple response sets are constructed from multiple variables in the data file. A multiple response set is a special construct within an SPSS-format data file. You can define and save multiple response sets in an SPSS-format data file, but you cannot import or export multiple response sets from/to other file formats. (You can copy multiple response sets from other SPSS data files using Copy Data Properties, which is accessed from the Data menu in the Data Editor window. For more information, see "Copying Data Properties" on p. 147.)

Defining Multiple Response Sets

To define multiple response sets:

▶ From the menus, choose:
Data
  Define Multiple Response Sets...

Figure 7-6
Define Multiple Response Sets dialog box

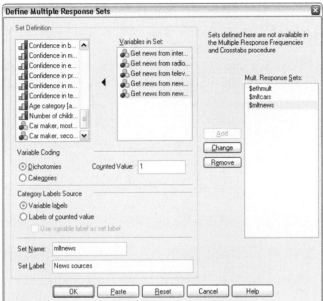

▶ Select two or more variables. If your variables are coded as dichotomies, indicate which value you want to have counted.

▶ Enter a unique name for each multiple response set. The name can be up to 63 bytes long. A dollar sign is automatically added to the beginning of the set name.

▶ Enter a descriptive label for the set. (This is optional.)

▶ Click Add to add the multiple response set to the list of defined sets.

Dichotomies

A multiple dichotomy set typically consists of multiple dichotomous variables: variables with only two possible values of a yes/no, present/absent, checked/not checked nature. Although the variables may not be strictly dichotomous, all of the variables in the set are coded the same way, and the Counted Value represents the positive/present/checked condition.

For example, a survey asks the question, "Which of the following sources do you rely on for news?" and provides five possible responses. The respondent can indicate multiple choices by checking a box next to each choice. The five responses become five variables in the data file, coded 0 for *No* (not checked) and 1 for *Yes* (checked). In the multiple dichotomy set, the Counted Value is 1.

The sample data file *survey_sample.sav* (in the *tutorial/sample_files* directory of the installation directory) already has three defined multiple response sets. *$mltnews* is a multiple dichotomy set.

▶ Select (click) *$mltnews* in the Mult. Response Sets list.

This displays the variables and settings used to define this multiple response set.

- ■ The Variables in Set list displays the five variables used to construct the multiple response set.

- ■ The Variable Coding group indicates that the variables are dichotomous.

- ■ The Counted Value is 1.

▶ Select (click) one of the variables in the Variables in Set list.

▶ Right-click the variable and select Variable Information from the pop-up context menu.

▶ In the Variable Information window, click the arrow on the Value Labels drop-down list to display the entire list of defined value labels.

Figure 7-7
Variable information for multiple dichotomy source variable

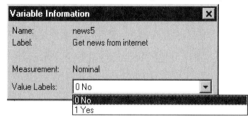

The value labels indicate that the variable is a dichotomy with values of 0 and 1, representing *No* and *Yes*, respectively. All five variables in the list are coded the same way, and the value of 1 (the code for *Yes*) is the counted value for the multiple dichotomy set.

Categories

A multiple category set consists of multiple variables, all coded the same way, often with many possible response categories. For example, a survey item states, "Name up to three nationalities that best describe your ethnic heritage." There may be hundreds of possible responses, but for coding purposes the list is limited to the 40 most common nationalities, with everything else relegated to an "other" category. In the data file, the three choices become three variables, each with 41 categories (40 coded nationalities and one "other" category).

In the sample data file, *$ethmult* and *$mltcars* are multiple category sets.

Category Label Source

For multiple dichotomies, you can control how sets are labeled.

■ **Variable labels.** Uses the defined variable labels (or variable names for variables without defined variable labels) as the set category labels. For example, if all of the variables in the set have the same value label (or no defined value labels) for the counted value (for example, *Yes*), then you should use the variable labels as the set category labels.

■ **Labels of counted values.** Uses the defined value labels of the counted values as set category labels. Select this option only if all variables have a defined value label for the counted value and the value label for the counted value is different for each variable.

■ **Use variable label as set label.** If you select Label of counted values, you can also use the variable label for the first variable in the set with a defined variable label as the set label. If none of the variables in the set have defined variable labels, the name of the first variable in the set is used as the set label.

Copying Data Properties

The Copy Data Properties Wizard provides the ability to use an external SPSS data file as a template for defining file and variable properties in the active dataset. You can also use variables in the active dataset as templates for other variables in the active dataset. You can:

■ Copy selected file properties from an external data file or open dataset to the active dataset. File properties include documents, file labels, multiple response sets, variable sets, and weighting.

- Copy selected variable properties from an external data file or open dataset to matching variables in the active dataset. Variable properties include value labels, missing values, level of measurement, variable labels, print and write formats, alignment, and column width (in the Data Editor).

- Copy selected variable properties from one variable in either an external data file, open dataset, or the active dataset to many variables in the active dataset.

- Create new variables in the active dataset based on selected variables in an external data file or open dataset.

When copying data properties, the following general rules apply:

- If you use an external data file as the source data file, it must be an SPSS-format data file.

- If you use the active dataset as the source data file, it must contain at least one variable. You cannot use a completely blank active dataset as the source data file.

- Undefined (empty) properties in the source dataset do not overwrite defined properties in the active dataset.

- Variable properties are copied from the source variable only to target variables of a matching type—string (alphanumeric) or numeric (including numeric, date, and currency).

Note: Copy Data Properties replaces Apply Data Dictionary, formerly available on the File menu.

To Copy Data Properties

▶ From the menus in the Data Editor window choose:
Data
  Copy Data Properties...

Figure 7-8
Copy Data Properties Wizard: Step 1

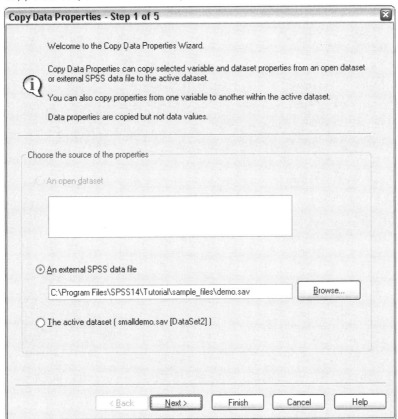

▶ Select the data file with the file and/or variable properties that you want to copy. This can be a currently open dataset, an external SPSS-format data file, or the active dataset.

▶ Follow the step-by-step instructions in the Copy Data Properties Wizard.

Selecting Source and Target Variables

In this step, you can specify the source variables containing the variable properties that you want to copy and the target variables that will receive those variable properties.

Figure 7-9
Copy Data Properties Wizard: Step 2

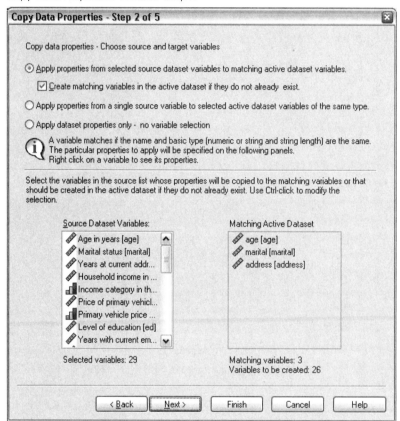

Apply properties from selected source dataset variables to matching active dataset variables. Variable properties are copied from one or more selected source variables to matching variables in the active dataset. Variables "match" if both the variable name and type (string or numeric) are the same. For string variables, the defined length must also be the same. By default, only matching variables are displayed in the two variable lists.

■ **Create matching variables in the active dataset if they do not already exist.** This updates the source list to display all variables in the source data file. If you select source variables that do not exist in the active dataset (based on variable name), new variables will be created in the active dataset with the variable names and properties from the source data file.

If the active dataset contains no variables (a blank, new dataset), all variables in the source data file are displayed and new variables based on the selected source variables are automatically created in the active dataset.

Apply properties from a single source variable to selected active dataset variables of the same type. Variable properties from a single selected variable in the source list can be applied to one or more selected variables in the active dataset list. Only variables of the same type (numeric or string) as the selected variable in the source list are displayed in the active dataset list. For string variables, only strings of the same defined length as the source variable are displayed. This option is not available if the active dataset contains no variables.

Note: You cannot create new variables in the active dataset with this option.

Apply dataset properties only—no variable selection. Only file properties (for example, documents, file label, weight) will be applied to the active dataset. No variable properties will be applied. This option is not available if the active dataset is also the source data file.

Choosing Variable Properties to Copy

You can copy selected variable properties from the source variables to the target variables. Undefined (empty) properties in the source variables do not overwrite defined properties in the target variables.

Figure 7-10
Copy Data Properties Wizard: Step 3

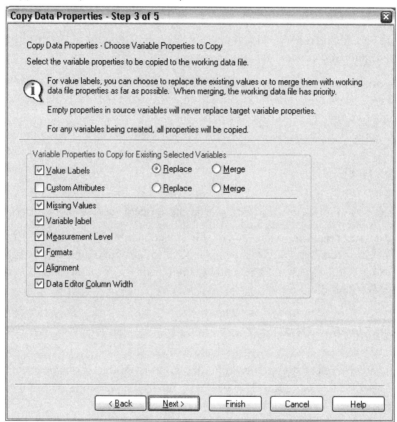

Value Labels. Value labels are descriptive labels associated with data values. Value labels are often used when numeric data values are used to represent non-numeric categories (for example, codes of 1 and 2 for *Male* and *Female*). You can replace or merge value labels in the target variables.

■ Replace deletes any defined value labels for the target variable and replaces them with the defined value labels from the source variable.

■ Merge merges the defined value labels from the source variable with any existing defined value label for the target variable. If the same value has a defined value label in both the source and target variables, the value label in the target variable is unchanged.

Custom Attributes. User-defined custom variable attributes. For more information, see "Custom Variable Attributes" in Chapter 5 on p. 116.

- Replace deletes any custom attributes for the target variable and replaces them with the defined attributes from the source variable.

- Merge merges the defined attributes from the source variable with any existing defined attributes for the target variable.

Missing Values. Missing values are values identified as representing missing data (for example, 98 for *Do not know* and 99 for *Not applicable*). Typically, these values also have defined value labels that describe what the missing value codes stand for. Any existing defined missing values for the target variable are deleted and replaced with the defined missing values from the source variable.

Variable Label. Descriptive variable labels can contain spaces and reserved characters not allowed in variable names. If you're copying variable properties from a single source variable to multiple target variables, you might want to think twice before selecting this option.

Measurement Level. The measurement level can be nominal, ordinal, or scale. For those procedures that distinguish between different measurement levels, nominal and ordinal are both considered **categorical**.

Formats. For numeric variables, this controls numeric type (such as numeric, date, or currency), width (total number of displayed characters, including leading and trailing characters and decimal indicator), and number of decimal places displayed. This option is ignored for string variables.

Alignment. This affects only alignment (left, right, center) in Data View in the Data Editor.

Data Editor Column Width. This affects only column width in Data View in the Data Editor.

Copying Dataset (File) Properties

You can apply selected, global dataset properties from the source data file to the active dataset. (This is not available if the active dataset is the source data file.)

Figure 7-11
Copy Data Properties Wizard: Step 4

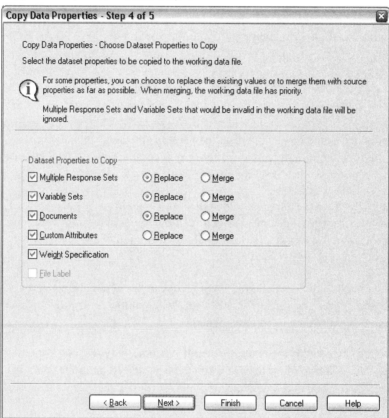

Multiple Response Sets. Applies multiple response set definitions from the source data file to the active dataset. (*Note*: Multiple response sets are currently used only by the Chart Builder and the Tables add-on component.)

■ Multiple response sets in the source data file that contain variables that do not exist in the active dataset are ignored unless those variables will be created based on specifications in step 2 (Selecting Source and Target Variables) in the Copy Data Properties Wizard.

■ Replace deletes all multiple response sets in the active dataset and replaces them with the multiple response sets from the source data file.

■ Merge adds multiple response sets from the source data file to the collection of multiple response sets in the active dataset. If a set with the same name exists in both files, the existing set in the active dataset is unchanged.

Variable Sets. Variable sets are used to control the list of variables that are displayed in dialog boxes. Variable sets are defined by selecting Define Sets from the Utilities menu.

■ Sets in the source data file that contain variables that do not exist in the active dataset are ignored unless those variables will be created based on specifications in step 2 (Selecting Source and Target Variables) in the Copy Data Properties Wizard.

■ Replace deletes any existing variable sets in the active dataset, replacing them with variable sets from the source data file.

■ Merge adds variable sets from the source data file to the collection of variable sets in the active dataset. If a set with the same name exists in both files, the existing set in the active dataset is unchanged.

Documents. Notes appended to the data file via the DOCUMENT command.

■ Replace deletes any existing documents in the active dataset, replacing them with the documents from the source data file.

■ Merge combines documents from the source and active dataset. Unique documents in the source file that do not exist in the active dataset are added to the active dataset. All documents are then sorted by date.

Custom Attributes. Custom data file attributes, typically created with the DATAFILE ATTRIBUTE command in command syntax.

■ Replace deletes any existing custom data file attributes in the active dataset, replacing them with the data file attributes from the source data file.

■ Merge combines data file attributes from the source and active dataset. Unique attribute names in the source file that do not exist in the active dataset are added to the active dataset. If the same attribute name exists in both data files, the named attribute in the active dataset is unchanged.

Weight Specification. Weights cases by the current weight variable in the source data file if there is a matching variable in the active dataset. This overrides any weighting currently in effect in the active dataset.

File Label. Descriptive label applied to a data file with the FILE LABEL command.

Results

Figure 7-12
Copy Data Properties Wizard: Step 5

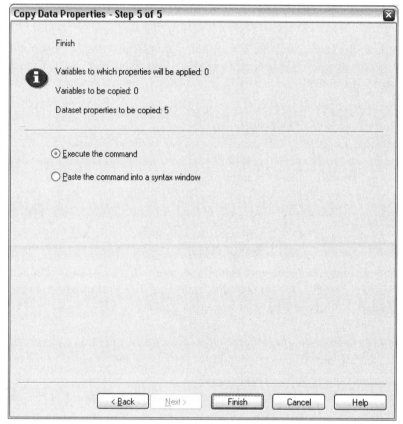

The last step in the Copy Data Properties Wizard provides information on the number of variables for which variable properties will be copied from the source data file, the number of new variables that will be created, and the number of dataset (file) properties that will be copied.

You can also choose to paste the generated command syntax into a syntax window and save the syntax for later use.

Identifying Duplicate Cases

"Duplicate" cases may occur in your data for many reasons, including:

■ Data entry errors in which the same case is accidentally entered more than once.

■ Multiple cases share a common primary ID value but have different secondary ID values, such as family members who all live in the same house.

■ Multiple cases represent the same case but with different values for variables other than those that identify the case, such as multiple purchases made by the same person or company for different products or at different times.

Identify Duplicate Cases allows you to define *duplicate* almost any way that you want and provides some control over the automatic determination of primary versus duplicate cases.

To Identify and Flag Duplicate Cases

▶ From the menus choose:
Data
  Identify Duplicate Cases...

▶ Select one or more variables that identify matching cases.

▶ Select one or more of the options in the Variables to Create group.

Optionally, you can:

▶ Select one or more variables to sort cases within groups defined by the selected matching cases variables. The sort order defined by these variables determines the "first" and "last" case in each group. Otherwise, the original file order is used.

▶ Automatically filter duplicate cases so that they won't be included in reports, charts, or calculations of statistics.

Figure 7-13
Identify Duplicate Cases dialog box

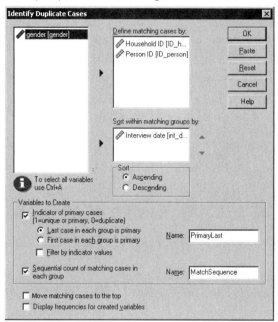

Define matching cases by. Cases are considered duplicates if their values match for *all* selected variables. If you want to identify only cases that are a 100% match in all respects, select all of the variables.

Sort within matching groups by. Cases are automatically sorted by the variables that define matching cases. You can select additional sorting variables that will determine the sequential order of cases in each matching group.

- For each sort variable, you can sort in ascending or descending order.

- If you select multiple sort variables, cases are sorted by each variable within categories of the preceding variable in the list. For example, if you select *date* as the first sorting variable and *amount* as the second sorting variable, cases will be sorted by amount within each date.

- Use the up and down arrow buttons to the right of the list to change the sort order of the variables.

- The sort order determines the "first" and "last" case within each matching group, which determines the value of the optional primary indicator variable. For example, if you want to filter out all but the most recent case in each matching group, you could sort cases within the group in ascending order of a date variable, which would make the most recent date the last date in the group.

Indicator of primary cases. Creates a variable with a value of 1 for all unique cases and the case identified as the primary case in each group of matching cases and a value of 0 for the nonprimary duplicates in each group.

- The primary case can be either the last or first case in each matching group, as determined by the sort order within the matching group. If you don't specify any sort variables, the original file order determines the order of cases within each group.

- You can use the indicator variable as a **filter variable** to exclude nonprimary duplicates from reports and analyses without deleting those cases from the data file.

Sequential count of matching cases in each group. Creates a variable with a sequential value from 1 to *n* for cases in each matching group. The sequence is based on the current order of cases in each group, which is either the original file order or the order determined by any specified sort variables.

Move matching cases to the top. Sorts the data file so that all groups of matching cases are at the top of the data file, making it easy to visually inspect the matching cases in the Data Editor.

Display frequencies for created variables. Frequency tables containing counts for each value of the created variables. For example, for the primary indicator variable, the table would show the number of cases with a value 0 for that variable, which indicates the number of duplicates, and the number of cases with a value of 1 for that variable, which indicates the number of unique and primary cases.

Missing Values. For numeric variables, the system-missing value is treated like any other value—cases with the system-missing value for an identifier variable are treated as having matching values for that variable. For string variables, cases with no value for an identifier variable are treated as having matching values for that variable.

Visual Binning

Visual Binning is designed to assist you in the process of creating new variables based on grouping contiguous values of existing variables into a limited number of distinct categories. You can use Visual Binning to:

- Create categorical variables from continuous scale variables. For example, you could use a scale income variable to create a new categorical variable that contains income ranges.

- Collapse a large number of ordinal categories into a smaller set of categories. For example, you could collapse a rating scale of nine down to three categories representing low, medium, and high.

In the first step, you:

▶ Select the numeric scale and/or ordinal variables for which you want to create new categorical (binned) variables.

Figure 7-14
Initial dialog box for selecting variables to bin

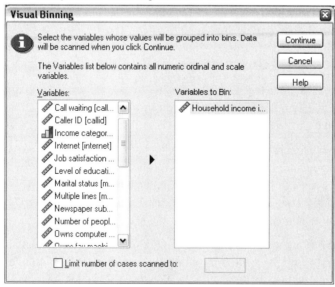

Optionally, you can limit the number of cases to scan. For data files with a large number of cases, limiting the number of cases scanned can save time, but you should avoid this if possible because it will affect the distribution of values used in subsequent calculations in Visual Binning.

Note: String variables and nominal numeric variables are not displayed in the source variable list. Visual Binning requires numeric variables, measured on either a scale or ordinal level, since it assumes that the data values represent some logical order that can be used to group values in a meaningful fashion. You can change the defined measurement level of a variable in Variable View in the Data Editor. For more information, see "Variable Measurement Level" in Chapter 5 on p. 107.

To Bin Variables

▶ From the menus in the Data Editor window choose:
Transform
  Visual Binning...

▶ Select the numeric scale and/or ordinal variables for which you want to create new categorical (binned) variables.

▶ Select a variable in the Scanned Variable List.

▶ Enter a name for the new binned variable. Variable names must be unique and must follow SPSS variable naming rules. For more information, see "Variable Names" in Chapter 5 on p. 106.

▶ Define the binning criteria for the new variable. For more information, see "Binning Variables" on p. 162.

▶ Click OK.

Binning Variables

Figure 7-15
Visual Binning, main dialog box

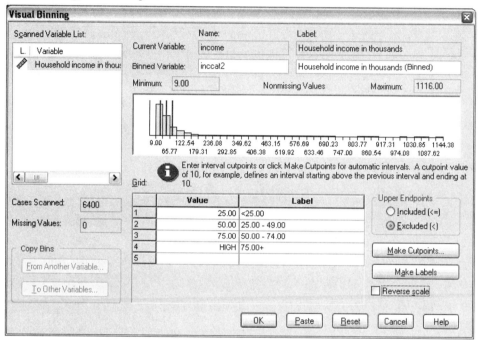

The Visual Binning main dialog box provides the following information for the scanned variables:

Scanned Variable List. Displays the variables you selected in the initial dialog box. You can sort the list by measurement level (scale or ordinal) or by variable label or name by clicking on the column headings.

Cases Scanned. Indicates the number of cases scanned. All scanned cases without user-missing or system-missing values for the selected variable are used to generate the distribution of values used in calculations in Visual Binning, including the histogram displayed in the main dialog box and cutpoints based on percentiles or standard deviation units.

Missing Values. Indicates the number of scanned cases with user-missing or system-missing values. Missing values are not included in any of the binned categories. For more information, see "User-Missing Values in Visual Binning" on p. 169.

Current Variable. The name and variable label (if any) for the currently selected variable that will be used as the basis for the new, binned variable.

Binned Variable. Name and optional variable label for the new, binned variable.

■ **Name.** You must enter a name for the new variable. Variable names must be unique and must follow SPSS variable naming rules. For more information, see "Variable Names" in Chapter 5 on p. 106.

■ **Label.** You can enter a descriptive variable label up to 255 characters long. The default variable label is the variable label (if any) or variable name of the source variable with *(Binned)* appended to the end of the label.

Minimum and Maximum. Minimum and maximum values for the currently selected variable, based on the scanned cases and not including values defined as user-missing.

Nonmissing Values. The histogram displays the distribution of nonmissing values for the currently selected variable, based on the scanned cases.

■ After you define bins for the new variable, vertical lines on the histogram are displayed to indicate the cutpoints that define bins.

■ You can click and drag the cutpoint lines to different locations on the histogram, changing the bin ranges.

■ You can remove bins by dragging cutpoint lines off the histogram.

Note: The histogram (displaying nonmissing values), the minimum, and the maximum are based on the scanned values. If you do not include all cases in the scan, the true distribution may not be accurately reflected, particularly if the data file has been sorted by the selected variable. If you scan zero cases, no information about the distribution of values is available.

Grid. Displays the values that define the upper endpoints of each bin and optional value labels for each bin.

■ **Value.** The values that define the upper endpoints of each bin. You can enter values or use Make Cutpoints to automatically create bins based on selected criteria. By default, a cutpoint with a value of *HIGH* is automatically included. This bin will contain any nonmissing values above the other cutpoints. The bin defined by the lowest cutpoint will include all nonmissing values lower than or equal to that value (or simply lower than that value, depending on how you define upper endpoints).

■ **Label.** Optional, descriptive labels for the values of the new, binned variable. Since the values of the new variable will simply be sequential integers from 1 to *n*, labels that describe what the values represent can be very useful. You can enter labels or use Make Labels to automatically create value labels.

To Delete a Bin from the Grid

▶ Right-click on the either the *Value* or *Label* cell for the bin.

▶ From the pop-up context menu, select Delete Row.

Note: If you delete the *HIGH* bin, any cases with values higher than the last specified cutpoint value will be assigned the system-missing value for the new variable.

To Delete All Labels or Delete All Defined Bins

▶ Right-click anywhere in the grid.

▶ From the pop-up context menu select either Delete All Labels or Delete All Cutpoints.

Upper Endpoints. Controls treatment of upper endpoint values entered in the *Value* column of the grid.

■ **Included (<=).** Cases with the value specified in the *Value* cell are included in the binned category. For example, if you specify values of 25, 50, and 75, cases with a value of exactly 25 will go in the first bin, since this will include all cases with values less than or equal to 25.

■ **Excluded (<).** Cases with the value specified in the *Value* cell are not included in the binned category. Instead, they are included in the next bin. For example, if you specify values of 25, 50, and 75, cases with a value of exactly 25 will go in the second bin rather than the first, since the first bin will contain only cases with values less than 25.

Make Cutpoints. Generates binned categories automatically for equal width intervals, intervals with the same number of cases, or intervals based on standard deviations. This is not available if you scanned zero cases. For more information, see "Automatically Generating Binned Categories" on p. 165.

Make Labels. Generates descriptive labels for the sequential integer values of the new, binned variable, based on the values in the grid and the specified treatment of upper endpoints (included or excluded).

Reverse scale. By default, values of the new, binned variable are ascending sequential integers from 1 to n. Reversing the scale makes the values descending sequential integers from n to 1.

Copy Bins. You can copy the binning specifications from another variable to the currently selected variable or from the selected variable to multiple other variables. For more information, see "Copying Binned Categories" on p. 168.

Automatically Generating Binned Categories

The Make Cutpoints dialog box allows you to auto-generate binned categories based on selected criteria.

To Use the Make Cutpoints Dialog Box

▶ Select (click) a variable in the Scanned Variable List.

▶ Click Make Cutpoints.

▶ Select the criteria for generating cutpoints that will define the binned categories.

▶ Click Apply.

Figure 7-16
Make Cutpoints dialog box

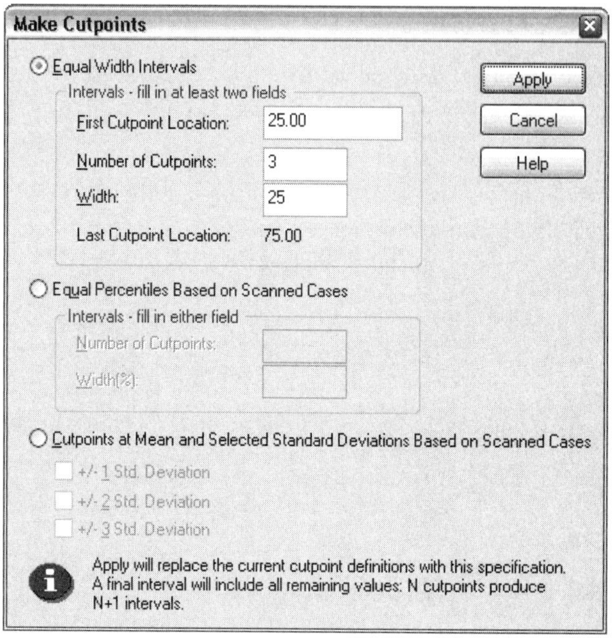

Note: The Make Cutpoints dialog box is not available if you scanned zero cases.

Equal Width Intervals. Generates binned categories of equal width (for example, 1–10, 11–20, and 21–30) based on any two of the following three criteria:

- **First Cutpoint Location.** The value that defines the upper end of the lowest binned category (for example, a value of 10 indicates a range that includes all values up to 10).

- **Number of Cutpoints.** The number of binned categories is the number of cutpoints *plus one*. For example, 9 cutpoints generate 10 binned categories.

- **Width.** The width of each interval. For example, a value of 10 would bin *age in years* into 10-year intervals.

Equal Percentiles Based on Scanned Cases. Generates binned categories with an equal number of cases in each bin (using the **aempirical algorithm** for percentiles), based on either of the following criteria:

- **Number of Cutpoints.** The number of binned categories is the number of cutpoints *plus one*. For example, three cutpoints generate four percentile bins (quartiles), each containing 25% of the cases.

- **Width (%).** Width of each interval, expressed as a percentage of the total number of cases. For example, a value of 33.3 would produce three binned categories (two cutpoints), each containing 33.3% of the cases.

If the source variable contains a relatively small number of distinct values or a large number of cases with the same value, you may get fewer bins than requested. If there are multiple identical values at a cutpoint, they will all go into the same interval; so the actual percentages may not always be exactly equal.

Cutpoints at Mean and Selected Standard Deviations Based on Scanned Cases. Generates binned categories based on the values of the mean and standard deviation of the distribution of the variable.

- If you don't select any of the standard deviation intervals, two binned categories will be created, with the mean as the cutpoint dividing the bins.

- You can select any combination of standard deviation intervals based on one, two, and/or three standard deviations. For example, selecting all three would result in eight binned categories—six bins in one standard deviation intervals and two bins for cases more than three standard deviations above and below the mean.

In a normal distribution, 68% of the cases fall within one standard deviation of the mean; 95%, within two standard deviations; and 99%, within three standard deviations. Creating binned categories based on standard deviations may result in some defined bins outside of the actual data range and even outside of the range of possible data values (for example, a negative salary range).

Note: Calculations of percentiles and standard deviations are based on the scanned cases. If you limit the number of cases scanned, the resulting bins may not contain the proportion of cases that you wanted in those bins, particularly if the data file is sorted by the source variable. For example, if you limit the scan to the first 100 cases of a data file with 1000 cases and the data file is sorted in ascending order of age of respondent, instead of four percentile age bins each containing 25% of the cases, you may find that the first three bins each contain only about 3.3% of the cases, and the last bin contains 90% of the cases.

Copying Binned Categories

When creating binned categories for one or more variables, you can copy the binning specifications from another variable to the currently selected variable or from the selected variable to multiple other variables.

Figure 7-17
Copying bins to or from the current variable

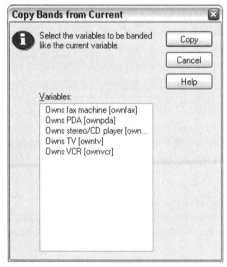

To Copy Binning Specifications

▶ Define binned categories for at least one variable—but do *not* click OK or Paste.

▶ Select (click) a variable in the Scanned Variable List for which you have defined binned categories.

▶ Click To Other Variables.

▶ Select the variables for which you want to create new variables with the same binned categories.

▶ Click Copy.

or

▶ Select (click) a variable in the Scanned Variable List to which you want to copy defined binned categories.

▶ Click From Another Variable.

▶ Select the variable with the defined binned categories that you want to copy.

▶ Click Copy.

If you have specified value labels for the variable from which you are copying the binning specifications, those are also copied.

Note: Once you click OK in the Visual Binning main dialog box to create new binned variables (or close the dialog box in any other way), you cannot use Visual Binning to copy those binned categories to other variables.

User-Missing Values in Visual Binning

Values defined as user-missing (values identified as codes for missing data) for the source variable are not included in the binned categories for the new variable. User-missing values for the source variables are copied as user-missing values for the new variable, and any defined value labels for missing value codes are also copied.

If a missing value code conflicts with one of the binned category values for the new variable, the missing value code for the new variable is recoded to a nonconflicting value by adding 100 to the highest binned category value. For example, if a value of 1 is defined as user-missing for the source variable and the new variable will have six binned categories, any cases with a value of 1 for the source variable will have a value of 106 for the new variable, and 106 will be defined as user-missing. If the user-missing value for the source variable had a defined value label, that label will be retained as the value label for the recoded value of the new variable.

Note: If the source variable has a defined range of user-missing values of the form *LO-n*, where *n* is a positive number, the corresponding user-missing values for the new variable will be negative numbers.

Data Transformations

In an ideal situation, your raw data are perfectly suitable for the type of analysis you want to perform, and any relationships between variables are either conveniently linear or neatly orthogonal. Unfortunately, this is rarely the case. Preliminary analysis may reveal inconvenient coding schemes or coding errors, or data transformations may be required in order to expose the true relationship between variables.

You can perform data transformations ranging from simple tasks, such as collapsing categories for analysis, to more advanced tasks, such as creating new variables based on complex equations and conditional statements.

Computing Variables

Use the Compute dialog box to compute values for a variable based on numeric transformations of other variables.

- You can compute values for numeric or string (alphanumeric) variables.

- You can create new variables or replace the values of existing variables. For new variables, you can also specify the variable type and label.

- You can compute values selectively for subsets of data based on logical conditions.

- You can use more than 70 built-in functions, including arithmetic functions, statistical functions, distribution functions, and string functions.

Figure 8-1
Compute Variable dialog box

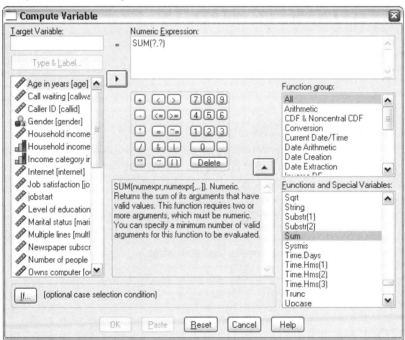

To Compute Variables

▶ From the menus choose:

Transform
    Compute Variable...

▶ Type the name of a single target variable. It can be an existing variable or a new variable to be added to the active dataset.

▶ To build an expression, either paste components into the Expression field or type directly in the Expression field.

  ■ You can paste functions or commonly used system variables by selecting a group from the Function group list and double-clicking the function or variable in the Functions and Special Variables list (or select the function or variable and click the arrow adjacent to the Function group list). Fill in any parameters indicated by question marks (only applies to functions). The function group labeled All provides a listing of all available functions and system variables. A brief description of the currently selected function or variable is displayed in a reserved area in the dialog box.

■ String constants must be enclosed in quotation marks or apostrophes.

■ If values contain decimals, a period (.) must be used as the decimal indicator.

■ For new string variables, you must also select Type & Label to specify the data type.

Compute Variable: If Cases

The If Cases dialog box allows you to apply data transformations to selected subsets of cases, using conditional expressions. A conditional expression returns a value of *true*, *false*, or *missing* for each case.

Figure 8-2
Compute Variable If Cases dialog box

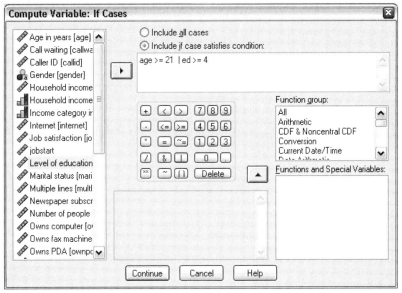

■ If the result of a conditional expression is *true*, the case is included in the selected subset.

■ If the result of a conditional expression is *false* or *missing*, the case is not included in the selected subset.

- Most conditional expressions use one or more of the six relational operators ($<$, $>$, $<=$, $>=$, $=$, and $\sim=$) on the calculator pad.

- Conditional expressions can include variable names, constants, arithmetic operators, numeric (and other) functions, logical variables, and relational operators.

Compute Variable: Type and Label

By default, new computed variables are numeric. To compute a new string variable, you must specify the data type and width.

Label. Optional, descriptive variable label up to 120 characters long. You can enter a label or use the first 110 characters of the compute expression as the label.

Type. Computed variables can be numeric or string (alphanumeric). String variables cannot be used in calculations.

Figure 8-3
Type and Label dialog box

Functions

Many types of functions are supported, including:

- Arithmetic functions.
- Statistical functions.
- String functions.
- Date and time functions.
- Distribution functions.
- Random variable functions.

- Missing value functions.

- Scoring functions (SPSS Server only).

For more information and a detailed description of each function, type functions on the Index tab of the Help system.

Missing Values in Functions

Functions and simple arithmetic expressions treat missing values in different ways. In the expression:

```
(var1+var2+var3)/3
```

the result is missing if a case has a missing value for any of the three variables.

In the expression:

```
MEAN(var1, var2, var3)
```

the result is missing only if the case has missing values for all three variables.

For statistical functions, you can specify the minimum number of arguments that must have nonmissing values. To do so, type a period and the minimum number after the function name, as in:

```
MEAN.2(var1, var2, var3)
```

Random Number Generators

The Random Number Generators dialog box allows you to select the random number generator and set the starting sequence value so you can reproduce a sequence of random numbers.

Active Generator. Two different random number generators are available:

- **SPSS 12 Compatible.** The random number generator used in SPSS 12 and previous releases. If you need to reproduce randomized results generated in previous releases based on a specified seed value, use this random number generator.

- **Mersenne Twister.** A newer random number generator that is more reliable for simulation purposes. If reproducing randomized results from SPSS 12 or earlier is not an issue, use this random number generator.

Active Generator Initialization. The random number seed changes each time a random number is generated for use in transformations (such as random distribution functions), random sampling, or case weighting. To replicate a sequence of random numbers, set the initialization starting point value prior to each analysis that uses the random numbers. The value must be a positive integer.

Figure 8-4
Random Number Generators dialog box

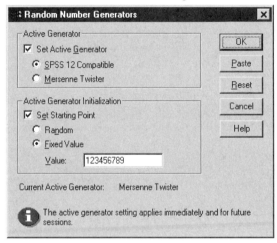

To select the random number generator and/or set the initialization value:

▶ From the menus choose:
Transform
    Random Number Generators

Count Occurrences of Values within Cases

This dialog box creates a variable that counts the occurrences of the same value(s) in a list of variables for each case. For example, a survey might contain a list of magazines with *yes/no* check boxes to indicate which magazines each respondent reads. You could count the number of *yes* responses for each respondent to create a new variable that contains the total number of magazines read.

Figure 8-5
Count Occurrences of Values within Cases dialog box

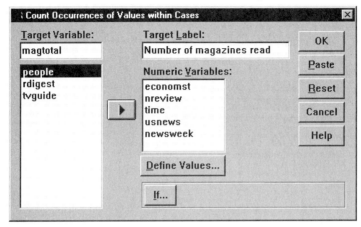

To Count Occurrences of Values within Cases

▶ From the menus choose:

Transform
 Count Values within Cases...

▶ Enter a target variable name.

▶ Select two or more variables of the same type (numeric or string).

▶ Click Define Values and specify which value or values should be counted.

Optionally, you can define a subset of cases for which to count occurrences of values.

Count Values within Cases: Values to Count

The value of the target variable (on the main dialog box) is incremented by 1 each time one of the selected variables matches a specification in the Values to Count list here. If a case matches several specifications for any variable, the target variable is incremented several times for that variable.

Value specifications can include individual values, missing or system-missing values, and ranges. Ranges include their endpoints and any user-missing values that fall within the range.

Figure 8-6
Values to Count dialog box

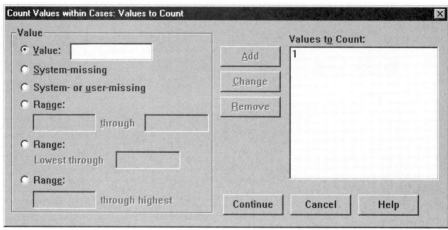

Count Occurrences: If Cases

The If Cases dialog box allows you to count occurrences of values for a selected subset of cases, using conditional expressions. A conditional expression returns a value of *true*, *false*, or *missing* for each case.

Figure 8-7
Count Occurrences If Cases dialog box

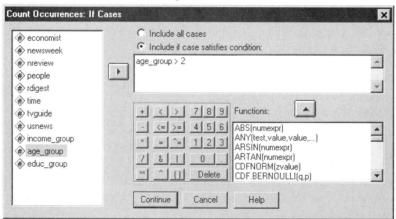

For general considerations on using the If Cases dialog box, see "Compute Variable: If Cases" on p. 173.

Recoding Values

You can modify data values by recoding them. This is particularly useful for collapsing or combining categories. You can recode the values within existing variables, or you can create new variables based on the recoded values of existing variables.

Recode into Same Variables

The Recode into Same Variables dialog box allows you to reassign the values of existing variables or collapse ranges of existing values into new values. For example, you could collapse salaries into salary range categories.

You can recode numeric and string variables. If you select multiple variables, they must all be the same type. You cannot recode numeric and string variables together.

Figure 8-8
Recode into Same Variables dialog box

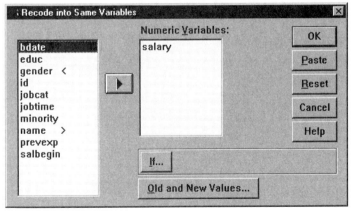

To Recode Values of a Variable

▶ From the menus choose:

Transform
  Recode Into Same Variables...

▶ Select the variables you want to recode. If you select multiple variables, they must be the same type (numeric or string).

▶ Click Old and New Values and specify how to recode values.

Optionally, you can define a subset of cases to recode. The If Cases dialog box for doing this is the same as the one described for Count Occurrences.

Recode into Same Variables: Old and New Values

You can define values to recode in this dialog box. All value specifications must be the same data type (numeric or string) as the variables selected in the main dialog box.

Old Value. The value(s) to be recoded. You can recode single values, ranges of values, and missing values. System-missing values and ranges cannot be selected for string variables because neither concept applies to string variables. Ranges include their endpoints and any user-missing values that fall within the range.

■ **Value.** Individual old value to be recoded into a new value. The value must be the same data type (numeric or string) as the variable(s) being recoded.

■ **System-missing.** Values assigned by the program when values in your data are undefined according to the format type you have specified, when a numeric field is blank, or when a value resulting from a transformation command is undefined. Numeric system-missing values are displayed as periods. String variables cannot have system-missing values, since any character is legal in a string variable.

■ **System- or user-missing.** Observations with values that either have been defined as user-missing values or are unknown and have been assigned the system-missing value, which is indicated with a period (.).

■ **Range.** Inclusive range of values. Not available for string variables. Any user-missing values within the range are included.

■ **All other values.** Any remaining values not included in one of the specifications on the Old-New list. This appears as ELSE on the Old-New list.

New Value. The single value into which each old value or range of values is recoded. You can enter a value or assign the system-missing value.

- **Value.** Value into which one or more old values will be recoded. The value must be the same data type (numeric or string) as the old value.

- **System-missing.** Recodes specified old values into the system-missing value. The system-missing value is not used in calculations, and cases with the system-missing value are excluded from many procedures. Not available for string variables.

Old–>New. The list of specifications that will be used to recode the variable(s). You can add, change, and remove specifications from the list. The list is automatically sorted, based on the old value specification, using the following order: single values, missing values, ranges, and all other values. If you change a recode specification on the list, the procedure automatically re-sorts the list, if necessary, to maintain this order.

Figure 8-9
Old and New Values dialog box

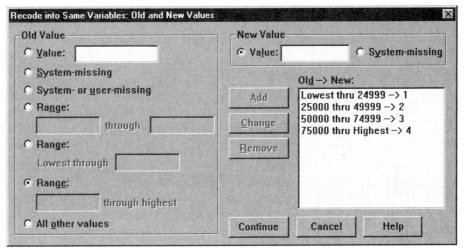

Recode into Different Variables

The Recode into Different Variables dialog box allows you to reassign the values of existing variables or collapse ranges of existing values into new values for a new variable. For example, you could collapse salaries into a new variable containing salary-range categories.

- You can recode numeric and string variables.

■ You can recode numeric variables into string variables and vice versa.

■ If you select multiple variables, they must all be the same type. You cannot recode numeric and string variables together.

Figure 8-10
Recode into Different Variables dialog box

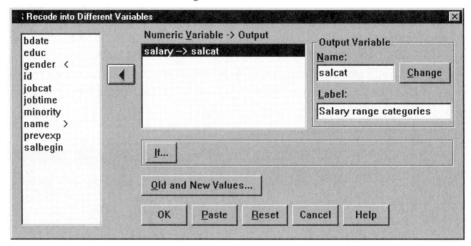

To Recode Values of a Variable into a New Variable

▶ From the menus choose:
Transform
  Recode Into Different Variables...

▶ Select the variables you want to recode. If you select multiple variables, they must be the same type (numeric or string).

▶ Enter an output (new) variable name for each new variable and click Change.

▶ Click Old and New Values and specify how to recode values.

Optionally, you can define a subset of cases to recode. The If Cases dialog box for doing this is the same as the one described for Count Occurrences.

Recode into Different Variables: Old and New Values

You can define values to recode in this dialog box.

Old Value. The value(s) to be recoded. You can recode single values, ranges of values, and missing values. System-missing values and ranges cannot be selected for string variables because neither concept applies to string variables. Old values must be the same data type (numeric or string) as the original variable. Ranges include their endpoints and any user-missing values that fall within the range.

- **Value.** Individual old value to be recoded into a new value. The value must be the same data type (numeric or string) as the variable(s) being recoded.

- **System-missing.** Values assigned by the program when values in your data are undefined according to the format type you have specified, when a numeric field is blank, or when a value resulting from a transformation command is undefined. Numeric system-missing values are displayed as periods. String variables cannot have system-missing values, since any character is legal in a string variable.

- **System- or user-missing.** Observations with values that either have been defined as user-missing values or are unknown and have been assigned the system-missing value, which is indicated with a period (.).

- **Range.** Inclusive range of values. Not available for string variables. Any user-missing values within the range are included.

- **All other values.** Any remaining values not included in one of the specifications on the Old-New list. This appears as ELSE on the Old-New list.

New Value. The single value into which each old value or range of values is recoded. New values can be numeric or string.

- **Value.** Value into which one or more old values will be recoded. The value must be the same data type (numeric or string) as the old value.

- **System-missing.** Recodes specified old values into the system-missing value. The system-missing value is not used in calculations, and cases with the system-missing value are excluded from many procedures. Not available for string variables.

- **Copy old values.** Retains the old value. If some values don't require recoding, use this to include the old values. Any old values that are not specified are not included in the new variable, and cases with those values will be assigned the system-missing value for the new variable.

Output variables are strings. Defines the new, recoded variable as a string (alphanumeric) variable. The old variable can be numeric or string.

Convert numeric strings to numbers. Converts string values containing numbers to numeric values. Strings containing anything other than numbers and an optional sign (+ or -) are assigned the system-missing value.

Old–>New. The list of specifications that will be used to recode the variable(s). You can add, change, and remove specifications from the list. The list is automatically sorted, based on the old value specification, using the following order: single values, missing values, ranges, and all other values. If you change a recode specification on the list, the procedure automatically re-sorts the list, if necessary, to maintain this order.

Figure 8-11
Old and New Values dialog box

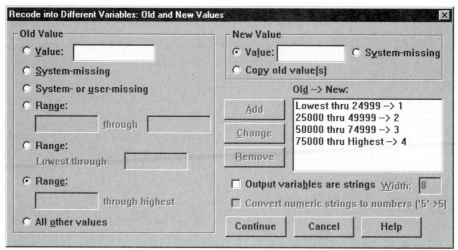

Rank Cases

The Rank Cases dialog box allows you to create new variables containing ranks, normal and Savage scores, and percentile values for numeric variables.

New variable names and descriptive variable labels are automatically generated, based on the original variable name and the selected measure(s). A summary table lists the original variables, the new variables, and the variable labels.

Optionally, you can:

■ Rank cases in ascending or descending order.

■ Organize rankings into subgroups by selecting one or more grouping variables for the By list. Ranks are computed within each group. Groups are defined by the combination of values of the grouping variables. For example, if you select *gender* and *minority* as grouping variables, ranks are computed for each combination of *gender* and *minority*.

Figure 8-12
Rank Cases dialog box

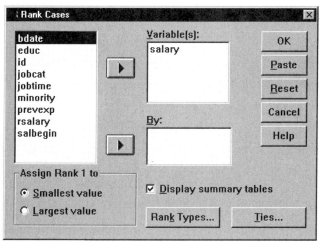

To Rank Cases

▶ From the menus choose:
Transform
  Rank Cases...

▶ Select one or more variables to rank. You can rank only numeric variables.

Optionally, you can rank cases in ascending or descending order and organize ranks into subgroups.

Rank Cases: Types

You can select multiple ranking methods. A separate ranking variable is created for each method. Ranking methods include simple ranks, Savage scores, fractional ranks, and percentiles. You can also create rankings based on proportion estimates and normal scores.

Rank. Simple rank. The value of the new variable equals its rank.

Savage score. The new variable contains Savage scores based on an exponential distribution.

Fractional rank. The value of the new variable equals rank divided by the sum of the weights of the nonmissing cases.

Fractional rank as percent. Each rank is divided by the number of cases with valid values and multiplied by 100.

Sum of case weights. The value of the new variable equals the sum of case weights. The new variable is a constant for all cases in the same group.

Ntiles. Ranks are based on percentile groups, with each group containing approximately the same number of cases. For example, 4 Ntiles would assign a rank of 1 to cases below the 25th percentile, 2 to cases between the 25th and 50th percentile, 3 to cases between the 50th and 75th percentile, and 4 to cases above the 75th percentile.

Proportion estimates. Estimates of the cumulative proportion of the distribution corresponding to a particular rank.

Normal scores. The z scores corresponding to the estimated cumulative proportion.

Proportion Estimation Formula. For proportion estimates and normal scores, you can select the proportion estimation formula: Blom, Tukey, Rankit, or Van der Waerden.

- **Blom.** Creates new ranking variable based on proportion estimates that uses the formula $(r-3/8) / (w+1/4)$, where w is the sum of the case weights and r is the rank.

- **Tukey.** Uses the formula $(r-1/3) / (w+1/3)$, where r is the rank and w is the sum of the case weights.

- **Rankit.** Uses the formula $(r-1/2) / w$, where w is the number of observations and r is the rank, ranging from 1 to w.

- **Van der Waerden.** Van der Waerden's transformation, defined by the formula $r/(w+1)$, where w is the sum of the case weights and r is the rank, ranging from 1 to w.

Figure 8-13
Rank Cases Types dialog box

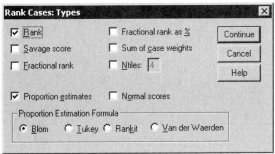

Rank Cases: Ties

This dialog box controls the method for assigning rankings to cases with the same value on the original variable.

Figure 8-14
Rank Cases Ties dialog box

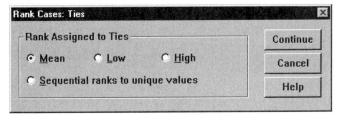

The following table shows how the different methods assign ranks to tied values:

Value	Mean	Low	High	Sequential
10	1	1	1	1
15	3	2	4	2
15	3	2	4	2
15	3	2	4	2
16	5	5	5	3
20	6	6	6	4

Automatic Recode

The Automatic Recode dialog box allows you to convert string and numeric values into consecutive integers. When category codes are not sequential, the resulting empty cells reduce performance and increase memory requirements for many procedures. Additionally, some procedures cannot use string variables, and some require consecutive integer values for factor levels.

Figure 8-15
Automatic Recode dialog box

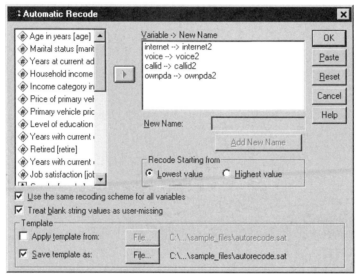

- The new variable(s) created by Automatic Recode retain any defined variable and value labels from the old variable. For any values without a defined value label, the original value is used as the label for the recoded value. A table displays the old and new values and value labels.

- String values are recoded in alphabetical order, with uppercase letters preceding their lowercase counterparts.

- Missing values are recoded into missing values higher than any nonmissing values, with their order preserved. For example, if the original variable has 10 nonmissing values, the lowest missing value would be recoded to 11, and the value 11 would be a missing value for the new variable.

Use the same recoding scheme for all variables. This option allows you to apply a single autorecoding scheme to all the selected variables, yielding a consistent coding scheme for all the new variables.

If you select this option, the following rules and limitations apply:

- All variables must be of the same type (numeric or string).

- All observed values for all selected variables are used to create a sorted order of values to recode into sequential integers.

- User-missing values for the new variables are based on the first variable in the list with defined user-missing values. All other values from other original variables, except for system-missing, are treated as valid.

Treat blank string values as user-missing. For string variables, blank or null values are not treated as **system-missing**. This option will autorecode blank strings into a **user-missing** value higher than the highest nonmissing value.

Templates

You can save the autorecoding scheme in a template file and then apply it to other variables and other data files.

For example, you may have a large number of alphanumeric product codes that you autorecode into integers every month, but some months new product codes are added that change the original autorecoding scheme. If you save the original scheme in a template and then apply it to the new data that contain the new set of codes, any new codes encountered in the data are autorecoded into values higher than the last value in the template, preserving the original autorecode scheme of the original product codes.

Save template as. Saves the autorecode scheme for the selected variables in an external template file.

- The template contains information that maps the original nonmissing values to the recoded values.

- Only information for nonmissing values is saved in the template. User-missing value information is not retained.

■ If you have selected multiple variables for recoding but you have not selected to use the same autorecoding scheme for all variables or you are not applying an existing template as part of the autorecoding, the template will be based on the first variable in the list.

■ If you have selected multiple variables for recoding and you have also selected Use the same recoding scheme for all variables and/or you have selected Apply template, then the template will contain the combined autorecoding scheme for all variables.

Apply template from. Applies a previously saved autorecode template to variables selected for recoding, appending any additional values found in the variables to the end of the scheme and preserving the relationship between the original and autorecoded values stored in the saved scheme.

■ All variables selected for recoding must be the same type (numeric or string), and that type must match the type defined in the template.

■ Templates do not contain any information on user-missing values. User-missing values for the target variables are based on the first variable in the original variable list with defined user-missing values. All other values from other original variables, except for system-missing, are treated as valid.

■ Value mappings from the template are applied first. All remaining values are recoded into values higher than the last value in the template, with user-missing values (based on the first variable in the list with defined user-missing values) recoded into values higher than the last valid value.

■ If you have selected multiple variables for autorecoding, the template is applied first, followed by a common, combined autorecoding for all additional values found in the selected variables, resulting in a single, common autorecoding scheme for all selected variables.

To Recode String or Numeric Values into Consecutive Integers

▶ From the menus choose:
Transform
  Automatic Recode...

▶ Select one or more variables to recode.

▶ For each selected variable, enter a name for the new variable and click New Name.

Date and Time Wizard

The Date and Time Wizard simplifies a number of common tasks associated with date and time variables.

To Use the Date and Time Wizard

▶ From the menus choose:

Transform
  Date/Time...

▶ Select the task you wish to accomplish and follow the steps to define the task.

Figure 8-16
Date and Time Wizard introduction screen

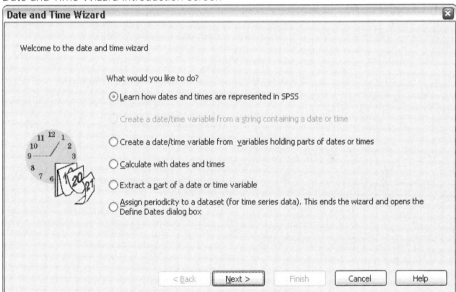

- **Learn how dates and times are represented in SPSS.** This choice leads to a screen that provides a brief overview of date/time variables in SPSS. By clicking on the Help button, it also provides a link to more detailed information.

- **Create a date/time variable from a string containing a date or time.** Use this option to create a date/time variable from a string variable. For example, you have a string variable representing dates in the form mm/dd/yyyy and want to create a date/time variable from this.

■ **Create a date/time variable from variables holding parts of dates or times.** This choice allows you to construct a date/time variable from a set of existing variables. For example, you have a variable that represents the month (as an integer), a second that represents the day of the month, and a third that represents the year. You can combine these three variables into a single date/time variable.

■ **Calculate with dates and times.** Use this option to add or subtract values from date/time variables. For example, you can calculate the duration of a process by subtracting a variable representing the start time of the process from another variable representing the end time of the process.

■ **Extract a part of a date or time variable.** This option allows you to extract part of a date/time variable, such as the day of the month from a date/time variable, which has the form mm/dd/yyyy.

■ **Assign periodicity to a dataset.** This choice takes you to the Define Dates dialog box, used to create date/time variables that consist of a set of sequential dates. This feature is typically used to associate dates with time series data.

Note: Tasks are disabled when the dataset lacks the types of variables required to accomplish the task. For instance, if the dataset contains no string variables, then the task to create a date/time variable from a string does not apply and is disabled.

Dates and Times in SPSS

Variables that represent dates and times in SPSS have a variable type of numeric, with display formats that correspond to the specific date/time formats. These variables are generally referred to as date/time variables. SPSS distinguishes between date/time variables that actually represent dates and those that represent a time duration that is independent of any date, such as 20 hours, 10 minutes, and 15 seconds. The latter are referred to as duration variables and the former as date or date/time variables. For a complete list of display formats, see "Date and Time" in the "Universals" section of the *SPSS Command Syntax Reference.*

Date and date/time variables. Date variables have a format representing a date, such as mm/dd/yyyy. Date/time variables have a format representing a date and time, such as dd-mmm-yyyy hh:mm:ss. Internally, date and date/time variables are stored as the number of seconds from October 14, 1582. Date and date/time variables are sometimes referred to as date-format variables.

- Both two- and four-digit year specifications are recognized. By default, two-digit years represent a range beginning 69 years prior to the current date and ending 30 years after the current date. This range is determined by your Options settings and is configurable (from the Edit menu, choose Options and click the Data tab).

- Dashes, periods, commas, slashes, or blanks can be used as delimiters in day-month-year formats.

- Months can be represented in digits, Roman numerals, or three-character abbreviations, and they can be fully spelled out. Three-letter abbreviations and fully spelled-out month names must be in English; month names in other languages are not recognized.

Duration Variables. Duration variables have a format representing a time duration, such as hh:mm. They are stored internally as seconds without reference to a particular date.

- In time specifications (applies to date/time and duration variables), colons can be used as delimiters between hours, minutes, and seconds. Hours and minutes are required, but seconds are optional. A period is required to separate seconds from fractional seconds. Hours can be of unlimited magnitude, but the maximum value for minutes is 59 and for seconds is 59.999....

Current Date and Time. The system variable *$TIME* holds the current date and time. It represents the number of seconds from October 14, 1582, to the date and time when the transformation command that uses it is executed.

Create a Date/Time Variable from a String

To create a date/time variable from a string variable:

▶ Select Create a date/time variable from a string containing a date or time on the introduction screen of the Date and Time Wizard.

Select String Variable to Convert to Date/Time Variable

Figure 8-17
Create date/time variable from string variable, step 1

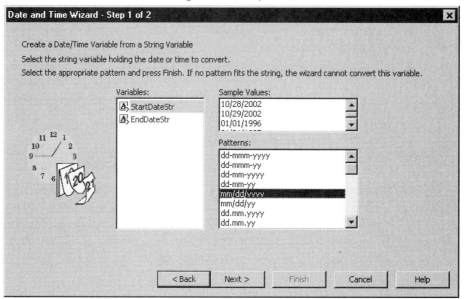

▶ Select the string variable to convert in the Variables list. Note that the list displays only string variables.

▶ Select the pattern from the Patterns list that matches how dates are represented by the string variable. The Sample Values list displays actual values of the selected variable in the data file. Values of the string variable that do not fit the selected pattern result in a value of system-missing for the new variable.

Specify Result of Converting String Variable to Date/Time Variable

Figure 8-18
Create date/time variable from string variable, step 2

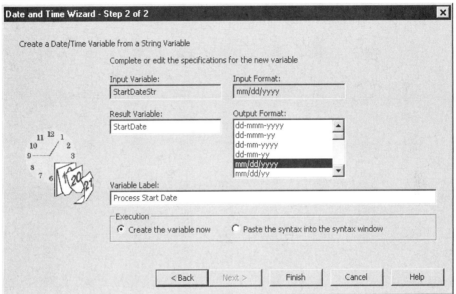

▶ Enter a name for the Result Variable. This cannot be the name of an existing variable.

Optionally, you can:

- Select a date/time format for the new variable from the Output Format list.

- Assign a descriptive variable label to the new variable.

Create a Date/Time Variable from a Set of Variables

To merge a set of existing variables into a single date/time variable:

▶ Select Create a date/time variable from variables holding parts of dates or times on the introduction screen of the Date and Time Wizard.

Select Variables to Merge into Single Date/Time Variable

Figure 8-19
Create date/time variable from variable set, step 1

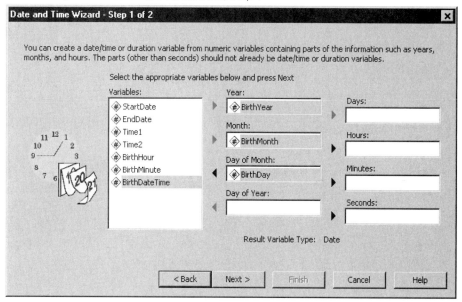

▶ Select the variables that represent the different parts of the date/time.

■ Some combinations of selections are not allowed. For instance, creating a date/time variable from Year and Day of Month is invalid because once Year is chosen, a full date is required.

■ You cannot use an existing date/time variable as one of the parts of the final date/time variable you're creating. Variables that make up the parts of the new date/time variable must be integers. The exception is the allowed use of an existing date/time variable as the Seconds part of the new variable. Since fractional seconds are allowed, the variable used for Seconds is not required to be an integer.

■ Values, for any part of the new variable, that are not within the allowed range result in a value of system-missing for the new variable. For instance, if you inadvertently use a variable representing day of month for Month, any cases with day of month values in the range 14–31 will be assigned the system-missing value for the new variable since the valid range for months in SPSS is 1–13

Specify Date/Time Variable Created by Merging Variables

Figure 8-20
Create date/time variable from variable set, step 2

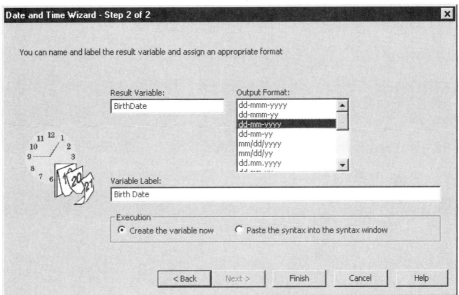

▶ Enter a name for the Result Variable. This cannot be the name of an existing variable.

▶ Select a date/time format from the Output Format list.

Optionally, you can:

■ Assign a descriptive variable label to the new variable.

Add or Subtract Values from Date/Time Variables

To add or subtract values from date/time variables:

▶ Select Calculate with dates and times on the introduction screen of the Date and Time Wizard.

Select Type of Calculation to Perform with Date/Time Variables

Figure 8-21
Add or subtract values from date/time variables, step 1

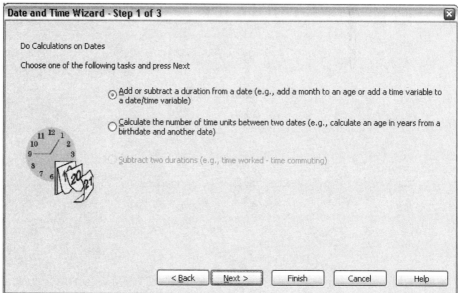

- **Add or subtract a duration from a date.** Use this option to add to or subtract from a date-format variable. You can add or subtract durations that are fixed values, such as 10 days, or the values from a numeric variable, such as a variable that represents years.

- **Calculate the number of time units between two dates.** Use this option to obtain the difference between two dates as measured in a chosen unit. For example, you can obtain the number of years or the number of days separating two dates.

- **Subtract two durations.** Use this option to obtain the difference between two variables that have formats of durations, such as hh:mm or hh:mm:ss.

Note: Tasks are disabled when the dataset lacks the types of variables required to accomplish the task. For instance, if the dataset lacks two variables with formats of durations, then the task to subtract two durations does not apply and is disabled.

Add or Subtract a Duration from a Date

To add or subtract a duration from a date-format variable:

▶ Select Add or subtract a duration from a date on the screen of the Date and Time Wizard labeled Do Calculations on Dates.

Select Date/Time Variable and Duration to Add or Subtract

Figure 8-22
Add or subtract duration, step 2

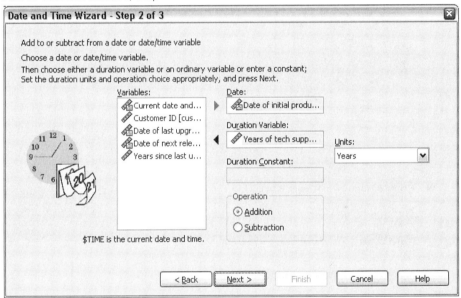

▶ Select a date (or time) variable.

▶ Select a duration variable or enter a value for Duration Constant. Variables used for durations cannot be date or date/time variables. They can be duration variables or simple numeric variables.

▶ Select the unit that the duration represents from the drop-down list. Select Duration if using a variable and the variable is in the form of a duration, such as hh:mm or hh:mm:ss.

Specify Result of Adding or Subtracting a Duration from a Date/Time Variable

Figure 8-23
Add or subtract duration, step 3

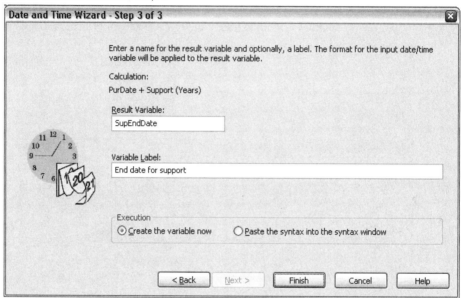

▶ Enter a name for Result Variable. This cannot be the name of an existing variable.

Optionally, you can:

■ Assign a descriptive variable label to the new variable.

Subtract Date-Format Variables

To subtract two date-format variables:

▶ Select Calculate the number of time units between two dates on the screen of the Date and Time Wizard labeled Do Calculations on Dates.

Select Date-Format Variables to Subtract

Figure 8-24
Subtract dates, step 2

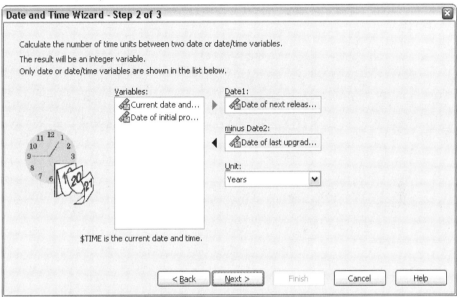

▶ Select the variables to subtract.

▶ Select the unit for the result from the drop-down list.

Specify Result of Subtracting Two Date-Format Variables

Figure 8-25
Subtract dates, step 3

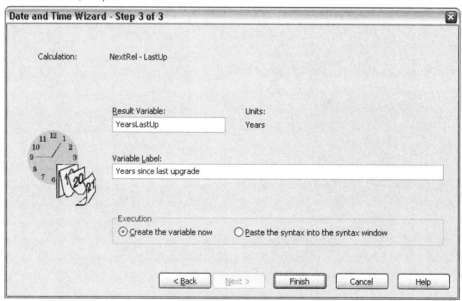

▶ Enter a name for Result Variable. This cannot be the name of an existing variable.

Optionally, you can:

■ Assign a descriptive variable label to the new variable.

Subtract Duration Variables

To subtract two duration variables:

▶ Select Subtract two durations on the screen of the Date and Time Wizard labeled Do Calculations on Dates.

Select Duration Variables to Subtract

Figure 8-26
Subtract two durations, step 2

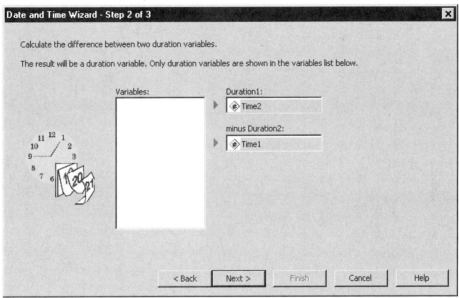

▶ Select the variables to subtract.

Specify Result of Subtracting Two Duration Variables

Figure 8-27
Subtract two durations, step 3

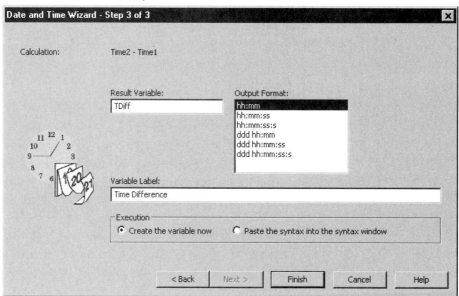

▶ Enter a name for Result Variable. This cannot be the name of an existing variable.

▶ Select a duration format from the Output Format list.

Optionally, you can:

■ Assign a descriptive variable label to the new variable.

Extract Part of a Date/Time Variable

To extract a component—such as the year—from a date/time variable:

▶ Select Extract a part of a date or time variable on the introduction screen of the Date and Time Wizard.

Select Component to Extract from Date/Time Variable

Figure 8-28
Get part of a date/time variable, step 1

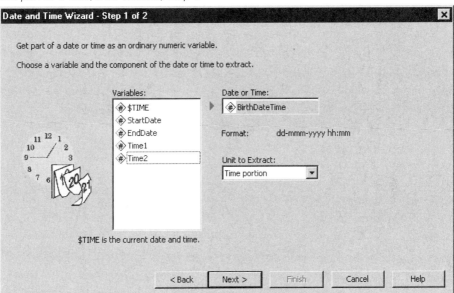

▶ Select the variable containing the date or time part to extract.

▶ Select the part of the variable to extract, from the drop-down list. You can extract information from dates that is not explicitly part of the display date, such as day of the week.

Specify Result of Extracting Component from Date/Time Variable

Figure 8-29
Get part of a date/time variable, step 2

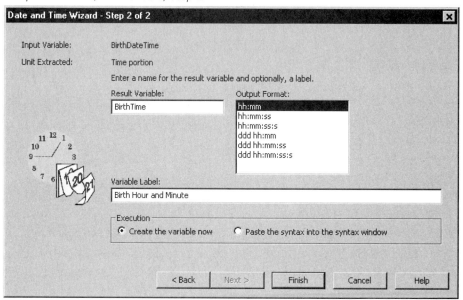

▶ Enter a name for Result Variable. This cannot be the name of an existing variable.

▶ If you're extracting the date or time part of a date/time variable, then you must select a format from the Output Format list. In cases where the output format is not required, the Output Format list will be disabled.

Optionally, you can:

■ Assign a descriptive variable label to the new variable.

Time Series Data Transformations

Several data transformations that are useful in time series analysis are provided:

■ Generate date variables to establish periodicity and to distinguish between historical, validation, and forecasting periods.

■ Create new time series variables as functions of existing time series variables.

■ Replace system- and user-missing values with estimates based on one of several methods.

A **time series** is obtained by measuring a variable (or set of variables) regularly over a period of time. Time series data transformations assume a data file structure in which each case (row) represents a set of observations at a different time, and the length of time between cases is uniform.

Define Dates

The Define Dates dialog box allows you to generate date variables that can be used to establish the periodicity of a **time series** and to label output from time series analysis.

Figure 8-30
Define Dates dialog box

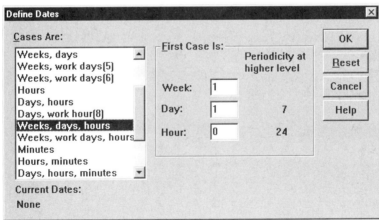

Cases Are. Defines the time interval used to generate dates.

■ Not dated removes any previously defined date variables. Any variables with the following names are deleted: *year_*, *quarter_*, *month_*, *week_*, *day_*, *hour_*, *minute_*, *second_*, and *date_*.

■ Custom indicates the presence of custom date variables created with command syntax (for example, a four-day workweek). This item merely reflects the current state of the active dataset. Selecting it from the list has no effect.

First Case Is. Defines the starting date value, which is assigned to the first case. Sequential values, based on the time interval, are assigned to subsequent cases.

Periodicity at higher level. Indicates the repetitive cyclical variation, such as the number of months in a year or the number of days in a week. The value displayed indicates the maximum value you can enter.

A new numeric variable is created for each component that is used to define the date. The new variable names end with an underscore. A descriptive string variable, *date_*, is also created from the components. For example, if you selected Weeks, days, hours, four new variables are created: *week_*, *day_*, *hour_*, and *date_*.

If date variables have already been defined, they are replaced when you define new date variables that will have the same names as the existing date variables.

To Define Dates for Time Series Data

▶ From the menus choose:
Data
  Define Dates...

▶ Select a time interval from the Cases Are list.

▶ Enter the value(s) that define the starting date for First Case Is, which determines the date assigned to the first case.

Date Variables versus Date Format Variables

Date variables created with Define Dates should not be confused with date format variables defined in the Variable view of the Data Editor. Date variables are used to establish periodicity for time series data. Date format variables represent dates and/or times displayed in various date/time formats. Date variables are simple integers representing the number of days, weeks, hours, and so on, from a user-specified starting point. Internally, most date format variables are stored as the number of seconds from October 14, 1582.

Create Time Series

The Create Time Series dialog box allows you to create new variables based on functions of existing numeric **time series** variables. These transformed values are useful in many time series analysis procedures.

Default new variable names are the first six characters of the existing variable used to create it, followed by an underscore and a sequential number. For example, for the variable *price*, the new variable name would be *price_1*. The new variables retain any defined value labels from the original variables.

Available functions for creating time series variables include differences, moving averages, running medians, lag, and lead functions.

Figure 8-31
Create Time Series dialog box

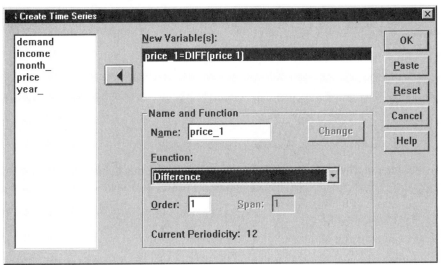

To Create New Time Series Variables

▶ From the menus choose:
Transform
  Create Time Series...

▶ Select the time series function that you want to use to transform the original variable(s).

▶ Select the variable(s) from which you want to create new time series variables. Only numeric variables can be used.

Optionally, you can:

■ Enter variable names to override the default new variable names.

■ Change the function for a selected variable.

Time Series Transformation Functions

Difference. Nonseasonal difference between successive values in the series. The order is the number of previous values used to calculate the difference. Because one observation is lost for each order of difference, system-missing values appear at the beginning of the series. For example, if the difference order is 2, the first two cases will have the system-missing value for the new variable.

Seasonal difference. Difference between series values a constant span apart. The span is based on the currently defined periodicity. To compute seasonal differences, you must have defined date variables (Data menu, Define Dates) that include a periodic component (such as months of the year). The order is the number of seasonal periods used to compute the difference. The number of cases with the system-missing value at the beginning of the series is equal to the periodicity multiplied by the order. For example, if the current periodicity is 12 and the order is 2, the first 24 cases will have the system-missing value for the new variable.

Centered moving average. Average of a span of series values surrounding and including the current value. The span is the number of series values used to compute the average. If the span is even, the moving average is computed by averaging each pair of uncentered means. The number of cases with the system-missing value at the beginning and at the end of the series for a span of *n* is equal to *n*/2 for even span values and for odd span values. For example, if the span is 5, the number of cases with the system-missing value at the beginning and at the end of the series is 2.

Prior moving average. Average of the span of series values preceding the current value. The span is the number of preceding series values used to compute the average. The number of cases with the system-missing value at the beginning of the series is equal to the span value.

Running median. Median of a span of series values surrounding and including the current value. The span is the number of series values used to compute the median. If the span is even, the median is computed by averaging each pair of uncentered medians. The number of cases with the system-missing value at the beginning and at the end of the series for a span of *n* is equal to *n*/2 for even span values and for odd span values. For example, if the span is 5, the number of cases with the system-missing value at the beginning and at the end of the series is 2.

Cumulative sum. Cumulative sum of series values up to and including the current value.

Lag. Value of a previous case, based on the specified lag order. The order is the number of cases prior to the current case from which the value is obtained. The number of cases with the system-missing value at the beginning of the series is equal to the order value.

Lead. Value of a subsequent case, based on the specified lead order. The order is the number of cases after the current case from which the value is obtained. The number of cases with the system-missing value at the end of the series is equal to the order value.

Smoothing. New series values based on a compound data smoother. The smoother starts with a running median of 4, which is centered by a running median of 2. It then resmoothes these values by applying a running median of 5, a running median of 3, and hanning (running weighted averages). Residuals are computed by subtracting the smoothed series from the original series. This whole process is then repeated on the computed residuals. Finally, the smoothed residuals are computed by subtracting the smoothed values obtained the first time through the process. This is sometimes referred to as T4253H smoothing.

Replace Missing Values

Missing observations can be problematic in analysis, and some time series measures cannot be computed if there are missing values in the series. Sometimes the value for a particular observation is simply not known. In addition, missing data can result from any of the following:

- Each degree of differencing reduces the length of a series by 1.

- Each degree of seasonal differencing reduces the length of a series by one season.

- If you create new series that contain forecasts beyond the end of the existing series (by clicking a Save button and making suitable choices), the original series and the generated residual series will have missing data for the new observations.

- Some transformations (for example, the log transformation) produce missing data for certain values of the original series.

Missing data at the beginning or end of a series pose no particular problem; they simply shorten the useful length of the series. Gaps in the middle of a series (*embedded missing data*) can be a much more serious problem. The extent of the problem depends on the analytical procedure you are using.

The Replace Missing Values dialog box allows you to create new **time series** variables from existing ones, replacing missing values with estimates computed with one of several methods. Default new variable names are the first six characters of the existing variable used to create it, followed by an underscore and a sequential number. For example, for the variable *price*, the new variable name would be *price_1*. The new variables retain any defined value labels from the original variables.

Figure 8-32
Replace Missing Values dialog box

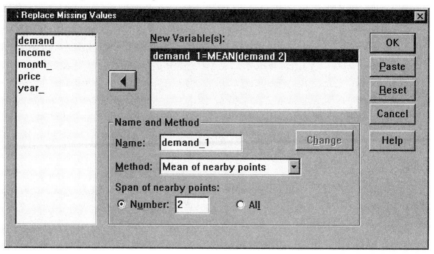

To Replace Missing Values for Time Series Variables

▶ From the menus choose:
Transform
  Replace Missing Values...

▶ Select the estimation method you want to use to replace missing values.

▶ Select the variable(s) for which you want to replace missing values.

Optionally, you can:

- Enter variable names to override the default new variable names.
- Change the estimation method for a selected variable.

Estimation Methods for Replacing Missing Values

Series mean. Replaces missing values with the mean for the entire series.

Mean of nearby points. Replaces missing values with the mean of valid surrounding values. The span of nearby points is the number of valid values above and below the missing value used to compute the mean.

Median of nearby points. Replaces missing values with the median of valid surrounding values. The span of nearby points is the number of valid values above and below the missing value used to compute the median.

Linear interpolation. Replaces missing values using a linear interpolation. The last valid value before the missing value and the first valid value after the missing value are used for the interpolation. If the first or last case in the series has a missing value, the missing value is not replaced.

Linear trend at point. Replaces missing values with the linear trend for that point. The existing series is regressed on an index variable scaled 1 to n. Missing values are replaced with their predicted values.

Scoring Data with Predictive Models

The process of applying a predictive model to a set of data is referred to as **scoring** the data. SPSS, Clementine, and AnswerTree have procedures for building predictive models such as regression, clustering, tree, and neural network models. Once a model has been built, the model specifications can be saved as an XML file containing all of the information necessary to reconstruct the model. The SPSS Server product then provides the means to read an XML model file and apply the model to a dataset.

Example. A credit application is rated for risk based on various aspects of the applicant and the loan in question. The credit score obtained from the risk model is used to accept or reject the loan application.

Scoring is treated as a transformation of the data. The model is expressed internally as a set of numeric transformations to be applied to a given set of variables—the predictor variables specified in the model—in order to obtain a predicted result. In this sense, the process of scoring data with a given model is inherently the same as applying any function, such as a square root function, to a set of data.

Scoring is only available with SPSS Server and can be done interactively by users working in distributed analysis mode. For scoring large data files you'll probably want to make use of the SPSS Batch Facility, a separate executable version of SPSS provided with SPSS Server. For information about using the SPSS Batch Facility, see the *SPSS Batch Facility User's Guide*, provided as a PDF file on the SPSS Server product CD.

The scoring process consists of the following:

► Loading a model from a file in XML (PMML) format.

► Computing your scores as a new variable, using the ApplyModel or StrApplyModel function available in the Compute Variable dialog box.

For details concerning the `ApplyModel` or `StrApplyModel` function, see Scoring Expressions in the Transformation Expressions section of the *SPSS Command Syntax Reference*.

The following table lists the SPSS procedures that support the export of model specifications to XML. The exported models can be used with SPSS Server to score new data as described above. The full list of model types that can be scored with SPSS Server is provided in the description of the `ApplyModel` function.

Procedure Name	Command Name	SPSS Option
Discriminant	DISCRIMINANT	Base
Linear Regression	REGRESSION	Base
TwoStep Cluster	TWOSTEP CLUSTER	Base
Generalized Linear Models	GENLIN	Advanced Models
Complex Samples General Linear Model	CSGLM	Complex Samples
Complex Samples Logistic Regression	CSLOGISTIC	Complex Samples
Complex Samples Ordinal Regression	CSORDINAL	Complex Samples
Logistic Regression	LOGISTIC REGRESSION	Regression

Procedure Name	Command Name	SPSS Option
Multinomial Logistic Regression	NOMREG	Regression
Classification Tree	TREE	Tree

Loading a Saved Model

The Load Model dialog box allows you to load predictive models saved in XML (PMML) format and is only available when working in distributed analysis mode. Loading models is a necessary first step to scoring data with them.

Figure 8-33
Load Model output

mregression

SPSS Variables	Model Variables					
	Name	Label	Type	Width	Role	Measurement
Age	Age	Age	Numeric	-1	Predictor	Continuous
Amount	Amount	Amount	Numeric	-1	Predictor	Continuous
Frequency	Frequency	Frequency	Numeric	7	Covariate	Continuous
Gender	Gender	Gender	String	6	Factor	Nominal
Has_Broadband	Has_Broadband	Has_Broadband	Numeric	10	Factor	Nominal
Has_Child	Has_Child	Has_Child	Numeric	10	Factor	Nominal
Income_Group	Income_Group	Income_Group	Numeric	10	Factor	Nominal
Recency	Recency	Recency	Numeric	10	Covariate	Continuous
Response	Response	Response	Numeric	10	Target	Nominal

To Load a Model

▶ From the menus choose:
Transform
  Prepare Model
    Load Model...

Figure 8-34
Prepare Model Load Model dialog box

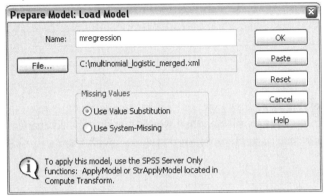

▶ Enter a name to associate with this model. Each loaded model must have a unique name.

▶ Click File and select a model file. The resulting Open File dialog box displays the files that are available in distributed analysis mode. This includes files on the machine where SPSS Server is installed and files on your local computer that reside in shared folders or on shared drives.

Note: When you score data, the model will be applied to variables in the active dataset with the same names as the variables from the model file. You can map variables in the original model to different variables in the active dataset with the use of command syntax (see the MODEL HANDLE command).

Name. A name used to identify this model. The rules for valid model names are the same as for SPSS variable names (see "Variable Names" in Chapter 5 on p. 106), with the addition of the $ character as an allowed first character. You will use this name to specify the model when scoring data with the ApplyModel or StrApplyModel functions.

File. The XML (PMML) file containing the model specifications.

Missing Values

This group of options controls the treatment of missing values, encountered during the scoring process, for the predictor variables defined in the model. A missing value in the context of scoring refers to one of the following:

■ A predictor variable contains no value. For numeric variables, this means the system-missing value. For string variables, this means a null string.

■ The value has been defined as user-missing, in the model, for the given predictor. Values defined as user-missing in the active dataset, but not in the model, are not treated as missing values in the scoring process.

■ The predictor variable is categorical and the value is not one of the categories defined in the model.

Use Value Substitution. Attempt to use value substitution when scoring cases with missing values. The method for determining a value to substitute for a missing value depends on the type of predictive model.

■ **SPSS Models.** For independent variables in linear regression and discriminant models, if mean value substitution for missing values was specified when building and saving the model, then this mean value is used in place of the missing value in the scoring computation, and scoring proceeds. If the mean value is not available, then the system-missing value is returned.

■ **AnswerTree models & SPSS TREE command models.** For the CHAID and Exhaustive CHAID models, the biggest child node is selected for a missing split variable. The biggest child node is the one with the largest population among the child nodes using learning sample cases. For C&RT and QUEST models, surrogate split variables (if any) are used first. (Surrogate splits are splits that attempt to match the original split as closely as possible using alternate predictors.) If no surrogate splits are specified or all surrogate split variables are missing, the biggest child node is used.

■ **Clementine models.** Linear regression models are handled as described under SPSS models. Logistic regression models are handled as described under Logistic Regression models. C&RT Tree models are handled as described for C&RT models under AnswerTree models.

■ **Logistic Regression models.** For covariates in logistic regression models, if a mean value of the predictor was included as part of the saved model, then this mean value is used in place of the missing value in the scoring computation, and scoring proceeds. If the predictor is categorical (for example, a factor in a logistic regression model), or if the mean value is not available, then the system-missing value is returned.

Use System-Missing. Return the system-missing value when scoring a case with a missing value.

Displaying a List of Loaded Models

You can obtain a list of the currently loaded models. From the menus (only available in distributed analysis mode) choose:

Transform
  Prepare Model
    List Model(s)

This will generate a Model Handles table. The table contains a list of all currently loaded models and includes the name (referred to as the model handle) assigned to the model, the type of model, the path to the model file, and the method for handling missing values.

Figure 8-35
Listing of loaded models

Model Handles

Name	Type	Input File Name	Validation handling	Missing handling
mregression	NOMREG	C:\models\mregression.xml	ON	SUBSTITUTE
tree	TREE	C:\models\tree.xml	ON	SUBSTITUTE

Additional Features Available with Command Syntax

From the Load Model dialog box, you can paste your selections into a syntax window and edit the resulting MODEL HANDLE command syntax. This allows you to:

■ Map variables in the original model to different variables in the active dataset (with the MAP subcommand). By default, the model is applied to variables in the active dataset with the same names as the variables from the model file.

See the *SPSS Command Syntax Reference* for complete syntax information.

File Handling and File Transformations

Data files are not always organized in the ideal form for your specific needs. You may want to combine data files, sort the data in a different order, select a subset of cases, or change the unit of analysis by grouping cases together. A wide range of file transformation capabilities is available, including the ability to:

Sort data. You can sort cases based on the value of one or more variables.

Transpose cases and variables. The SPSS data file format reads rows as cases and columns as variables. For data files in which this order is reversed, you can switch the rows and columns and read the data in the correct format.

Merge files. You can merge two or more data files. You can combine files with the same variables but different cases or the same cases but different variables.

Select subsets of cases. You can restrict your analysis to a subset of cases or perform simultaneous analyses on different subsets.

Aggregate data. You can change the unit of analysis by aggregating cases based on the value of one or more grouping variables.

Weight data. Weight cases for analysis based on the value of a weight variable.

Restructure data. You can restructure data to create a single case (record) from multiple cases or create multiple cases from a single case.

Sort Cases

This dialog box sorts cases (rows) of the data file based on the values of one or more sorting variables. You can sort cases in ascending or descending order.

- If you select multiple sort variables, cases are sorted by each variable within categories of the preceding variable on the Sort list. For example, if you select *gender* as the first sorting variable and *minority* as the second sorting variable, cases will be sorted by minority classification within each gender category.

- The sort sequence is based on the locale-defined order (and is not necessarily the same as the numerical order of the character codes). The default locale is the operating system locale. You can control the locale with the Language setting on the General tab of the Options dialog box (Edit menu).

Figure 9-1
Sort Cases dialog box

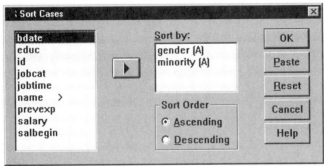

To Sort Cases

▶ From the menus choose:

Data
  Sort Cases...

▶ Select one or more sorting variables.

Transpose

Transpose creates a new data file in which the rows and columns in the original data file are transposed so that cases (rows) become variables and variables (columns) become cases. Transpose automatically creates new variable names and displays a list of the new variable names.

- A new string variable that contains the original variable name, *case_lbl*, is automatically created.

■ If the active dataset contains an ID or name variable with unique values, you can use it as the name variable, and its values will be used as variable names in the transposed data file. If it is a numeric variable, the variable names start with the letter *V*, followed by the numeric value.

■ User-missing values are converted to the system-missing value in the transposed data file. To retain any of these values, change the definition of missing values in the Variable View in the Data Editor.

To Transpose Variables and Cases

▶ From the menus choose:
Data
  Transpose...

▶ Select one or more variables to transpose into cases.

Merging Data Files

You can merge data from two files in two different ways. You can:

■ Merge the active dataset with another open dataset or SPSS-format data file containing the same variables but different cases.

■ Merge the active dataset with another open dataset or SPSS-format data file containing the same cases but different variables.

To Merge Files

▶ From the menus choose:
Data
  Merge Files

▶ Select Add Cases or Add Variables.

Figure 9-2
Selecting files to merge

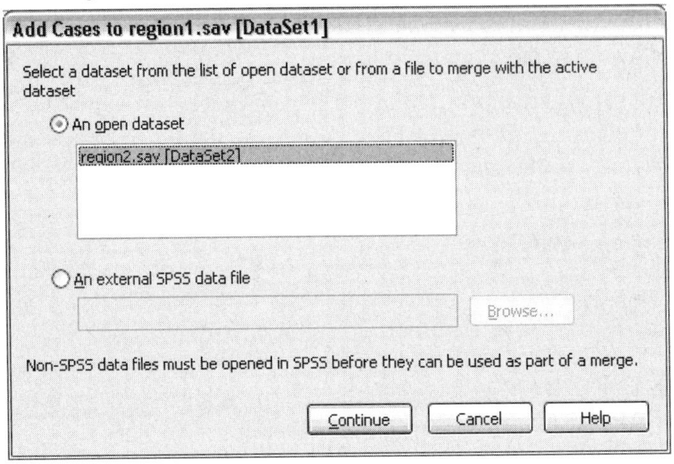

Add Cases

Add Cases merges the active dataset with a second dataset or SPSS-format data file that contains the same variables (columns) but different cases (rows). For example, you might record the same information for customers in two different sales regions and maintain the data for each region in separate files. The second dataset can be an external SPSS-format data file or a dataset available in the current session.

Figure 9-3
Add Cases dialog box

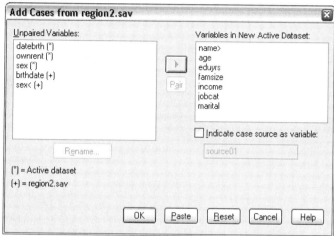

Unpaired Variables. Variables to be excluded from the new, merged data file. Variables from the active dataset are identified with an asterisk (*). Variables from the other dataset are identified with a plus sign (+). By default, this list contains:

- Variables from either data file that do not match a variable name in the other file. You can create pairs from unpaired variables and include them in the new, merged file.

- Variables defined as numeric data in one file and string data in the other file. Numeric variables cannot be merged with string variables.

- String variables of unequal width. The defined width of a string variable must be the same in both data files.

Variables in New Active Dataset. Variables to be included in the new, merged data file. By default, all of the variables that match both the name and the data type (numeric or string) are included on the list.

- You can remove variables from the list if you do not want them to be included in the merged file.

- Any unpaired variables included in the merged file will contain missing data for cases from the file that does not contain that variable.

Indicate case source as variable. Indicates the source data file for each case. This variable has a value of 0 for cases from the active dataset and a value of 1 for cases from the external data file.

To Merge Data Files with the Same Variables and Different Cases

▶ Open at least one of the data files that you want to merge. If you have multiple datasets open, make one of the datasets that you want to merge the active dataset. The cases from this file will appear first in the new, merged data file.

▶ From the menus choose:

Data
  Merge Files
    Add Cases...

▶ Select the dataset or SPSS-format data file to merge with the active dataset.

▶ Remove any variables that you do not want from the Variables in New Active Dataset list.

▶ Add any variable pairs from the Unpaired Variables list that represent the same information recorded under different variable names in the two files. For example, date of birth might have the variable name *brthdate* in one file and *datebrth* in the other file.

To Select a Pair of Unpaired Variables

▶ Click one of the variables on the Unpaired Variables list.

▶ Ctrl-click the other variable on the list. (Press the Ctrl key and click the left mouse button at the same time.)

▶ Click Pair to move the variable pair to the Variables in New Active Dataset list. (The variable name from the active dataset is used as the variable name in the merged file.)

Figure 9-4
Selecting pairs of variables with Ctrl-click

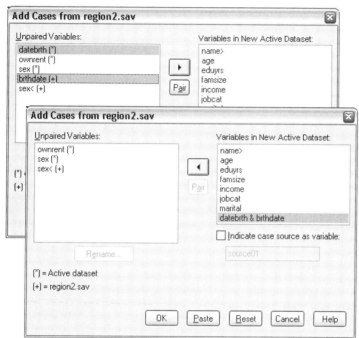

Add Cases: Rename

You can rename variables from either the active dataset or the other dataset before moving them from the unpaired list to the list of variables to be included in the merged data file. Renaming variables enables you to:

- Use the variable name from the other dataset rather than the name from the active dataset for variable pairs.

- Include two variables with the same name but of unmatched types or different string widths. For example, to include both the numeric variable *sex* from the active dataset and the string variable *sex* from the other dataset, one of them must be renamed first.

Add Cases: Dictionary Information

Any existing dictionary information (variable and value labels, user-missing values, display formats) in the active dataset is applied to the merged data file.

■ If any dictionary information for a variable is undefined in the active dataset, dictionary information from the other dataset is used.

■ If the active dataset contains any defined value labels or user-missing values for a variable, any additional value labels or user-missing values for that variable in the other dataset are ignored.

Merging More Than Two Data Sources

Using command syntax, you can merge up to 50 datasets and/or data files. For more information, see the ADD FILES command in the *SPSS Command Syntax Reference* (available from the Help menu).

Add Variables

Add Variables merges the active dataset with another open dataset or SPSS-format data file that contains the same cases (rows) but different variables (columns). For example, you might want to merge a data file that contains pre-test results with one that contains post-test results.

■ Cases must be sorted in the same order in both datasets.

■ If one or more key variables are used to match cases, the two datasets must be sorted by ascending order of the key variable(s).

■ Variable names in the second data file that duplicate variable names in the active dataset are excluded by default because Add Variables assumes that these variables contain duplicate information.

Indicate case source as variable. Indicates the source data file for each case. This variable has a value of 0 for cases from the active dataset and a value of 1 for cases from the external data file.

Figure 9-5
Add Variables dialog box

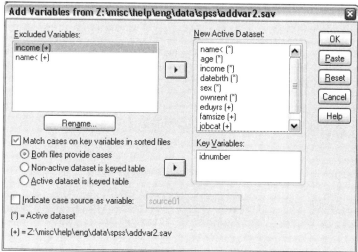

Excluded Variables. Variables to be excluded from the new, merged data file. By default, this list contains any variable names from the other dataset that duplicate variable names in the active dataset. Variables from the active dataset are identified with an asterisk (*). Variables from the other dataset are identified with a plus sign (+). If you want to include an excluded variable with a duplicate name in the merged file, you can rename it and add it to the list of variables to be included.

New Active Dataset. Variables to be included in the new, merged dataset. By default, all unique variable names in both datasets are included on the list.

Key Variables. If some cases in one dataset do not have matching cases in the other dataset (that is, some cases are missing in one dataset), use key variables to identify and correctly match cases from the two datasets. You can also use key variables with table lookup files.

- The key variables must have the same names in both datasets.

- Both datasets must be sorted by ascending order of the key variables, and the order of variables on the Key Variables list must be the same as their sort sequence.

- Cases that do not match on the key variables are included in the merged file but are not merged with cases from the other file. Unmatched cases contain values for only the variables in the file from which they are taken; variables from the other file contain the system-missing value.

Non-active or active dataset is keyed table. A keyed table, or **table lookup file**, is a file in which data for each "case" can be applied to multiple cases in the other data file. For example, if one file contains information on individual family members (such as sex, age, education) and the other file contains overall family information (such as total income, family size, location), you can use the file of family data as a table lookup file and apply the common family data to each individual family member in the merged data file.

To Merge Files with the Same Cases but Different Variables

▶ Open at least one of the data files that you want to merge. If you have multiple datasets open, make one of the datasets that you want to merge the active dataset.

▶ From the menus choose:
Data
  Merge Files
    Add Variables...

▶ Select the dataset or SPSS-format data file to merge with the active dataset.

To Select Key Variables

▶ Select the variables from the external file variables (+) on the Excluded Variables list.

▶ Select Match cases on key variables in sorted files.

▶ Add the variables to the Key Variables list.

The key variables must exist in both the active dataset and the other dataset. Both datasets must be sorted by ascending order of the key variables, and the order of variables on the Key Variables list must be the same as their sort sequence.

Add Variables: Rename

You can rename variables from either the active dataset or the other dataset before moving them to the list of variables to be included in the merged data file. This is primarily useful if you want to include two variables with the same name that contain different information in the two files.

Merging More Than Two Data Sources

Using command syntax, you can merge up to 50 datasets and/or data files. For more information, see the MATCH FILES command in the *SPSS Command Syntax Reference* (available from the Help menu).

Aggregate Data

Aggregate Data aggregates groups of cases in the active dataset into single cases and creates a new, aggregated file or creates new variables in the active dataset that contain aggregated data. Cases are aggregated based on the value of one or more break (grouping) variables.

- If you create a new, aggregated data file, the new data file contains one case for each group defined by the break variables. For example, if there is one break variable with two values, the new data file will contain only two cases.

- If you add aggregate variables to the active dataset, the data file itself is not aggregated. Each case with the same value(s) of the break variable(s) receives the same values for the new aggregate variables. For example, if *gender* is the only break variable, all males would receive the same value for a new aggregate variable that represents average age.

Figure 9-6
Aggregate Data dialog box

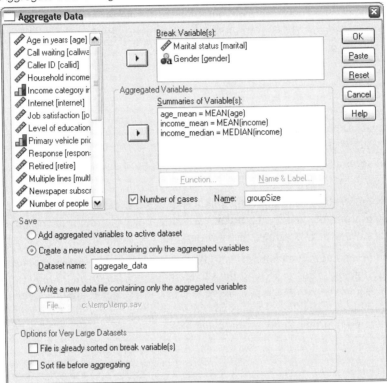

Break Variable(s). Cases are grouped together based on the values of the break variables. Each unique combination of break variable values defines a group. When creating a new, aggregated data file, all break variables are saved in the new file with their existing names and dictionary information. The break variable can be either numeric or string.

Aggregated Variables. Source variables are used with aggregate functions to create new aggregate variables. The aggregate variable name is followed by an optional variable label in quotes, the name of the aggregate function, and the source variable name in parentheses. Source variables for aggregate functions must be numeric.

You can override the default aggregate variable names with new variable names, provide descriptive variable labels, and change the functions used to compute the aggregated data values. You can also create a variable that contains the number of cases in each break group.

To Aggregate a Data File

▶ From the menus choose:
Data
  Aggregate...

▶ Select one or more break variables that define how cases are grouped to create aggregated data.

▶ Select one or more aggregate variables.

▶ Select an aggregate function for each aggregate variable.

Saving Aggregated Results

You can add aggregate variables to the active dataset or create a new, aggregated data file.

■ **Add aggregated variables to active dataset.** New variables based on aggregate functions are added to the active dataset. The data file itself is not aggregated. Each case with the same value(s) of the break variable(s) receives the same values for the new aggregate variables.

■ **Create a new dataset containing only the aggregated variables.** Saves aggregated data to a new dataset in the current session. The dataset includes the break variables that define the aggregated cases and all aggregate variables defined by aggregate functions. The active dataset is unaffected.

■ **Write a new data file containing only the aggregated variables.** Saves aggregated data to an external data file. The file includes the break variables that define the aggregated cases and all aggregate variables defined by aggregate functions. The active dataset is unaffected.

Sorting Options for Large Data Files

For very large data files, it may be more efficient to aggregate presorted data.

File is already sorted on break variable(s). If the data have already been sorted by values of the break variables, this option enables the procedure to run more quickly and use less memory. Use this option with caution.

■ Data must by sorted by values of the break variables in the same order as the break variables specified for the Aggregate Data procedure.

■ If you are adding variables to the active dataset, select this option only if the data are sorted by ascending values of the break variables.

Sort file before aggregating. In very rare instances with large data files, you may find it necessary to sort the data file by values of the break variables prior to aggregating. This option is not recommended unless you encounter memory or performance problems.

Aggregate Data: Aggregate Function

This dialog box specifies the function to use to calculate aggregated data values for selected variables on the Aggregate Variables list in the Aggregate Data dialog box. Aggregate functions include:

■ Summary functions, including mean, median, standard deviation, and sum

■ Number of cases, including unweighted, weighted, nonmissing, and missing

■ Percentage or fraction of values above or below a specified value

■ Percentage or fraction of values inside or outside of a specified range

Figure 9-7
Aggregate Function dialog box

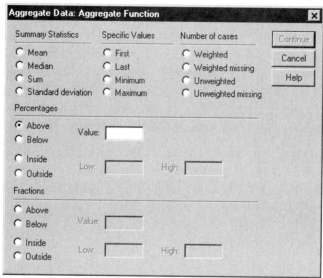

Aggregate Data: Variable Name and Label

Aggregate Data assigns default variable names for the aggregated variables in the new data file. This dialog box enables you to change the variable name for the selected variable on the Aggregate Variables list and provide a descriptive variable label. For more information, see "Variable Names" in Chapter 5 on p. 106.

Figure 9-8
Variable Name and Label dialog box

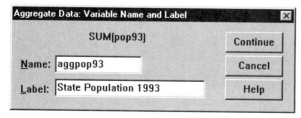

Split File

Split File splits the data file into separate groups for analysis based on the values of one or more grouping variables. If you select multiple grouping variables, cases are grouped by each variable within categories of the preceding variable on the Groups Based On list. For example, if you select *gender* as the first grouping variable and *minority* as the second grouping variable, cases will be grouped by minority classification within each gender category.

■   You can specify up to eight grouping variables.

■   Each eight characters of a long string variable (string variables longer than eight characters) counts as a variable toward the limit of eight grouping variables.

■   Cases should be sorted by values of the grouping variables and in the same order that variables are listed in the Groups Based On list. If the data file isn't already sorted, select Sort the file by grouping variables.

Figure 9-9
Split File dialog box

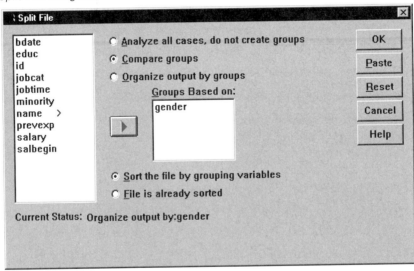

Compare groups. Split-file groups are presented together for comparison purposes. For pivot tables, a single pivot table is created and each split-file variable can be moved between table dimensions. For charts, a separate chart is created for each split-file group and the charts are displayed together in the Viewer.

Organize output by groups. All results from each procedure are displayed separately for each split-file group.

To Split a Data File for Analysis

▶ From the menus choose:
Data
  Split File...

▶ Select Compare groups or Organize output by groups.

▶ Select one or more grouping variables.

Select Cases

Select Cases provides several methods for selecting a subgroup of cases based on criteria that include variables and complex expressions. You can also select a random sample of cases. The criteria used to define a subgroup can include:

■ Variable values and ranges

■ Date and time ranges

■ Case (row) numbers

■ Arithmetic expressions

■ Logical expressions

■ Functions

Figure 9-10
Select Cases dialog box

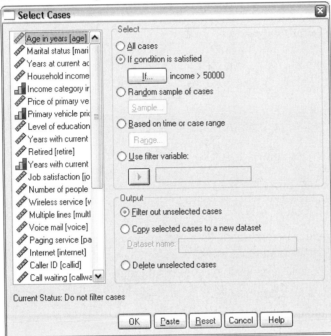

All cases. Turns case filtering off and uses all cases.

If condition is satisfied. Use a conditional expression to select cases. If the result of the conditional expression is true, the case is selected. If the result is false or missing, the case is not selected.

Random sample of cases. Selects a random sample based on an approximate percentage or an exact number of cases.

Based on time or case range. Selects cases based on a range of case numbers or a range of dates/times.

Use filter variable. Use the selected numeric variable from the data file as the filter variable. Cases with any value other than 0 or missing for the filter variable are selected.

Output

This section controls the treatment of unselected cases. You can choose one of the following alternatives for the treatment of unselected cases:

■ **Filter out unselected cases.** Unselected cases are not included in the analysis but remain in the dataset. You can use the unselected cases later in the session if you turn filtering off. If you select a random sample or if you select cases based on a conditional expression, this generates a variable named *filter_$* with a value of 1 for selected cases and a value of 0 for unselected cases.

■ **Copy selected cases to a new dataset.** Selected cases are copied to a new dataset, leaving the original dataset unaffected. Unselected cases are not included in the new dataset and are left in their original state in the original dataset.

■ **Delete unselected cases.** Unselected cases are deleted from the dataset. Deleted cases can be recovered only by exiting from the file without saving any changes and then reopening the file. The deletion of cases is permanent if you save the changes to the data file. *Note*: If you delete unselected cases and save the file, the cases cannot be recovered.

To Select a Subset of Cases

▶ From the menus choose:
Data
 Select Cases...

▶ Select one of the methods for selecting cases.

▶ Specify the criteria for selecting cases.

Select Cases: If

This dialog box allows you to select subsets of cases using conditional expressions. A conditional expression returns a value of *true*, *false*, or *missing* for each case.

Figure 9-11
Select Cases If dialog box

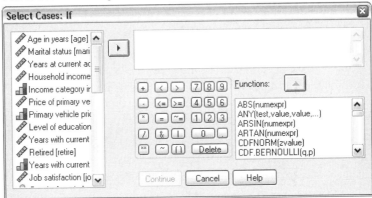

- If the result of a conditional expression is *true*, the case is included in the selected subset.

- If the result of a conditional expression is *false* or *missing*, the case is not included in the selected subset.

- Most conditional expressions use one or more of the six relational operators (<, >, <=, >=, =, and ~=) on the calculator pad.

- Conditional expressions can include variable names, constants, arithmetic operators, numeric (and other) functions, logical variables, and relational operators.

Select Cases: Random Sample

This dialog box allows you to select a random sample based on an approximate percentage or an exact number of cases. Sampling is performed without replacement; so, the same case cannot be selected more than once.

Figure 9-12
Select Cases Random Sample dialog box

Approximately. Generates a random sample of approximately the specified percentage of cases. Since this routine makes an independent pseudo-random decision for each case, the percentage of cases selected can only approximate the specified percentage. The more cases there are in the data file, the closer the percentage of cases selected is to the specified percentage.

Exactly. A user-specified number of cases. You must also specify the number of cases from which to generate the sample. This second number should be less than or equal to the total number of cases in the data file. If the number exceeds the total number of cases in the data file, the sample will contain proportionally fewer cases than the requested number.

Select Cases: Range

This dialog box selects cases based on a range of case numbers or a range of dates or times.

■ Case ranges are based on row number as displayed in the Data Editor.

■ Date and time ranges are available only for **time series data** with defined date variables (Data menu, Define Dates).

Figure 9-13
Select Cases Range dialog box for range of cases (no defined date variables)

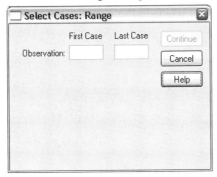

Figure 9-14
Select Cases Range dialog box for time series data with defined date variables

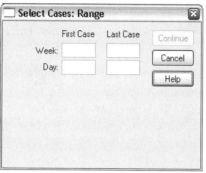

Weight Cases

Weight Cases gives cases different weights (by simulated replication) for statistical analysis.

■ The values of the weighting variable should indicate the number of observations represented by single cases in your data file.

■ Cases with zero, negative, or missing values for the weighting variable are excluded from analysis.

■ Fractional values are valid; they are used exactly where this is meaningful and most likely where cases are tabulated.

Figure 9-15
Weight Cases dialog box

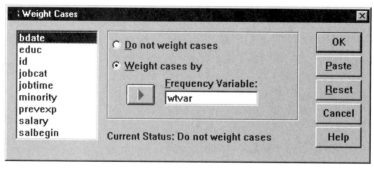

Once you apply a weight variable, it remains in effect until you select another weight variable or turn off weighting. If you save a weighted data file, weighting information is saved with the data file. You can turn off weighting at any time, even after the file has been saved in weighted form.

Weights in Crosstabs. The Crosstabs procedure has several options for handling case weights. For more information, see "Crosstabs Cell Display" in Chapter 17 on p. 377.

Weights in scatterplots and histograms. Scatterplots and histograms have an option for turning case weights on and off, but this does not affect cases with a zero, negative, or missing value for the weight variable. These cases remain excluded from the chart even if you turn weighting off from within the chart.

To Weight Cases

▶ From the menus choose:
Data
  Weight Cases...

▶ Select Weight cases by.

▶ Select a frequency variable.

The values of the frequency variable are used as case weights. For example, a case with a value of 3 for the frequency variable will represent three cases in the weighted data file.

Restructuring Data

Use the Restructure Data Wizard to restructure your data for the SPSS procedure that you want to use. The wizard replaces the current file with a new, restructured file. The wizard can:

- Restructure selected variables into cases.
- Restructure selected cases into variables.
- Transpose all data.

To Restructure Data

▶ From the menus choose:
  Data
    Restructure...

▶ Select the type of restructuring that you want to do.

▶ Select the data to restructure.

  Optionally, you can:

  ■ Create identification variables, which allow you to trace a value in the new file back to a value in the original file.

  ■ Sort the data prior to restructuring.

  ■ Define options for the new file.

  ■ Paste the command syntax into a syntax window.

Restructure Data Wizard: Select Type

Use the Restructure Data Wizard to restructure your data. In the first dialog box, select the type of restructuring that you want to do.

Figure 9-16
Restructure Data Wizard

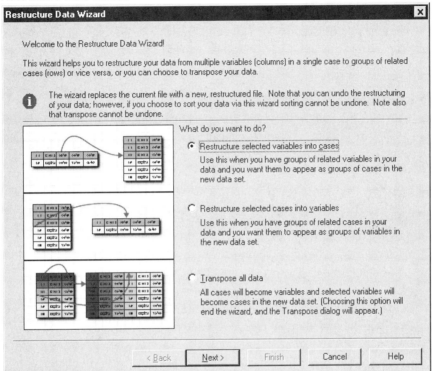

- **Restructure selected variables into cases.** Choose this when you have groups of related columns in your data and you want them to appear in groups of rows in the new data file. If you choose this, the wizard will display the steps for **Variables to Cases**.

- **Restructure selected cases into variables.** Choose this when you have groups of related rows in your data and you want them to appear in groups of columns in the new data file. If you choose this, the wizard will display the steps for **Cases to Variables**.

- **Transpose all data.** Choose this when you want to transpose your data. All rows will become columns and all columns will become rows in the new data. This choice closes the Restructure Data Wizard and opens the Transpose Data dialog box.

Deciding How to Restructure the Data

A **variable** contains information that you want to analyze—for example, a measurement or a score. A **case** is an observation—for example, an individual. In a *simple* data structure, each variable is a single column in your data and each case is a single row. So, for example, if you were measuring test scores for all students in a class, all score values would appear in only one column, and there would be a row for each student.

When you analyze data, you are often analyzing how a variable varies according to some condition. The condition can be a specific experimental treatment, a demographic, a point in time, or something else. In data analysis, conditions of interest are often referred to as **factors**. When you analyze factors, you have a *complex* data structure. You may have information about a variable in more than one column in your data (for example, a column for each level of a factor), or you may have information about a case in more than one row (for example, a row for each level of a factor). The Restructure Data Wizard helps you to restructure files with a complex data structure.

The structure of the current file and the structure that you want in the new file determine the choices that you make in the wizard.

How are the data arranged in the current file? The current data may be arranged so that factors are recorded in a *separate* variable (in groups of cases) or *with* the variable (in groups of variables).

■ **Groups of cases.** Does the current file have variables and conditions recorded in separate columns? For example:

var	factor
8	1
9	1
3	2
1	2

In this example, the first two rows are a **case group** because they are related. They contain data for the same factor level. In SPSS data analysis, the factor is often referred to as a **grouping variable** when the data are structured this way.

■ **Groups of columns.** Does the current file have variables and conditions recorded in the same column? For example:

var_1	var_2
8	3
9	1

In this example, the two columns are a **variable group** because they are related. They contain data for the same variable—*var_1* for factor level 1 and *var_2* for factor level 2. In SPSS data analysis, the factor is often referred to as a **repeated measure** when the data are structured this way.

How should the data be arranged in the new file? This is usually determined by the procedure that you want to use to analyze your data.

■ **Procedures that require groups of cases.** Your data must be structured in case groups to do analyses that require a grouping variable. Examples are *univariate*, *multivariate*, and *variance components* with General Linear Model, Mixed Models, and OLAP Cubes; and *independent samples* with T Test or Nonparametric Tests. If your current data structure is variable groups and you want to do these analyses, select Restructure selected variables into cases.

■ **Procedures that require groups of variables.** Your data must be structured in variable groups to analyze repeated measures. Examples are *repeated measures* with General Linear Model, *time-dependent covariate* analysis with Cox Regression Analysis, *paired samples* with T Test, or *related samples* with Nonparametric Tests. If your current data structure is case groups and you want to do these analyses, select Restructure selected cases into variables.

Example of Variables to Cases

In this example, test scores are recorded in separate columns for each factor, *A* and *B*.

Figure 9-17
Current data for variables to cases

	score_a	score_b
1	1014.00	864.00
2	684.00	636.00
3	810.00	638.00

You want to do an independent-samples *t* test. You have a column group consisting of *score_a* and *score_b*, but you don't have the **grouping variable** that the procedure requires. Select Restructure selected variables into cases in the Restructure Data Wizard, restructure one variable group into a new variable named *score*, and create an index named *group*. The new data file is shown in the following figure.

Figure 9-18
New, restructured data for variables to cases

	group	score
1	SCORE_A	1014.00
2	SCORE_B	864.00
3	SCORE_A	684.00
4	SCORE_B	636.00
5	SCORE_A	810.00
6	SCORE_B	638.00

When you run the independent-samples *t* test, you can now use *group* as the grouping variable.

Example of Cases to Variables

In this example, test scores are recorded twice for each subject—before and after a treatment.

Figure 9-19
Current data for cases to variables

	id	scor	time
1	1	1014.00	bef
2	1	864.00	aft
3	2	684.00	bef
4	2	636.00	aft

You want to do a paired-samples *t* test. Your data structure is case groups, but you don't have the **repeated measures** for the paired variables that the procedure requires. Select Restructure selected cases into variables in the Restructure Data Wizard, use *id* to identify the row groups in the current data, and use *time* to create the variable group in the new file.

Figure 9-20
New, restructured data for cases to variables

	id	aft	bef
1	1	864.00	1014.00
2	2	636.00	684.00

When you run the paired-samples *t* test, you can now use *bef* and *aft* as the variable pair.

Restructure Data Wizard (Variables to Cases): Number of Variable Groups

Note: The wizard presents this step if you choose to restructure variable groups into rows.

In this step, decide how many variable groups in the current file that you want to restructure in the new file.

How many variable groups are in the current file? Think about how many variable groups exist in the current data. A group of related columns, called a **variable group**, records repeated measures of the same variable in separate columns. For example, if you have three columns in the current data—*w1*, *w2*, and *w3*—that record **width**, you have one variable group. If you have an additional three columns—*h1*, *h2*, and *h3*—that record **height**, you have two variable groups.

How many variable groups should be in the new file? Consider how many variable groups you want to have represented in the new data file. You do not have to restructure all variable groups into the new file.

Figure 9-21
Restructure Data Wizard: Number of Variable Groups, Step 2

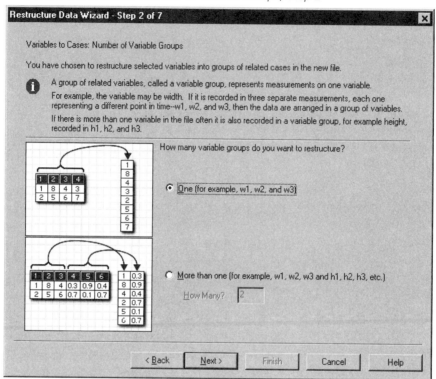

- **One.** The wizard will create a single restructured variable in the new file from one variable group in the current file.

- **More than one.** The wizard will create multiple restructured variables in the new file. The number that you specify affects the next step, in which the wizard automatically creates the specified number of new variables.

Restructure Data Wizard (Variables to Cases): Select Variables

Note: The wizard presents this step if you choose to restructure variable groups into rows.

In this step, provide information about how the variables in the current file should be used in the new file. You can also create a variable that identifies the rows in the new file.

Figure 9-22
Restructure Data Wizard: Select Variables, Step 3

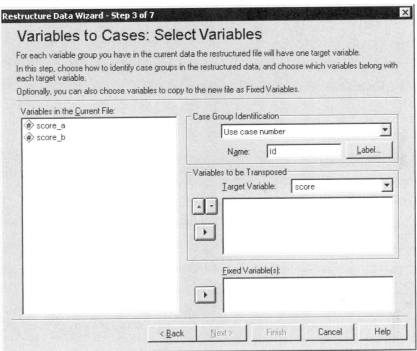

How should the new rows be identified? You can create a variable in the new data file that identifies the row in the current data file that was used to create a group of new rows. The identifier can be a sequential case number or it can be the values of the variable. Use the controls in Case Group Identification to define the identification variable in the new file. Click a cell to change the default variable name and provide a descriptive variable label for the identification variable.

What should be restructured in the new file? In the previous step, you told the wizard how many variable groups you want to restructure. The wizard created one new variable for each group. The values for the variable group will appear in that variable in the new file. Use the controls in Variables to be Transposed to define the restructured variable in the new file.

To Specify One Restructured Variable

▶ Put the variables that make up the variable group that you want to transform into the Variables to be Transposed list. All of the variables in the group must be of the same type (numeric or string).

You can include the same variable more than once in the variable group (variables are copied rather than moved from the source variable list); its values are repeated in the new file.

To Specify Multiple Restructured Variables

▶ Select the first target variable that you want to define from the Target Variable drop-down list.

▶ Put the variables that make up the variable group that you want to transform into the Variables to be Transposed list. All of the variables in the group must be of the same type (numeric or string). You can include the same variable more than once in the variable group. (A variable is copied rather than moved from the source variable list, and its values are repeated in the new file.)

▶ Select the next target variable that you want to define, and repeat the variable selection process for all available target variables.

- Although you can include the same variable more than once in the same target variable group, you cannot include the same variable in more than one target variable group.

- Each target variable group list must contain the same number of variables. (Variables that are listed more than once are included in the count.)

- The number of target variable groups is determined by the number of variable groups that you specified in the previous step. You can change the default variable names here, but you must return to the previous step to change the number of variable groups to restructure.

- You must define variable groups (by selecting variables in the source list) for all available target variables before you can proceed to the next step.

What should be copied into the new file? Variables that aren't restructured can be copied into the new file. Their values will be propagated in the new rows. Move variables that you want to copy into the new file into the Fixed Variable(s) list.

Restructure Data Wizard (Variables to Cases): Create Index Variables

Note: The wizard presents this step if you choose to restructure variable groups into rows.

In this step, decide whether to create index variables. An index is a new variable that sequentially identifies a row group based on the original variable from which the new row was created.

Figure 9-23
Restructure Data Wizard: Create Index Variables, Step 4

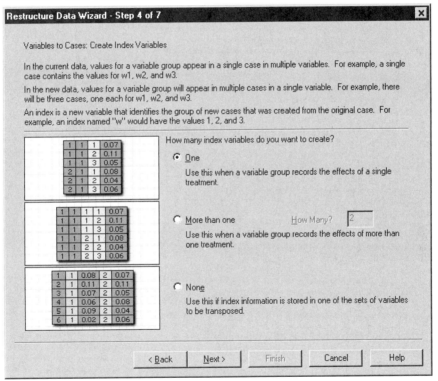

How many index variables should be in the new file? Index variables can be used as grouping variables in SPSS procedures. In most cases, a single index variable is sufficient; however, if the variable groups in your current file reflect multiple factor levels, multiple indices may be appropriate.

■ **One.** The wizard will create a single index variable.

■ **More than one.** The wizard will create multiple indices and enter the number of indices that you want to create. The number that you specify affects the next step, in which the wizard automatically creates the specified number of indices.

■ **None.** Select this if you do not want to create index variables in the new file.

Example of One Index for Variables to Cases

In the current data, there is one variable group, *width*, and one factor, *time*. Width was measured three times and recorded in *w1*, *w2*, and *w3*.

Figure 9-24
Current data for one index

	subject	w1	w2	w3
1	1	6.70	4.30	5.70
2	2	7.10	5.90	5.60

We'll restructure the variable group into a single variable, *width*, and create a single numeric index. The new data are shown in the following table.

Figure 9-25
New, restructured data with one index

	subject	index	width
1	1	1	6.70
2	1	2	4.30
3	1	3	5.70
4	2	1	7.10
5	2	2	5.90
6	2	3	5.60

Index starts with 1 and increments for each variable in the group. It restarts each time a new row is encountered in the original file. We can now use *index* in SPSS procedures that require a grouping variable.

Example of Two Indices for Variables to Cases

When a variable group records more than one factor, you can create more than one index; however, the current data must be arranged so that the levels of the first factor are a primary index within which the levels of subsequent factors cycle. In the current data, there is one variable group, *width*, and two factors, *A* and *B*. The data are arranged so that levels of factor *B* cycle within levels of factor *A*.

Figure 9-26
Current data for two indices

	subject	w_a1b1	w_a1b2	w_a2b1	w_a2b2
1	1	5.50	6.40	5.80	5.90
2	2	7.40	7.10	5.60	6.70

We'll restructure the variable group into a single variable, *width*, and create two indices. The new data are shown in the following table.

Figure 9-27
New, restructured data with two indices

	subject	index_a	index_b	width
1	1	1	1	5.50
2	1	1	2	6.40
3	1	2	1	5.80
4	1	2	2	5.90
5	2	1	1	7.40
6	2	1	2	7.10
7	2	2	1	5.60
8	2	2	2	6.70

Restructure Data Wizard (Variables to Cases): Create One Index Variable

Note: The wizard presents this step if you choose to restructure variable groups into rows and create one index variable.

In this step, decide what values you want for the index variable. The values can be sequential numbers or the names of the variables in an original variable group. You can also specify a name and a label for the new index variable.

Figure 9-28
Restructure Data Wizard: Create One Index Variable, Step 5

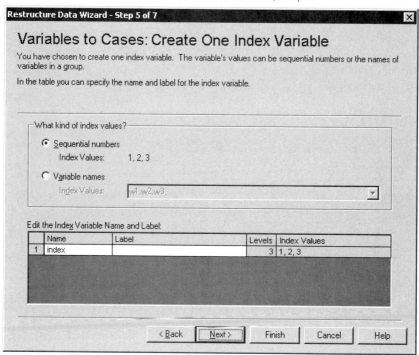

For more information, see "Example of One Index for Variables to Cases" on p. 252.

■ **Sequential numbers.** The wizard will automatically assign sequential numbers as index values.

■ **Variable names.** The wizard will use the names of the selected variable group as index values. Choose a variable group from the list.

■ **Names and labels.** Click a cell to change the default variable name and provide a descriptive variable label for the index variable.

Restructure Data Wizard (Variables to Cases): Create Multiple Index Variables

Note: The wizard presents this step if you choose to restructure variable groups into rows and create multiple index variables.

In this step, specify the number of levels for each index variable. You can also specify a name and a label for the new index variable.

Figure 9-29
Restructure Data Wizard: Create Multiple Index Variables, Step 5

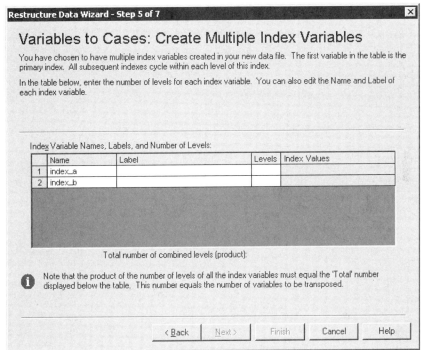

For more information, see "Example of Two Indices for Variables to Cases" on p. 252.

How many levels are recorded in the current file? Consider how many factor levels are recorded in the current data. A **level** defines a group of cases that experienced identical conditions. If there are multiple factors, the current data must be arranged so that the levels of the first factor are a primary index within which the levels of subsequent factors cycle.

How many levels should be in the new file? Enter the number of levels for each index. The values for multiple index variables are always sequential numbers. The values start at 1 and increment for each level. The first index increments the slowest, and the last index increments the fastest.

Total combined levels. You cannot create more levels than exist in the current data. Because the restructured data will contain one row for each combination of treatments, the wizard checks the number of levels that you create. It will compare the product of the levels that you create to the number of variables in your variable groups. They must match.

Names and labels. Click a cell to change the default variable name and provide a descriptive variable label for the index variables.

Restructure Data Wizard (Variables to Cases): Options

Note: The wizard presents this step if you choose to restructure variable groups into rows.

In this step, specify options for the new, restructured file.

Figure 9-30
Restructure Data Wizard: Options, Step 6

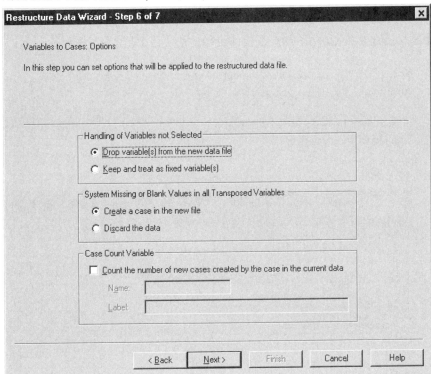

Drop unselected variables? In the Select Variables step (step 3), you selected variable groups to be restructured, variables to be copied, and an identification variable from the current data. The data from the selected variables will appear in the new file. If there are other variables in the current data, you can choose to discard or keep them.

Keep missing data? The wizard checks each potential new row for null values. A **null value** is a system-missing or blank value. You can choose to keep or discard rows that contain only null values.

Create a count variable? The wizard can create a **count variable** in the new file. It contains the number of new rows generated by a row in the current data. A count variable may be useful if you choose to discard null values from the new file because that makes it possible to generate a different number of new rows for a given row in the current data. Click a cell to change the default variable name and provide a descriptive variable label for the count variable.

Restructure Data Wizard (Cases to Variables): Select Variables

Note: The wizard presents this step if you choose to restructure case groups into columns.

In this step, provide information about how the variables in the current file should be used in the new file.

Figure 9-31
Restructure Data Wizard: Select Variables, Step 2

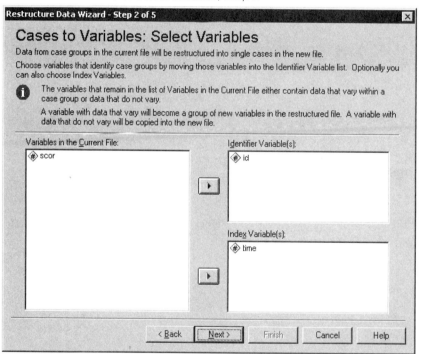

What identifies case groups in the current data? A **case group** is a group of rows that are related because they measure the same observational unit—for example, an individual or an institution. The wizard needs to know which variables in the current file identify the case groups so that it can consolidate each group into a single row in the new file. Move variables that identify case groups in the current file into the Identifier Variable(s) list. Variables that are used to split the current data file are automatically used to identify case groups. Each time a new combination of identification values is encountered, the wizard will create a new row, so cases in the current file should be sorted by values of the identification variables, in the same order that variables are listed in the Identifier Variable(s) list. If the current data file isn't already sorted, you can sort it in the next step.

How should the new variable groups be created in the new file? In the original data, a variable appears in a single column. In the new data file, that variable will appear in multiple new columns. **Index variables** are variables in the current data that the wizard should use to create the new columns. The restructured data will contain one new variable for each unique value in these columns. Move the variables that should be used to form the new variable groups to the Index Variable(s) list. When the wizard presents options, you can also choose to order the new columns by index.

What happens to the other columns? The wizard automatically decides what to do with the variables that remain in the Current File list. It checks each variable to see if the data values vary within a case group. If they do, the wizard restructures the values into a variable group in the new file. If they don't, the wizard copies the values into the new file.

Restructure Data Wizard (Cases to Variables): Sort Data

Note: The wizard presents this step if you choose to restructure case groups into columns.

In this step, decide whether to sort the current file before restructuring it. Each time the wizard encounters a new combination of identification values, a new row is created, so it is important that the data are sorted by the variables that identify case groups.

Figure 9-32
Restructure Data Wizard: Sorting Data, Step 3

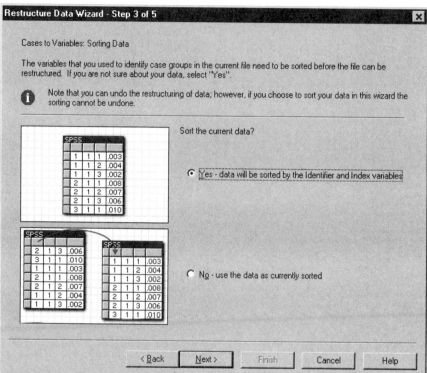

How are the rows ordered in the current file? Consider how the current data are sorted and which variables you are using to identify case groups (specified in the previous step).

- **Yes.** The wizard will automatically sort the current data by the identification variables, in the same order that variables are listed in the Identifier Variable(s) list in the previous step. Choose this when the data aren't sorted by the identification variables or when you aren't sure. This choice requires a separate pass of the data, but it guarantees that the rows are correctly ordered for restructuring.

- **No.** The wizard will not sort the current data. Choose this when you are sure that the current data are sorted by the variables that identify case groups.

Restructure Data Wizard (Cases to Variables): Options

Note: The wizard presents this step if you choose to restructure case groups into columns.

In this step, specify options for the new, restructured file.

Figure 9-33
Restructure Data Wizard: Options, Step 4

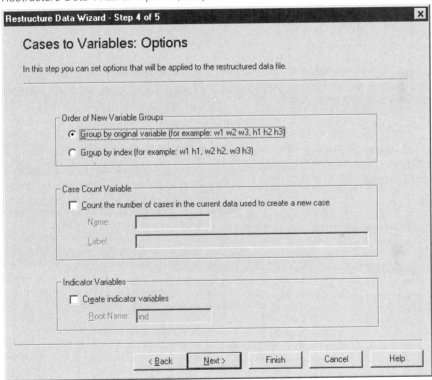

How should the new variable groups be ordered in the new file?

■ **By variable.** The wizard groups the new variables created from an original variable together.

■ **By index.** The wizard groups the variables according to the values of the index variables.

Example. The variables to be restructured are *w* and *h*, and the index is *month*:

w   h   month

Grouping by variable results in:

w.jan   w.feb   h.jan

Grouping by index results in:

w.jan   h.jan   w.feb

Create a count variable? The wizard can create a count variable in the new file. It contains the number of rows in the current data that were used to create a row in the new data file.

Create indicator variables? The wizard can use the index variables to create **indicator variables** in the new data file. It creates one new variable for each unique value of the index variable. The indicator variables signal the presence or absence of a value for a case. An indicator variable has the value of 1 if the case has a value; otherwise, it is 0.

Example. The index variable is *product*. It records the products that a customer purchased. The original data are:

customer	product
1	chick
1	eggs
2	eggs
3	chick

Creating an indicator variable results in one new variable for each unique value of *product*. The restructured data are:

customer	indchick	indeggs
1	1	1
2	0	1
3	1	0

In this example, the restructured data could be used to get frequency counts of the products that customers buy.

Restructure Data Wizard: Finish

This is the final step of the Restructure Data Wizard. Decide what to do with your specifications.

Figure 9-34
Restructure Data Wizard: Finish

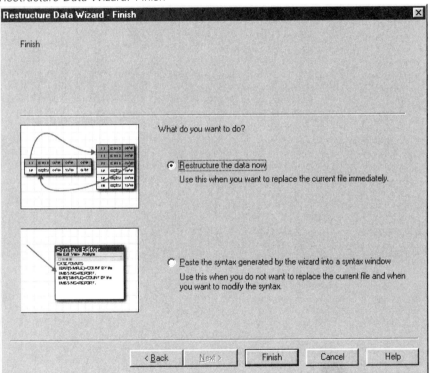

- **Restructure now.** The wizard will create the new, restructured file. Choose this if you want to replace the current file immediately. *Note*: If original data are weighted, the new data will be weighted unless the variable that is used as the weight is restructured or dropped from the new file.

- **Paste syntax**. The wizard will paste the syntax it generates into a syntax window. Choose this when you are not ready to replace the current file, when you want to modify the syntax, or when you want to save it for future use.

Working with Output

When you run a procedure, the results are displayed in a window called the Viewer. In this window, you can easily navigate to the output that you want to see. You can also manipulate the output and create a document that contains precisely the output that you want.

Viewer

Results are displayed in the Viewer. You can use the Viewer to:

- Browse results.
- Show or hide selected tables and charts.
- Change the display order of results by moving selected items.
- Move items between the Viewer and other applications.

Figure 10-1
Viewer

Double-click
a book icon to
show or hide
an item

Click to expand
or collapse the
outline view

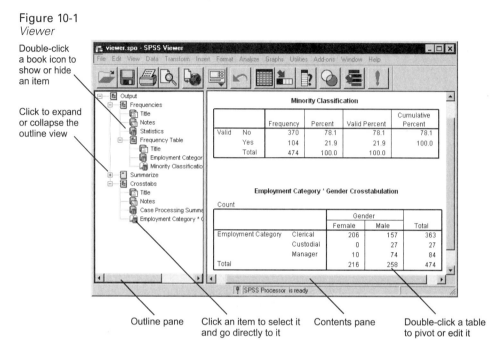

Outline pane   Click an item to select it   Contents pane   Double-click a table
and go directly to it   to pivot or edit it

The Viewer is divided into two panes:

■ The left pane of the Viewer contains an outline view of the contents.

■ The right pane contains statistical tables, charts, and text output.

You can use the scroll bars to browse the results, or you can click an item in the outline to go directly to the corresponding table or chart. You can click and drag the right border of the outline pane to change the width of the outline pane.

Using the Draft Viewer

If you prefer simple text output rather than interactive pivot tables, you can use the Draft Viewer.

To Use the Draft Viewer

▶ In any window, from the menus choose:
Edit
  Options...

▶ On the General tab, click Draft for the output viewer type.

▶ To change the format options for Draft Viewer output, click the Draft Viewer tab.

For more information, see "Draft Viewer" in Chapter 11 on p. 301. You can also search the Help facility to learn more:

▶ In any window, from the menus choose:
Help
  Topics

▶ Click the Index tab in the Help Topics window.

▶ Type draft viewer, and then double-click the index entry.

Showing and Hiding Results

In the Viewer, you can selectively show and hide individual tables or results from an entire procedure. This process is useful when you want to shorten the amount of visible output in the contents pane.

To Hide Tables and Charts

▶ Double-click the item's book icon in the outline pane of the Viewer.

or

▶ Click the item to select it.

▶ From the menus choose:
View
  Hide

or

▶ Click the closed book (Hide) icon on the Outlining toolbar.

The open book (Show) icon becomes the active icon, indicating that the item is now hidden.

To Hide Procedure Results

▶ Click the box to the left of the procedure name in the outline pane.

This hides all results from the procedure and collapses the outline view.

Moving, Deleting, and Copying Output

You can rearrange the results by copying, moving, or deleting an item or a group of items.

To Move Output in the Viewer

▶ Click an item in the outline or contents pane. (Shift-click to select multiple items, or Ctrl-click to select noncontiguous items.)

▶ Click and drag selected items (hold down the mouse button while dragging).

▶ Release the mouse button on the item just above the location where you want to drop the moved items.

You can also move items by using Cut and Paste After on the Edit menu.

To Delete Output in the Viewer

▶ Click an item in the outline or contents pane. (Shift-click to select multiple items, or Ctrl-click to select noncontiguous items.)

▶ Press Delete.

or

▶ From the menus choose:
Edit
Delete

To Copy Output in the Viewer

▶ Click an item in the outline or contents pane. (Shift-click to select multiple items, or Ctrl-click to select noncontiguous items.)

▶ Hold down the Ctrl key while you click and drag selected items (hold down the mouse button while dragging).

▶ Release the mouse button to drop the items where you want them.

You can also copy items by using Copy and Paste After on the Edit menu.

Changing Initial Alignment

By default, all results are initially left-aligned. You can change the initial alignment (choose Options on the Edit menu, then click the Viewer tab) at any time.

Changing Alignment of Output Items

▶ In the outline or contents pane, click the items that you want to align (Shift-click or Ctrl-click to select multiple items).

▶ From the menus choose:
Format
  Align Left

Other alignment options include Center and Align Right.

Note: All results are displayed left-aligned in the Viewer. Only the alignment of printed results is affected by the alignment settings. Centered and right-aligned items are identified by a small symbol above and to the left of the item.

Viewer Outline

The outline pane provides a table of contents of the Viewer document. You can use the outline pane to navigate through your results and control the display. Most actions in the outline pane have a corresponding effect on the contents pane.

■ Selecting an item in the outline pane displays the corresponding item in the contents pane.

■ Moving an item in the outline pane moves the corresponding item in the contents pane.

■ Collapsing the outline view hides the results from all items in the collapsed levels.

Figure 10-2
Collapsed outline view and hidden results

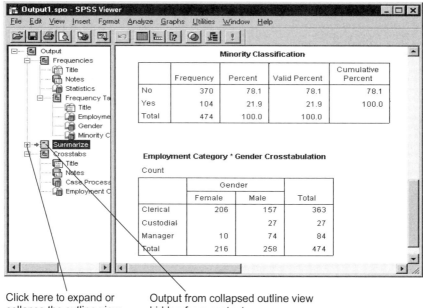

Click here to expand or
collapse the outline view

Output from collapsed outline view
hidden from contents pane

Controlling the outline display. To control the outline display, you can:

■ Expand and collapse the outline view.

■ Change the outline level for selected items.

■ Change the size of items in the outline display.

■ Change the font that is used in the outline display.

To Collapse and Expand the Outline View

▶ Click the box to the left of the outline item that you want to collapse or expand.

or

▶ Click the item in the outline.

▶ From the menus choose:
View
　Collapse

or

View
  Expand

To Change the Outline Level

▶ Click the item in the outline pane.

▶ Click the left arrow on the Outlining toolbar to promote the item (move the item to the left).

▶ Click the right arrow on the Outlining toolbar to demote the item (move the item to the right).

or

From the menus choose:
Edit
  Outline
    Promote

or

Edit
  Outline
    Demote

Changing the outline level is particularly useful after you move items in the outline level. Moving items can change the outline level of the items, and you can use the left- and right-arrow buttons on the Outlining toolbar to restore the original outline level.

To Change the Size of Outline Items

▶ From the menus choose:
View
  Outline Size
    Small

Other options include Medium and Large.

The icons and their associated text change size.

To Change the Font in the Outline

▶ From the menus choose:
View
  Outline Font...

▶ Select a font.

Adding Items to the Viewer

In the Viewer, you can add items such as titles, new text, charts, or material from other applications.

To Add a Title or Text

Text items that are not connected to a table or chart can be added to the Viewer.

▶ Click the table, chart, or other object that will precede the title or text.

▶ From the menus choose:
Insert
  New Title

or

Insert
  New Text

▶ Double-click the new object.

▶ Enter the text that you want at this location.

To Add a Text File

▶ In the outline pane or contents pane of the Viewer, click the table, chart, or other object that will precede the text.

▶ From the menus choose:
Insert
  Text File...

▶ Select a text file.

To edit the text, double-click it.

Using Output in Other Applications

Pivot tables and charts can be copied and pasted into another Windows application, such as a word-processing program or a spreadsheet. You can paste the pivot tables or charts in various formats, including the following formats:

Embedded object. For applications that support ActiveX objects, you can embed pivot tables and interactive charts. After you paste the table, it can be activated in place by double-clicking it, and then it can be edited as if in the Viewer.

Picture (metafile). You can paste pivot tables, text output, and charts as metafile pictures. The picture format can be resized in the other application, and sometimes a limited amount of editing can be done with the facilities of the other application. Pivot tables that are pasted as pictures retain all borders and font characteristics.

RTF (rich text format). Pivot tables can be pasted into other applications in RTF format. In most applications, this means that the pivot table is pasted as a table that can then be edited in the other application. (*Note*: Microsoft Word may not display extremely wide tables properly.)

Bitmap. Charts can be pasted into other applications as bitmaps.

BIFF. The contents of a table can be pasted into a spreadsheet and retain numeric precision.

Text. The contents of a table can be copied and pasted as text. This process can be useful for applications such as e-mail, where the application can accept or transmit only text.

To Copy a Table or Chart

▶ Select the table or chart to be copied.

▶ From the menus choose:
Edit
　Copy

To Copy and Paste Results into Another Application

▶ Copy the results in the Viewer.

▶ From the menus in the target application choose:
Edit
  Paste

or

Edit
  Paste Special...

Paste. Output is copied to the Clipboard in a number of formats. Each application determines the "best" format to use for Paste. In many applications, Paste will paste results as a picture (metafile). For word-processing applications, Paste will paste pivot tables in RTF format, which pastes the pivot table as a table. For spreadsheet applications, Paste will paste pivot tables in BIFF format. Charts are pasted as metafiles.

Paste Special. Results are copied to the Clipboard in multiple formats. Paste Special allows you to select the format that you want from the list of formats that are available to the target application.

To Embed a Table in Another Application

You can embed pivot tables and interactive charts in other applications in ActiveX format. An embedded object can be activated in place by double-clicking it, and the object can then be edited and pivoted as if in the Viewer.

If you have applications that support ActiveX objects:

▶ Double-click the file *objs-on.bat*, which is located in the directory in which SPSS is installed.

This action turns on ActiveX embedding for pivot tables. The file *objs-off.bat* turns ActiveX embedding off.

To embed a pivot table or interactive chart in another application:

▶ In the Viewer, copy the table.

▶ From the menus in the target application choose:
Edit
  Paste Special...

▶ From the list, select SPSS Pivot Table Object *or* SPSS Graphics Control Object.

The target application must support ActiveX objects. See the application's documentation for information about ActiveX support. Some applications that do not support ActiveX may initially accept ActiveX pivot tables but may then exhibit unstable behavior. Do not rely on embedded objects until you have tested the application's stability with embedded ActiveX objects.

To Paste a Pivot Table or Chart as a Picture (Metafile)

▶ In the Viewer, copy the table or chart.

▶ From the menus in the target application choose:
Edit
  Paste Special...

▶ From the list, select Picture.

The item is pasted as a metafile. Only the layer and columns that were visible when the item was copied are available in the metafile. Other layers or hidden columns are not available.

To Paste a Pivot Table as a Table (RTF)

▶ In the Viewer, copy the pivot table.

▶ From the menus in the target application choose:
Edit
  Paste Special...

▶ From the list, select Formatted Text (RTF) or Rich Text Format.

The pivot table is pasted as a table. Only the layer and columns that were visible when the item was copied are pasted into the table. Other layers or hidden columns are not available. You can copy and paste only one pivot table at a time in this format. (*Note*: Microsoft Word may not display extremely wide tables properly.)

To Paste a Pivot Table as Text

▶ In the Viewer, copy the table.

▶ From the menus in the target application choose:
  Edit
    Paste Special...

▶ From the list, select Unformatted Text.

Unformatted pivot table text contains tabs between columns. You can align columns by adjusting the tab stops in the other application.

To Copy and Paste Multiple Items into Another Application

▶ Select the tables and/or charts to be copied. (Shift-click or Ctrl-click to select multiple items.)

▶ From the menus choose:
  Edit
    Copy Objects

▶ In the target application, from the menus choose:
  Edit
    Paste

Note: Use Copy Objects only to copy multiple items from the Viewer to another application. For copying and pasting within Viewer documents (for example, between two Viewer windows), use Copy on the Edit menu.

Pasting Objects into the Viewer

Objects from other applications can be pasted into the Viewer. You can use either Paste After or Paste Special. Either type of pasting puts the new object after the currently selected object in the Viewer. Use Paste Special when you want to choose the format of the pasted object.

Paste Special

Paste Special allows you to select the format of a copied object that is pasted into the Viewer. The possible file types for the object are listed in the Paste Special dialog box. The object will be inserted in the Viewer following the currently selected object.

Figure 10-3
Paste Special dialog box

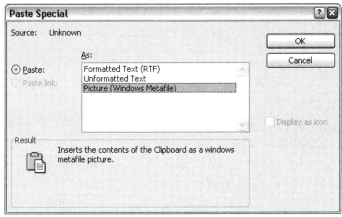

Pasting Objects from Other Applications into the Viewer

▶ In the other application, copy the object.

▶ In the outline pane or contents pane of the Viewer, click the table, chart, or other object that will precede the object.

▶ From the menus choose:
Edit
  Paste Special...

▶ From the list, select the format for the object.

Export Output

Export Output saves Viewer output in HTML, text, Word/RTF, Excel, PowerPoint (requires PowerPoint 97 or later), and PDF formats. Charts can also be exported in a number of different graphics formats. (*Note*: Export to PowerPoint is not available in the Student Version.)

Output Document. Exports any combination of pivot tables, text output, and charts.

■ For HTML and text formats, charts are exported in the currently selected chart export format. For HTML document format, charts are embedded by reference, and you should export charts in a suitable format for inclusion in HTML documents. For text document format, a line is inserted in the text file for each chart, indicating the filename of the exported chart.

■ For Word/RTF format, charts are included in the Word file in metafile format.

■ Charts are not included in Excel documents.

■ For PowerPoint format, charts are exported in TIFF format and embedded in the PowerPoint file.

■ For PDF format, charts are included in the PDF file in metafile format.

Output Document (No Charts). Exports pivot tables and text output. Any charts in the Viewer are ignored.

Charts Only. Available export formats include: Windows metafile (WMF), enhanced metafile (EMF), Windows bitmap (BMP), encapsulated PostScript (EPS), JPEG, TIFF, PNG, and Macintosh PICT.

Export What. You can export all objects in the Viewer, all visible objects, or only selected objects.

Export Format. For output documents, the available options are HTML, text, Excel, Word/RTF, PowerPoint, and PDF; for HTML and text format, charts are exported in the currently selected chart format in the Options dialog box for the selected format. For Charts Only, choose a chart export format from the drop-down list. For output documents, pivot tables and text are exported in the following manner:

■ **HTML file (*.htm).** Pivot tables are exported as HTML tables. Text output is exported as preformatted HTML.

■ **Text file (*.txt).** Pivot tables can be exported in tab-separated or space-separated format. All text output is exported in space-separated format.

■ **Excel file (*.xls).** Pivot table rows, columns, and cells are exported as Excel rows, columns, and cells, with all formatting attributes intact—for example, cell borders, font styles, and background colors. Text output is exported with all font attributes intact. Each line in the text output is a row in the Excel file, with the entire contents of the line contained in a single cell.

■ **Word/RTF file (*.doc).** Pivot tables are exported as Word tables with all formatting attributes intact—for example, cell borders, font styles, background colors, and so on. Text output is exported as formatted RTF. Text output in SPSS is always displayed in a fixed-pitch (monospaced) font and is exported with the same font attributes. A fixed-pitch font is required for proper alignment of space-separated text output. (*Note*: Microsoft Word may not display extremely wide tables properly.)

■ **PowerPoint file (*.ppt).** Pivot tables are exported as Word tables and are embedded on separate slides in the PowerPoint file, with one slide for each pivot table. All formatting attributes of the pivot table are retained—for example, cell borders, font styles, and background colors. Text output is exported as formatted RTF. Text output in SPSS is always displayed in a fixed-pitch (monospaced) font and is exported with the same font attributes. A fixed-pitch font is required for proper alignment of space-separated text output. (*Note*: Export to PowerPoint is not available in the Student Version.)

■ **Portable Document Format (*.pdf).** All output is exported as it appears in Print Preview, with all formatting attributes intact.

Output Management System. You can also automatically export all output or user-specified types of output as text, HTML, XML, or SPSS-format data files. For more information, see "Output Management System" in Chapter 49 on p. 713.

To Export Output

▶ Make the Viewer the active window (click anywhere in the window).

▶ From the menus choose:
File
  Export...

▶ Enter a filename (or prefix for charts) and select an export format.

Figure 10-4
Export Output dialog box

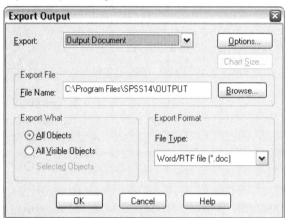

Figure 10-5
Output exported in Word/RTF format

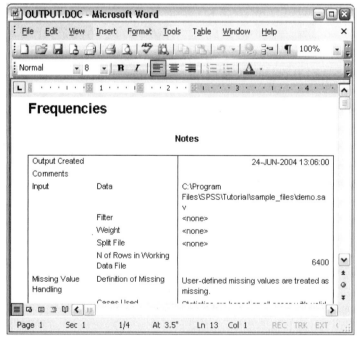

HTML, Word/RTF, and Excel Options

When you click Options in the Export Output dialog box, a dialog box is displayed, which controls the inclusion of footnotes and captions for documents that are exported in HTML, Word/RTF, and Excel formats, controls the chart export options for HTML documents, and controls the handling of multilayer pivot tables.

Image Format. Controls the chart export format and optional settings, including chart size for HTML documents. For Word/RTF, all charts are exported in Windows metafile (WMF) format. For Excel, charts are not included.

Export Footnotes and Caption. Check this box to include any footnotes and captions along with the export of pivot tables.

Export All Layers. Check this box to export all layers of a multilayer pivot table. If the box is left unchecked, only the top layer is exported.

To Set HTML, Word/RTF, and Excel Export Options

▶ Make the Viewer the active window (click anywhere in the window).

▶ From the menus choose:
File
  Export...

▶ Select HTML file or Word/RTF file or Excel file as the export format.

▶ Click Options.

PowerPoint Options

PowerPoint Options controls the inclusion of titles for slides, the inclusion of footnotes and captions for pivot tables, the handling of multilayer pivot tables, and the options for charts exported to PowerPoint. (*Note*: Export to PowerPoint is not available in the Student Version.)

Include Title on Slide. Check this box to include a title on each slide that is created by the export. Each slide contains a single item that is exported from the Viewer. The title is formed from the outline entry for the item in the outline pane of the Viewer.

Export Footnotes and Caption. Check this box to include any footnotes and captions along with the export of pivot tables.

Export All Layers. Check this box to export all layers of a multilayer pivot table; each layer is placed on a separate slide, and all layers have the same title. If this box is left unchecked, only the top layer is exported.

To Set PowerPoint Export Options

▶ Make the Viewer the active window (click anywhere in the window).

▶ From the menus choose:
File
  Export...

▶ Select PowerPoint file as the export format.

▶ Click Options.

Figure 10-6
PowerPoint Options dialog box

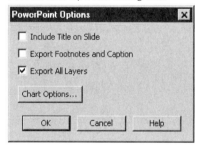

PDF Options

PDF Options controls the options for exporting Viewer output in portable document format (PDF).

Optimize for fast Web viewing. This option provides faster access and viewing when downloading the file from the Internet by downloading the document a page at a time (byte serving), allowing users to view and scroll the document while it is downloading.

Embed bookmarks. This option includes bookmarks in the PDF document that correspond to the Viewer outline entries. Like the Viewer outline pane, bookmarks can make it much easier to navigate documents with a large number of output objects.

Embed fonts. Embedding fonts ensures that the PDF document will look the same on all computers. Otherwise, if some fonts used in the document are not available on the computer being used to view (or print) the PDF document, font substitution may yield suboptimal results.

- **Only characters that are used in the document.** This option embeds only the font information necessary to display the PDF file consistently, regardless of the fonts installed on a particular computer.

- **All characters.** This option embeds all characters from all fonts used in the document. This is probably useful only if you need the ability to edit the PDF document and it is important for all of the edited material to use the same fonts as the original. For double-byte character sets (for example, Asian languages), this may result in a substantially larger file size.

Exporting Multilayer Pivot Tables. There are three options for exporting multilayer pivot tables in PDF format:

- **Honor Print Layer setting.** This uses the Table Properties settings for each table to determine if all layers or only the currently visible layer should be included. (With the default TableLook setting, only the currently visible layer is included.) For more information, see "Table Properties: Printing" in Chapter 12 on p. 331.

- **Export all layers.** All layers of all pivot tables are included, regardless of the Table Properties settings.

- **Export visible layer only.** Only the visible layer is included for all pivot tables, regardless of the Table Properties settings.

To Set PDF Export Options

▶ Make the Viewer the active window (click anywhere in the window).

▶ From the menus choose:
File
    Export...

▶ Select Portable Document Format as the export format.

▶ Click Options.

Figure 10-7
PDF Options dialog box

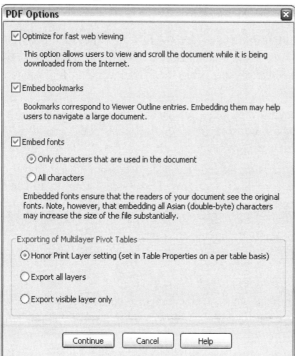

Other Settings That Affect PDF Output

Page Setup. Page size, orientation, margins, content and display of page headers and footers, and printed chart size in PDF documents are controlled by page setup options (File menu, Page Setup in the Viewer window). For more information, see "Page Setup" on p. 294.

Table Properties/TableLooks. Scaling of wide and/or long tables and printing of table layers are controlled by table properties for each table. These properties can also be saved in TableLooks. For more information, see "Table Properties: Printing" in Chapter 12 on p. 331.

Default/Current Printer. The resolution (DPI) of the PDF document is the current resolution setting for the default or currently selected printer (which can be changed using Page Setup). The maximum resolution is 1200 DPI. If the printer setting is higher, the PDF document resolution will be 1200 DPI. *Note*: High-resolution documents may yield poor results when printed on lower-resolution printers.

Text Options

Text Options controls pivot table, text output, and chart format options and the inclusion of footnotes and captions for documents that are exported in text format.

Pivot tables can be exported in tab-separated or space-separated format. For tab-separated format, if a cell is not empty, its contents and a tab character are printed. If a cell is empty, a tab character is printed.

All text output is exported in space-separated format. All space-separated output requires a fixed-pitch (monospaced) font for proper alignment.

Cell Formatting. For space-separated pivot tables, by default all line wrapping is removed, and each column is set to the width of the longest label or value in the column. To limit the width of columns and wrap long labels, specify a number of characters for the column width. This setting affects only pivot tables.

Cell Separators. For space-separated pivot tables, you can specify the characters that are used to create cell borders.

Image Format. Controls the chart export format and optional settings, including chart size.

Insert page break between tables. Inserts a form feed/page break between each table. For multilayer pivot tables, this option inserts a page break between each layer.

To Set Text Export Options

▶ Make the Viewer the active window (click anywhere in the window).

▶ From the menus choose:
File
  Export...

▶ Select Text file as the export format.

▶ Click Options.

Figure 10-8
Text Options dialog box

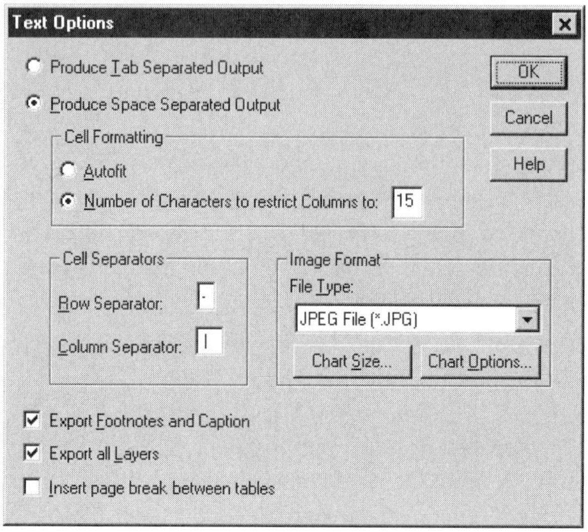

Chart Size Options

Chart Size controls the size of exported charts. The custom percentage specification allows you to decrease or increase the size of the exported chart up to 200%.

To Set the Size for Exported Charts

▶ Make the Viewer the active window (click anywhere in the window).

▶ From the menus choose:
File
  Export...

▶ For output documents, click Options, select the export format, and then click Chart Size.

▶ For Charts Only, select the export format, and then click Chart Size.

Figure 10-9
Export Chart Size dialog box

JPEG Chart Export Options

Color Depth. JPEG charts can be exported as true color (24 bit) or 256 grayscale.

Color Space. Refers to the way that colors are encoded in the image. The YUV color model is one form of color encoding, commonly used for digital video and MPEG transmission. The acronym stands for Y-signal, U-signal, V-signal. The Y component specifies grayscale or luminance, and the U and V components correspond to the chrominance (color information).

The ratios represent the sampling rates for each component. Reducing the U and V sampling rates reduces file size (and also quality). Color Space determines the degree of "lossiness" for colors in the exported image. YUV 4:4:4 is lossless, while YUV 4:2:2 and YUV 4:1:1 represent the decreasing trade-off between file size (disk space) and quality of the colors that are represented.

Progressive Encoding. Enables the image to load in stages, initially displaying at low resolution and then increasing in quality as the image continues to load.

Compression Quality Setting. Controls the ratio of compression to image quality. The higher the image quality, the larger the exported file size.

Color Operations. The following operations are available:

- **Invert.** Each pixel is saved as the inverse of the original color.
- **Gamma correction.** Adjusts the intensity of colors in the exported chart by changing the gamma constant that is used to map the intensity values. Basically, this feature can be used to lighten or darken the bitmapped image. The value can range from 0.10 (darkest) to 6.5 (lightest).

BMP and PICT Chart Export Options

Color Depth. Determines the number of colors in the exported chart. A chart that is saved under any depth will have a minimum of the number of colors that are actually used and a maximum of the number of colors that are allowed by the depth. For example, if the chart contains three colors—red, white, and black—and you save it as 16 colors, the chart will remain as three colors.

- If the number of colors in the chart exceeds the number of colors for that depth, the colors will be dithered to replicate the colors in the chart.

- Current screen depth is the number of colors that are currently displayed on your computer monitor.

Color Operations. The following operations are available:

- **Invert.** Each pixel is saved as the inverse of the original color.

- **Gamma correction.** Adjusts the intensity of colors in the exported chart by changing the gamma constant that is used to map the intensity values. Basically, this feature can be used to lighten or darken the bitmapped image. The value can range from 0.10 (darkest) to 6.5 (lightest).

Use RLE compression. (BMP only.) A lossless compression technique supported by common Windows file formats. Lossless compression means that image quality has not been sacrificed for the sake of smaller files.

PNG and TIFF Chart Export Options

Color Depth. Determines the number of colors in the exported chart. A chart that is saved under any depth will have a minimum of the number of colors that are actually used and a maximum of the number of colors that are allowed by the depth. For example, if the chart contains three colors—red, white, and black—and you save it as 16 colors, the chart will remain as three colors.

- If the number of colors in the chart exceeds the number of colors for that depth, the colors will be dithered to replicate the colors in the chart.

- Current screen depth is the number of colors currently displayed on your computer monitor.

Color Operations. The following operations are available:

- **Invert.** Each pixel is saved as the inverse of the original color.

- **Gamma correction.** Adjusts the intensity of colors in the exported chart by changing the gamma constant that is used to map the intensity values. Basically, this feature can be used to lighten or darken the bitmapped image. The value can range from 0.10 (darkest) to 6.5 (lightest).

Transparency. Allows you to select a color that will appear transparent in the exported chart. The feature is available only with 32-bit true color export. Enter integer values between 0 and 255 for each color. The default value for each color is 255, creating a default transparent color of white.

Format. (TIFF only.) Allows you to set the color space and compress the exported chart. All color depths are available with RGB color. Only 24- and 32-bit true color is available with CMYK. With the YCbCr option, only 24-bit true color is available.

EPS Chart Export Options

Trees, Maps, and Interactive Charts

For trees (Classification Trees option), maps (Maps option), and interactive charts (Graphs menu, Interactive submenu), the following EPS options are available:

Image Preview. Allows you to save a preview image within the EPS image. A preview image is used mainly when an EPS file is placed within another document. Many applications cannot display an EPS image on screen but can display the preview that is saved with the image. The preview image can be either WMF (smaller and more scalable) or TIFF (more portable and supported by other platforms). Check the application in which you want to include the EPS graphic to see what preview format the application supports.

Fonts. Controls the treatment of TrueType fonts in EPS images.

- **Convert to PostScript fonts.** Converts TrueType fonts to PostScript (Type 1) fonts based on font family. For example, Times New Roman is converted to Times, and Arial is converted to Helvetica. *Note*: This format is not recommended for interactive graphics that use the SPSS marker font (for example, scatterplots), because there are no meaningful PostScript equivalents for the SPSS TrueType marker symbols.

- **Replace fonts with curves.** Turns TrueType fonts into PostScript curve data. The text itself is no longer editable as text in applications that can edit EPS graphics. There is also a loss of quality, but this option is useful if you have a PostScript printer that doesn't support Type 42 fonts, and you need to preserve special TrueType symbols, such as the markers that are used in interactive scatterplots.

Other Charts

For all other charts, the following EPS options are available:

Include TIFF preview. Saves a preview with the EPS image in TIFF format for display in applications that cannot display EPS images on screen.

Fonts. Controls the treatment of fonts in EPS images.

- **Replace fonts with curves.** Turns fonts into PostScript curve data. The text itself is no longer editable as text in applications that can edit EPS graphics. This option is useful if the fonts that are used in the chart are not available on the output device.

- **Use font references.** If the fonts that are used in the chart are available on the output device, the fonts are used. Otherwise, the output device uses alternate fonts.

WMF Chart Export Options

Aldus placeable. Provides a degree of device independence (same physical size when opened at 96 dpi versus 120 dpi), but not all applications support this format.

Standard Windows. Supported by most applications that can display Windows metafiles.

To Set Chart Export Options

▶ Make the Viewer the active window (click anywhere in the window).

▶ From the menus choose:
  File
    Export...

▶ For output documents, click Options, select the export format, and then click Chart Options.

▶ For Charts Only, select the export format, and then click Options.

Viewer Printing

There are two options for printing the contents of the Viewer window:

All visible output. Prints only items that are currently displayed in the contents pane. Hidden items (items with a closed book icon in the outline pane or hidden in collapsed outline layers) are not printed.

Selection. Prints only items that are currently selected in the outline and/or contents panes.

Figure 10-10
Viewer Print dialog box

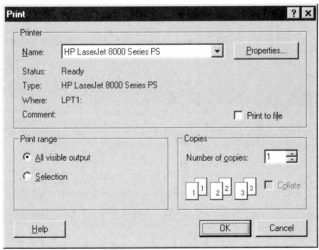

To Print Output and Charts

▶ Make the Viewer the active window (click anywhere in the window).

▶ From the menus choose:
File
  Print...

▶ Select the print settings that you want.

▶ Click OK to print.

Print Preview

Print Preview shows you what will print on each page for Viewer documents. It is a good idea to check Print Preview before actually printing a Viewer document, because Print Preview shows you items that may not be visible by looking at the contents pane of the Viewer, including:

- Page breaks
- Hidden layers of pivot tables
- Breaks in wide tables
- Complete output from large tables
- Headers and footers that are printed on each page

Figure 10-11
Print Preview

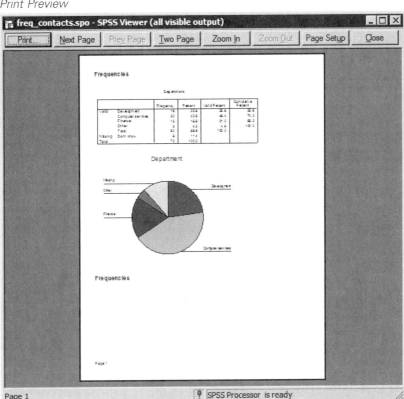

If any output is currently selected in the Viewer, the preview displays only the selected output. To view a preview for all output, make sure nothing is selected in the Viewer.

To View a Print Preview

▶ Make the Viewer the active window (click anywhere in the window).

▶ From the menus choose:
File
  Print Preview

Page Setup

With Page Setup, you can control:

■ Paper size and orientation

■ Page margins

■ Page headers and footers

■ Page numbering

■ Printed size for charts

Figure 10-12
Page Setup dialog box

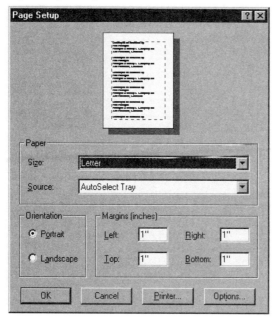

Page Setup settings are saved with the Viewer document. Page Setup affects settings for printing Viewer documents only. These settings have no effect on printing data from the Data Editor or printing syntax from a syntax window.

To Change Page Setup

▶ Make the Viewer the active window (click anywhere in the window).

▶ From the menus choose:

File
  Page Setup...

▶ Change the settings and click OK.

Page Setup Options: Headers and Footers

Headers and footers are the information that is printed at the top and bottom of each page. You can enter any text that you want to use as headers and footers. You can also use the toolbar in the middle of the dialog box to insert:

■ Date and time

■ Page numbers

■ Viewer filename

■ Outline heading labels

■ Page titles and subtitles

Figure 10-13
Page Setup Options dialog box, Header/Footer tab

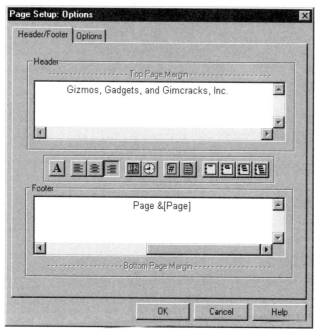

Outline heading labels indicate the first-, second-, third-, and/or fourth-level outline heading for the first item on each page.

Page titles and subtitles print the current page titles and subtitles. Page titles and subtitles are created with New Page Title on the Viewer Insert menu or the TITLE and SUBTITLE commands in command syntax. If you have not specified any page titles or subtitles, this setting is ignored.

Note: Font characteristics for new page titles and subtitles are controlled on the Viewer tab of the Options dialog box (accessed by choosing Options on the Edit menu). Font characteristics for existing page titles and subtitles can be changed by editing the titles in the Viewer.

To see how your headers and footers will look on the printed page, choose Print Preview from the File menu.

Page Setup Options: Options

This dialog box controls the printed chart size, the space between printed output items, and page numbering.

Printed Chart Size. Controls the size of the printed chart relative to the defined page size. The chart's aspect ratio (width-to-height ratio) is not affected by the printed chart size. The overall printed size of a chart is limited by both its height and width. When the outer borders of a chart reach the left and right borders of the page, the chart size cannot increase further to fill additional page height.

Space between items. Controls the space between printed items. Each pivot table, chart, and text object is a separate item. This setting does not affect the display of items in the Viewer.

Number pages starting with. Numbers pages sequentially, starting with the specified number.

Figure 10-14
Page Setup Options dialog box, Options tab

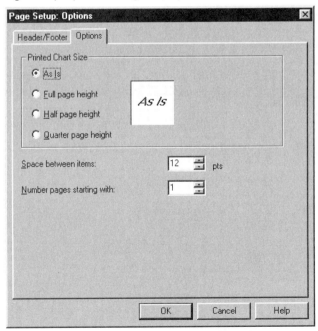

Saving Output

The contents of the Viewer can be saved to a Viewer document. The saved document includes both panes of the Viewer window (the outline and the contents).

To Save a Viewer Document

▶ From the Viewer window menus choose:
File
  Save

▶ Enter the name of the document, and then click Save.

To save results in external formats (for example, HTML or text), use Export on the File menu. (This feature is not available in the standalone SmartViewer.)

Save With Password Option

Save With Password allows you to password-protect your Viewer files.

Password. The password is case-sensitive and can be up to 16 characters. If you assign a password, the file cannot be viewed without entering the password.

OEM Code. Leave this field blank unless you have a contractual agreement with SPSS Inc. to redistribute the SmartViewer. The OEM license code is provided with the contract.

To Save Viewer Files with a Password

▶ From the Viewer window menus choose:
File
  Save with Password...

▶ Enter the password.

▶ Reenter the password to confirm it, and then click OK.

▶ Enter a filename in the Save As dialog box.

▶ Click Save.

Note: Leave the OEM Code field blank unless you have a contractual agreement with SPSS Inc. to redistribute the SmartViewer.

Draft Viewer

The Draft Viewer provides results in draft form, including:

- Simple text output (instead of pivot tables)
- Charts as metafile pictures (instead of chart objects)

Text output in the Draft Viewer can be edited, charts can be resized, and both text output and charts can be pasted into other applications. However, charts cannot be edited, and the interactive features of pivot tables and charts are not available.

Figure 11-1
Draft Viewer window

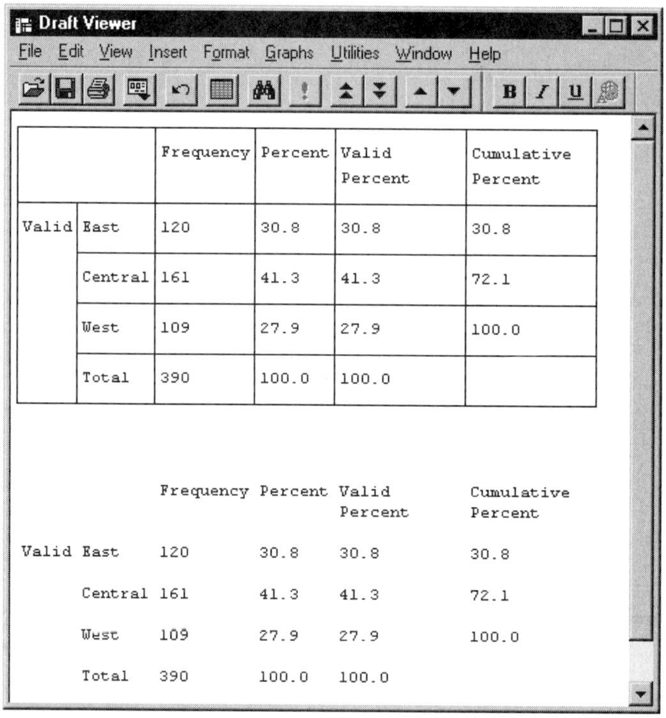

To Create Draft Output

▶ From the menus choose:
File
  New
    Draft Output

▶ To make draft output the default output type, from the menus choose:
Edit
  Options...

▶ Click the General tab.

▶ Select Draft under Viewer Type at Startup.

Note: New output is always displayed in the designated Viewer window. If you have both a Viewer and a Draft Viewer window open, the **designated window** is the one opened most recently or the one designated with the Designate Window button (the plus sign icon) on the toolbar.

Controlling Draft Output Format

Output that would be displayed as pivot tables in the Viewer is converted to text output for the Draft Viewer. The default settings for converted pivot table output include the following:

■ Each column is set to the width of the column label, and labels are not wrapped to multiple lines.

■ Alignment is controlled by spaces (instead of tabs).

■ Box characters from the SPSS Marker Set font are used as row and column separators.

■ If box characters are turned off, vertical line characters (|) are used as column separators and dashes (–) are used as row separators.

You can control the format of new draft output using Draft Viewer Options (Edit menu, Options, Draft Viewer tab).

Figure 11-2
Draft Viewer Options

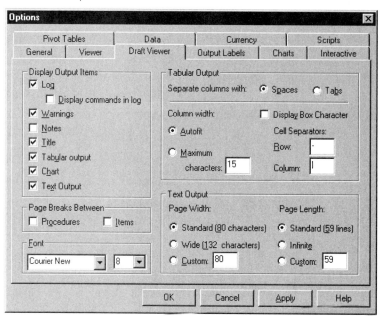

Column width. To reduce the width of tables that contain long labels, select Maximum characters under Column width. Labels longer than the specified width are wrapped to fit the maximum width.

Figure 11-3
Draft output before and after setting maximum column width

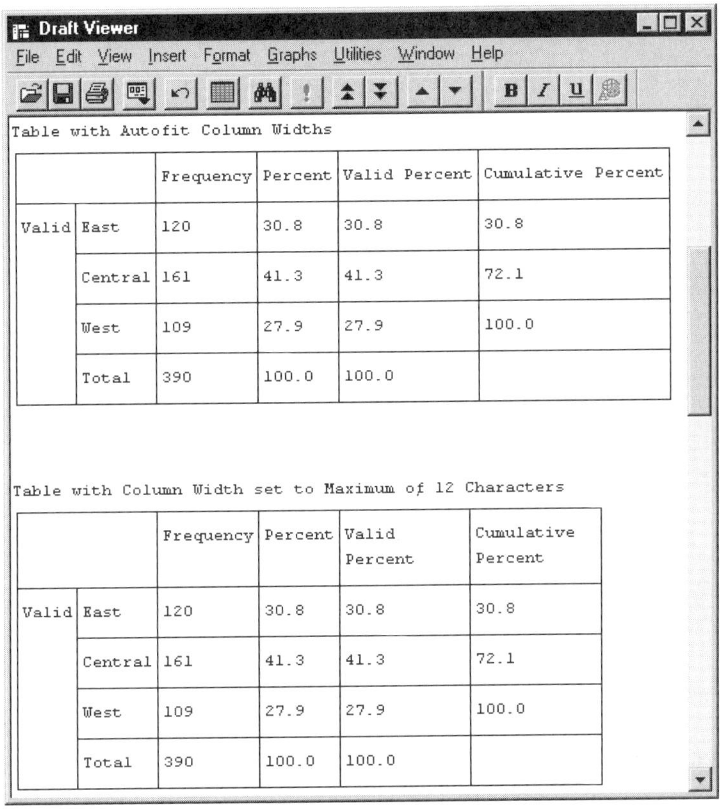

Table with Autofit Column Widths

		Frequency	Percent	Valid Percent	Cumulative Percent
Valid	East	120	30.8	30.8	30.8
	Central	161	41.3	41.3	72.1
	West	109	27.9	27.9	100.0
	Total	390	100.0	100.0	

Table with Column Width set to Maximum of 12 Characters

		Frequency	Percent	Valid Percent	Cumulative Percent
Valid	East	120	30.8	30.8	30.8
	Central	161	41.3	41.3	72.1
	West	109	27.9	27.9	100.0
	Total	390	100.0	100.0	

Row and column separators. As an alternative to box characters for row and column borders, you can use the Cell Separators settings to control the row and column separators displayed in new draft output. You can specify different cell separators or enter blank spaces if you don't want any characters used to mark rows and columns. You must deselect Display Box Character to specify cell separators.

Figure 11-4
Draft output before and after setting cell separators

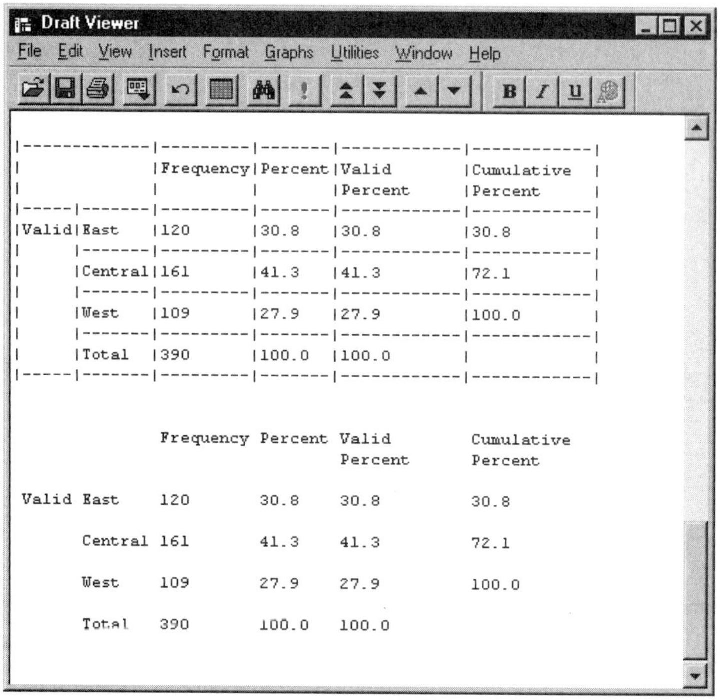

Space-separated versus tab-separated columns. The Draft Viewer is designed to display space-separated output in a fixed-pitch (monospaced) font. If you want to paste draft output into another application, you must use a fixed-pitch font to align space-separated columns properly. If you select Tabs for the column separator, you can use any font that you want in the other application and set the tabs to align output properly. However, tab-separated output will not align properly in the Draft Viewer.

Figure 11-5
Tab-separated output in the Draft Viewer and formatted in a word processor

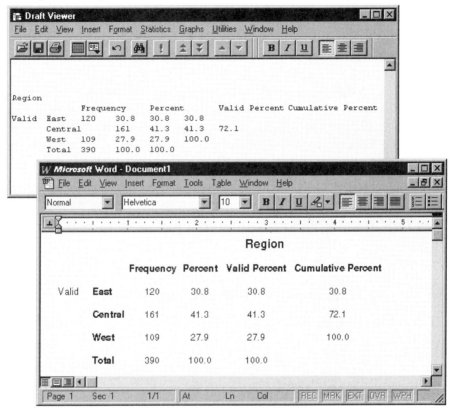

To Set Draft Viewer Options

▶ From the menus choose:
Edit
  Options...

▶ Click the Draft Viewer tab.

▶ Select the settings that you want.

▶ Click OK or Apply.

Draft Viewer output display options affect only new output produced after you change the settings. Output already displayed in the Draft Viewer is not affected by changes in these settings.

Fonts in Draft Output

You can modify the font attributes (such as font, size, and style) of text output in the Draft Viewer. However, if you use box characters for row and column borders, proper column alignment for space-separated text requires a fixed-pitch (monospaced) font, such as Courier. Additionally, other font changes, such as size and style (for example, bold and italic), applied to only part of a table can affect column alignment.

Row and column borders. The default solid-line row and column borders use the SPSS Marker Set font. The line-drawing characters used to draw the borders are not supported by other fonts.

To Change Fonts in Draft Viewer

▶ Select the text to which you want to apply the font change.

▶ From the Draft Viewer menus choose:
Format
  Font...

▶ Select the font attributes that you want to apply to the selected text.

To Print Draft Output

▶ From the Draft Viewer menus choose:
File
  Print...

To Print Only a Selected Portion of the Draft Output

▶ Select the output that you want to print.

▶ From the menus choose:
File
  Print...

▶ Select Selection.

Draft Viewer Print Preview

Print Preview shows you what will print on each page for draft documents. It is usually a good idea to check Print Preview before actually printing a Viewer document because Print Preview shows you items that may not fit on the page, including:

- Long tables
- Wide tables produced by converted pivot table output without column-width control
- Text output created with the Wide page-width option (Draft Viewer Options) with the printer set to Portrait mode

Output that is too wide for the page is truncated, not printed on another page. There are several things that you can do to prevent wide output from being truncated:

- Use a smaller font size (Format menu, Fonts).
- Select Landscape for the page orientation (File menu, Page Setup).
- For new output, specify a narrow maximum column width (Edit menu, Options, Draft Viewer tab).

For long tables, use page breaks (Insert menu, Page Break) to control where the table breaks between pages.

To View Draft Viewer Print Preview

▶ From the Draft Viewer menus choose:
File
   Print Preview

To Save Draft Viewer Output

▶ From the Draft Viewer menus choose:
File
   Save

Draft Viewer output is saved in rich text format (RTF).

To Save Draft Output as Text

▶ From the Draft Viewer menus choose:
File
  Export...

You can export all text or just the selected text. Only text output (converted pivot table output and text output) is saved in the exported files; charts are not included.

Pivot Tables

Many of the results in the Viewer are presented in tables that can be pivoted interactively. That is, you can rearrange the rows, columns, and layers.

Manipulating a Pivot Table

Options for manipulating a pivot table include:

- Transposing rows and columns
- Moving rows and columns
- Creating multidimensional layers
- Grouping and ungrouping rows and columns
- Showing and hiding cells
- Rotating row and column labels
- Finding definitions of terms

To Edit a Pivot Table

▶ Double-click the table.

This activates the Pivot Table Editor.

To Edit Two or More Pivot Tables at a Time

▶ Right-click the pivot table.

▶ From the context menu choose:
SPSS Pivot Table Object
　Open

▶ Repeat the process for each pivot table that you want to edit.

Each pivot table is ready to edit in its own separate window.

To Pivot a Table by Using Icons

▶ Activate the pivot table.

▶ From the Pivot Table menus choose:
Pivot
  Pivoting Trays

▶ Place the pointer over each icon to see a ToolTip that tells you which table dimension the icon represents.

▶ Drag an icon from one tray to another tray.

Figure 12-1
Pivoting trays

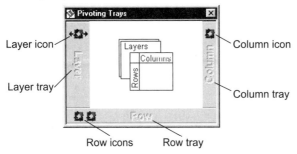

This changes the arrangement of the table. For example, suppose that the icon represents a variable with categories Yes and No. If you drag the icon from the Row tray to the Column tray, Yes and No become column labels. Before the move, Yes and No were row labels.

To Identify Pivot Table Dimensions

▶ Activate the pivot table.

▶ If pivoting trays are not on, from the Pivot Table menus choose:
Pivot
  Pivoting Trays

▶ Click and hold down the mouse button on an icon.

This highlights the dimension labels in the pivot table.

To Transpose Rows and Columns

▶ From the Pivot Table menus choose:
Pivot
  Transpose Rows and Columns

This has the same effect as dragging all of the row icons into the Column tray and dragging all of the column icons into the Row tray.

To Change Display Order

The order of pivot icons in a dimension tray reflects the display order of elements in the pivot table. To change the display order of elements in a dimension:

▶ Activate the pivot table.

▶ If pivoting trays are not already on, from the Pivot Table menus choose:
Pivot
  Pivoting Trays

▶ Drag the icons in each tray to the order that you want (left to right or top to bottom).

To Move Rows and Columns in a Pivot Table

▶ Activate the pivot table.

▶ Click the label for the row or column you want to move.

▶ Drag the label to the new position.

▶ From the context menu choose Insert Before or Swap.

Note: Make sure that Drag to Copy on the Edit menu is *not* enabled (checked). If Drag to Copy is enabled, deselect it.

To Group Rows or Columns and Insert Group Labels

▶ Activate the pivot table.

▶ Select the labels for the rows or columns that you want to group together (click and drag or Shift-click to select multiple labels).

▶ From the menus choose:
Edit
  Group

A group label is automatically inserted. Double-click the group label to edit the label text.

Figure 12-2
Row and column groups and labels

		Column Group Label		Total
		Female	Male	
Row Group Label	Clerical	206	157	363
	Custodial		27	27
	Manager	10	74	84

Note: To add rows or columns to an existing group, you must first ungroup the items that are currently in the group; then you can create a new group that includes the additional items.

To Ungroup Rows or Columns and Remove Group Labels

▶ Activate the pivot table.

▶ Click anywhere in the group label for the rows or columns that you want to ungroup.

▶ From the menus choose:
Edit
  Ungroup

Ungrouping automatically deletes the group label.

To Rotate Pivot Table Labels

▶ Activate the pivot table.

▶ From the menus choose:

Format
  Rotate InnerColumn Labels

or

Rotate OuterRow Labels

Figure 12-3
Rotated column labels

	Frequency	Percent	Valid Percent	Cumulative Percent
Clerical	363	76.6	76.6	76.6
Custodial	27	5.7	5.7	82.3
Manager				
Total				

	Frequency	Percent	Valid Percent	Cumulative Percent
Clerical	363	76.6	76.6	76.6
Custodial	27	5.7	5.7	82.3
Manager	84	17.7	17.7	100.0
Total	474	100.0	100.0	

Only the innermost column labels and the outermost row labels can be rotated.

To Reset Pivots to Defaults

After performing one or more pivoting operations, you can return to the original arrangement of the pivot table.

▶ From the Pivot menu choose Reset Pivots to Defaults.

This resets only changes that are the result of pivoting row, column, and layer elements between dimensions. The action does not affect changes such as grouping or ungrouping or moving rows and columns.

To Find a Definition of a Pivot Table Label

You can obtain context-sensitive Help on cell labels in pivot tables. For example, if *Mean* appears as a label, you can obtain a definition of the mean.

▶ Right-click a label cell.

▶ From the context menu choose What's This?.

You must click your right mouse button on the label cell itself, rather than on the data cells in the row or column. Context-sensitive Help is not available for user-defined labels, such as variable names or value labels.

Working with Layers

You can display a separate two-dimensional table for each category or combination of categories. The table can be thought of as stacked in layers, with only the top layer visible.

To Create and Display Layers

▶ Activate the pivot table.

▶ From the Pivot menu choose Pivoting Trays (if it is not already selected).

▶ Drag an icon from the Row tray or the Column tray into the Layer tray.

Figure 12-4
Moving categories into layers

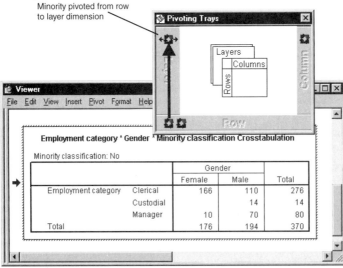

Each layer icon has left and right arrows. The visible table is the table for the top layer.

Figure 12-5
Categories in separate layers

Minority classification: Yes
Minority classification: No

To Change Layers

▶ Click one of the layer icon arrows.

or

▶ Choose a category from the drop-down list of layers.

Figure 12-6
Selecting layers from drop-down lists

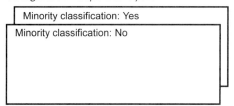

Layered Reports

Region:	Total							

Division		Sum	% of Total Sum	Mean	N	% of Total N
Business Products		89,707,150	61.9	$425,152	211	54.1
Consumer Products		$55,331,100	38.1	$309,112	179	45.9
Total		$145,038,250	100.0	$371,893	390	100.0

Drop-down list in header shows: Total, East, Central, West

Go to Layer Category

Go to Layer Category allows you to change layers in a pivot table. This dialog box is particularly useful when there are a large number of layers or if one layer has many categories.

Chapter 12

To Go to a Table Layer

▶ From the Pivot Table menus choose:

Pivot
  Go to Layer...

Figure 12-7
Go to Layer Category dialog box

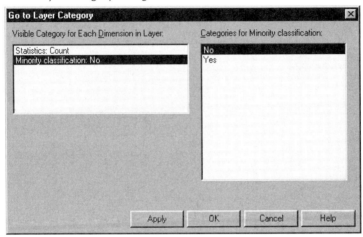

▶ In the Visible Category list, select a layer dimension. The Categories list will display all categories for the selected dimension.

▶ In the Categories list, select the category that you want, and then click OK. This changes the layer and closes the dialog box.

To view another layer without closing the dialog box:

▶ Select the category and click Apply.

To Move Layers to Rows or Columns

If the table that you are viewing is stacked in layers, with only the top layer showing, you can display all of the layers at once, either down the rows or across the columns. There must be at least one icon in the Layer tray.

▶ From the Pivot menu choose Move Layers to Rows.

or

▶ From the Pivot menu choose Move Layers to Columns.

You can also move layers to rows or columns by dragging their icons between the Layer, Row, and Column pivoting trays.

Bookmarks

Bookmarks allow you to save different views of a pivot table. Bookmarks save:

■ Placement of elements in row, column, and layer dimensions

■ Display order of elements in each dimension

■ Currently displayed layer for each layer element

To Bookmark Pivot Table Views

▶ Activate the pivot table.

▶ Pivot the table to the view that you want to bookmark.

▶ From the menus choose:
Pivot
    Bookmarks...

▶ Enter a name for the bookmark. (Bookmark names are not case sensitive.)

▶ Click Add.

Each pivot table has its own set of bookmarks. Within a pivot table, each bookmark name must be unique, but you can use duplicate bookmark names in different pivot tables.

To Display a Bookmarked Pivot Table View

▶ Activate the pivot table.

▶ From the menus choose:

Pivot
  Bookmarks...

▶ Click the name of the bookmark in the list.

▶ Click Go To.

To Rename a Pivot Table Bookmark

▶ Activate the pivot table.

▶ From the menus choose:

Pivot
  Bookmarks...

▶ Click the name of the bookmark in the list.

▶ Click Rename.

▶ Enter the new bookmark name.

▶ Click OK.

Showing and Hiding Cells

Many types of cells can be hidden, including:

- Dimension labels
- Categories, including the label cell and data cells in a row or column
- Category labels (without hiding the data cells)
- Footnotes, titles, and captions

To Hide Rows and Columns in a Table

▶ Ctrl-Alt-click the category label of the row or column to be hidden.

▶ From the Pivot Table menus choose:
  View
    Hide

or

▶ Right-click the row or column.

▶ From the context menu choose Hide Category.

To Show Hidden Rows and Columns in a Table

▶ Select another label that is in the same dimension as the hidden row or column.

For example, if the *Female* category of the Gender dimension is hidden, click the *Male* category.

▶ From the Pivot Table menus choose:
  View
    Show All Categories in dimension name

For example, choose Show All Categories in Gender.

or

▶ From the Pivot Table menus choose:
  View
    Show All

This displays all hidden cells in the table. (If Hide empty rows and columns is selected in Table Properties for this table, a completely empty row or column remains hidden.)

To Hide or Show a Dimension Label

▶ Activate the pivot table.

▶ Select the dimension label or any category label within the dimension.

▶ From the menus choose:
  View
    Hide (or Show) Dimension Label

To Hide or Show a Footnote in a Table

▶ Select a footnote.

▶ From the menus choose:
  View
    Hide (or Show)

To Hide or Show a Caption or Title in a Table

▶ Select a caption or title.

▶ From the menus choose:
  View
    Hide (or Show)

Editing Results

The appearance and contents of each table or text output item can be edited. You can:

- Apply a TableLook.
- Change the properties of the current table.
- Change the properties of cells in the table.
- Modify text.
- Add footnotes and captions to tables.
- Add items to the Viewer.
- Copy and paste results into other applications.
- Modify data cell widths.

Changing the Appearance of Tables

You can change the appearance of a table either by editing table properties or by applying a TableLook. Each TableLook consists of a collection of table properties, including general appearance, footnote properties, cell properties, and borders. You can select one of the preset TableLooks, or you can create and save a custom TableLook.

TableLooks

A TableLook is a set of properties that define the appearance of a table. You can select a previously defined TableLook or create your own TableLook.

Before or after a TableLook is applied, you can change cell formats for individual cells or groups of cells by using cell properties. The edited cell formats will remain intact, even when you apply a new TableLook.

For example, you might start by applying TableLook *9POINT,* then select a data column, and from the Cell Formats dialog box, change to a bold font for that column. Later, you change the TableLook to *BOXED*. The previously selected column retains the bold font while the rest of the characteristics are applied from the *BOXED* TableLook.

Optionally, you can reset all cells to the cell formats that are defined by the current TableLook. This resets any cells that have been edited. If As Displayed is selected in the TableLook Files list, any edited cells are reset to the current table properties.

To Apply or Save a TableLook

▶ Activate a pivot table.

▶ From the menus choose:
Format
  TableLooks...

Figure 12-8
TableLooks dialog box

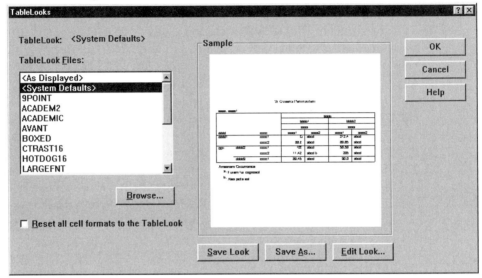

▶ Select a TableLook from the list of files. To select a file from another directory, click Browse.

▶ Click OK to apply the TableLook to the selected pivot table.

To Edit or Create a TableLook

▶ In the TableLooks dialog box, select a TableLook from the list of files.

▶ Click Edit Look.

▶ Adjust the table properties for the attributes that you want, and then click OK.

▶ Click Save Look to save the edited TableLook, or click Save As to save it as a new TableLook.

Editing a TableLook affects only the selected pivot table. An edited TableLook is not applied to any other tables that use that TableLook unless you select those tables and reapply the TableLook.

Table Properties

The Table Properties dialog box allows you to set general properties of a table, set cell styles for various parts of a table, and save a set of those properties as a TableLook. Using the tabs on this dialog box, you can:

■ Control general properties, such as hiding empty rows or columns and adjusting printing properties.

■ Control the format and position of footnote markers.

■ Determine specific formats for cells in the data area, for row and column labels, and for other areas of the table.

■ Control the width and color of the lines that form the borders of each area of the table.

To Change Pivot Table Properties

▶ Double-click anywhere in the pivot table to activate it.

▶ From the Pivot Table menus choose:
Format
  Table Properties...

▶ Select a tab (General, Footnotes, Cell Formats, Borders, or Printing).

▶ Select the options that you want.

▶ Click OK or Apply.

The new properties are applied to the selected pivot table. To apply new table properties to a TableLook instead of just the selected table, edit the TableLook (Format menu, TableLooks).

Table Properties: General

Several properties apply to the table as a whole. You can:

■ Show or hide empty rows and columns. (An empty row or column has nothing in any of the data cells.)

- Control the placement of row labels, which can be in the upper left corner or nested.

- Control maximum and minimum column width (expressed in points).

Figure 12-9
Table Properties dialog box, General tab

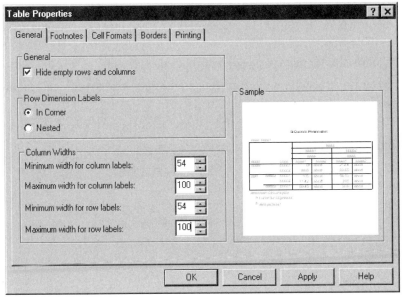

To Change General Table Properties

▶ Click the General tab.

▶ Select the options that you want.

▶ Click OK or Apply.

Table Properties: Footnotes

The properties of footnote markers include style and position in relation to text.

- The style of footnote markers is either numbers (1, 2, 3, ...) or letters (a, b, c, ...).

- The footnote markers can be attached to text as superscripts or subscripts.

Figure 12-10
Table Properties dialog box, Footnotes tab

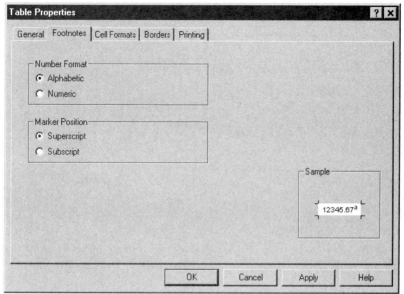

To Change Footnote Marker Properties

▶ Click the Footnotes tab.

▶ Select a footnote number format.

▶ Select a marker position.

▶ Click OK or Apply.

Table Properties: Cell Formats

For formatting, a table is divided into areas: title, layers, corner labels, row labels, column labels, data, caption, and footnotes. For each area of a table, you can modify the associated cell formats. Cell formats include text characteristics (such as font, size, color, and style), horizontal and vertical alignment, cell shading, foreground and background colors, and inner cell margins.

Figure 12-11
Areas of a table

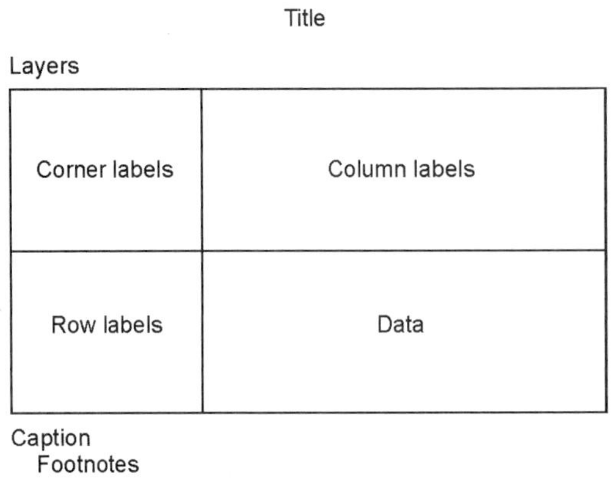

Cell formats are applied to areas (categories of information). They are not characteristics of individual cells. This distinction is an important consideration when pivoting a table.

For example,

- If you specify a bold font as a cell format of column labels, the column labels will appear bold no matter what information is currently displayed in the column dimension—and if you move an item from the column dimension to another dimension, it does not retain the bold characteristic of the column labels.

- If you make column labels bold simply by highlighting the cells in an activated pivot table and clicking the Bold button on the toolbar, the contents of those cells will remain bold no matter what dimension you move them to, and the column labels will not retain the bold characteristic for other items moved into the column dimension.

Figure 12-12
Table Properties dialog box, Cell Formats tab

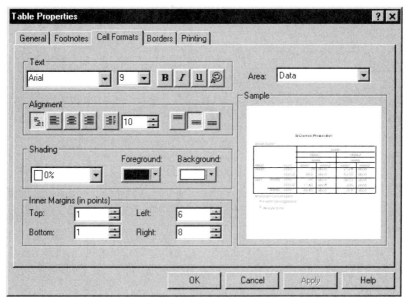

To Change Cell Formats

▶ Select the Cell Formats tab.

▶ Select an area from the drop-down list or click an area of the sample.

▶ Select characteristics for the area. Your selections are reflected in the sample.

▶ Click OK or Apply.

Table Properties: Borders

For each border location in a table, you can select a line style and a color. If you select None as the style, there will be no line at the selected location.

Figure 12-13
Table Properties dialog box, Borders tab

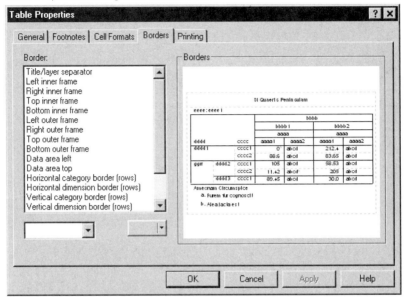

To Change Borders in a Table

▶ Click the Borders tab.

▶ Select a border location, either by clicking its name in the list or by clicking a line in the Sample area. (Shift-click to select multiple names, or Ctrl-click to select noncontiguous names.)

▶ Select a line style or select None.

▶ Select a color.

▶ Click OK or Apply.

To Display Hidden Borders in a Pivot Table

For tables without many visible borders, you can display the hidden borders. This can simplify tasks like changing column widths. The hidden borders (gridlines) are displayed in the Viewer but are not printed.

▶ Double-click anywhere in the pivot table to activate it.

▶ From the menus choose:
  View
    Gridlines

Table Properties: Printing

You can control the following properties for printed pivot tables:

■ Print all layers or only the top layer of the table, and print each layer on a separate page. (This setting affects only printing, not the display of layers in the Viewer.)

■ Shrink a table horizontally or vertically to fit the page for printing.

■ Control widow/orphan lines by controlling the minimum number of rows and columns that will be contained in any printed section of a table if the table is too wide and/or too long for the defined page size. (*Note*: If a table is too long to fit on the current page because there is other output above it, but it will fit within the defined page length, the table is automatically printed on a new page, regardless of the widow/orphan setting.)

■ Include continuation text for tables that don't fit on a single page. You can display continuation text at the bottom of each page and at the top of each page. If neither option is selected, the continuation text will not be displayed.

To Control Pivot Table Printing

▶ Click the Printing tab.

▶ Select the printing options that you want.

▶ Click OK or Apply.

Font

A TableLook allows you to specify font characteristics for different areas of the table. You can also change the font for any individual cell. Options for the font in a cell include the font face, style, size, and color. You can also hide the text or underline it.

If you specify font properties in a cell, they apply in all of the table layers that have the same cell.

Figure 12-14
Font dialog box

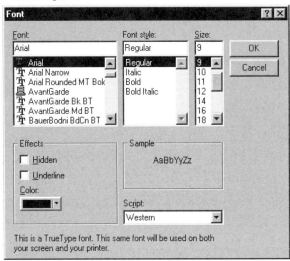

To Change the Font in a Cell

▶ Activate the pivot table and select the text that you want to change.

▶ From the Pivot Table menus choose:

Format
  Font...

Optionally, you can select a font, font style, and point size, and you can choose a color and a script style, as well as whether you want the text to be hidden or underlined.

Data Cell Widths

Set Data Cell Width is used to set all data cells to the same width.

Figure 12-15
Set Data Cell Width dialog box

To Change Data Cell Widths

▶ Activate the pivot table.

▶ From the menus choose:
Format
  Set Data Cell Widths...

▶ Enter a value for the cell width.

To Change the Width of a Pivot Table Column

▶ Double-click anywhere in the table to activate it.

▶ Move the pointer through the category labels until it is on the right border of the column that you want to change. (The pointer changes to an arrow with points on both ends.)

▶ Hold down the mouse button while you drag the border to its new position.

Figure 12-16
Changing the width of a column

Drag column border

		Gender		Total
		Female	Male	
Employment Category	Clerical	206	157	363
	Custodial	0	27	27
	Manager	10	74	84
Total		216	258	474

You can change vertical category and dimension borders in the row labels area, regardless of whether they are showing.

▶ Move the mouse pointer through the row labels until you see the double-pointed arrow.

▶ Drag the border to the new width.

Cell Properties

Cell properties are applied to a selected cell. You can change the value format, alignment, margins, and shading. Cell properties override table properties; therefore, if you change table properties, you do not change any individually applied cell properties.

To Change Cell Properties

▶ Activate a table and select a cell in the table.

▶ From the menus choose:
Format
  Cell Properties...

Cell Properties: Value

This dialog box tab controls the value format for a cell. You can select formats for number, date, time, or currency, and you can adjust the number of decimal digits that are displayed.

Figure 12-17
Cell Properties dialog box, Value tab

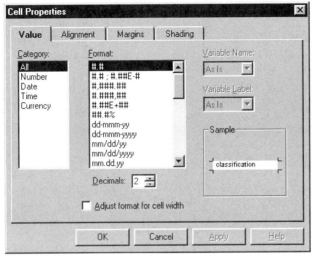

To Change Value Formats in a Cell

▶ Click the Value tab.

▶ Select a category and a format.

▶ Select the number of decimal places.

To Change Value Formats for a Column

▶ Ctrl-Alt-click the column label.

▶ Right-click the highlighted column.

▶ From the context menu choose Cell Properties.

▶ Click the Value tab.

▶ Select the format that you want to apply to the column.

You can use this method to suppress or add percentage signs and dollar signs, change the number of decimals that are displayed, and switch between scientific notation and regular numeric display.

Cell Properties: Alignment

This dialog box tab sets horizontal and vertical alignment and text direction for a cell. If you select Mixed, contents of the cell are aligned according to its type (number, date, or text).

Figure 12-18
Cell Properties dialog box, Alignment tab

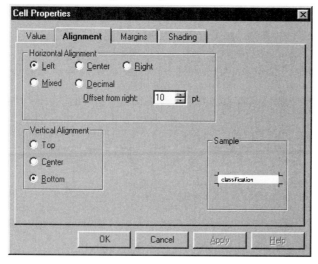

To Change Alignment in Cells

▶ Select a cell in the table.

▶ From the Pivot Table menus choose:
Format
  Cell Properties...

▶ Click the Alignment tab.

As you select the alignment properties for the cell, the properties are illustrated in the Sample area.

Cell Properties: Margins

This dialog box tab specifies the inset at each edge of a cell.

Figure 12-19
Cell Properties dialog box, Margins tab

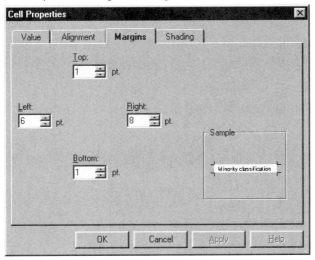

To Change Margins in Cells

▶ Click the Margins tab.

▶ Select the inset for each of the four margins.

Cell Properties: Shading

This dialog box tab specifies the percentage of shading and specifies foreground and background colors for a selected cell area. This setting does not change the color of the text.

Figure 12-20
Cell Properties dialog box, Shading tab

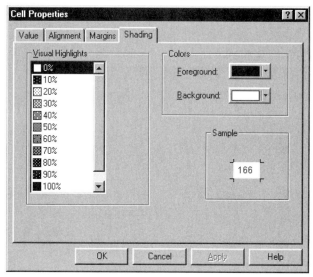

To Change Shading in Cells

▶ Click the Shading tab.

▶ Select the highlights and colors for the cell.

Footnote Marker

Footnote Marker changes the character(s) that can be used to mark a footnote.

Figure 12-21
Footnote Marker dialog box

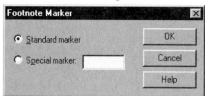

To Change Footnote Marker Characters

▶ Select a footnote.

▶ From the Pivot Table menus choose:
Format
  Footnote Marker...

▶ Enter one or two characters.

To Renumber Footnotes

When you have pivoted a table by switching rows, columns, and layers, the footnotes may be out of order. To renumber the footnotes:

▶ Activate the pivot table.

▶ From the menus choose:
Format
  Renumber Footnotes

Selecting Rows and Columns in Pivot Tables

In pivot tables, there are some constraints on how you select entire rows and columns, and the visual highlight that indicates the selected row or column may span noncontiguous areas of the table.

To Select a Row or Column in a Pivot Table

▶ Double-click anywhere in the pivot table to activate it.

▶ Click a row or column label.

▶ From the menus choose:
Edit
  Select
    Data and Label Cells

or

▶ Ctrl-Alt-click the row or column label.

If the table contains more than one dimension in the row or column area, the highlighted selection may span multiple noncontiguous cells.

Modifying Pivot Table Results

Text appears in the Viewer in many items. You can edit the text or add new text. Pivot tables can be modified by:

■ Editing text within pivot table cells

■ Adding captions and footnotes

To Modify Text in a Cell

▶ Activate the pivot table.

▶ Double-click the cell.

▶ Edit the text.

▶ Press Enter to record your changes, or press Esc to revert to the previous contents of the cell.

To Add Captions to a Table

▶ From the Pivot Table menus choose:
Insert
  Caption

The words Table Caption are displayed at the bottom of the table.

▶ Double-click the words Table Caption and replace them with your caption text.

To Add a Footnote to a Table

A footnote can be attached to any item in a table.

▶ Click a title, cell, or caption within an activated pivot table.

▶ From the Pivot Table menus choose:

Insert
 Footnote...

▶ Double-click the word Footnote and replace it with the footnote text.

Printing Pivot Tables

Several factors can affect the way that printed pivot tables look, and these factors can be controlled by changing pivot table attributes.

■ For multidimensional pivot tables (tables with layers), you can either print all layers or print only the top (visible) layer.

■ For long or wide pivot tables, you can automatically resize the table to fit the page or control the location of table breaks and page breaks.

Use Print Preview on the File menu to see how printed pivot tables will look.

To Print Hidden Layers of a Pivot Table

▶ Double-click anywhere in the table to activate it.

▶ From the menus choose:

Format
 Table Properties...

▶ On the Printing tab, select Print all layers.

You can also print each layer of a pivot table on a separate page.

Controlling Table Breaks for Wide and Long Tables

Pivot tables that are either too wide or too long to print within the defined page size are automatically split and printed in multiple sections. (For wide tables, multiple sections will print on the same page if there is room.) You can:

■ Control the row and column locations where large tables are split.

■ Specify rows and columns that should be kept together when tables are split.

■ Rescale large tables to fit the defined page size.

To Specify Row and Column Breaks for Printed Pivot Tables

▶ Activate the pivot table.

▶ Click the column label to the left of where you want to insert the break, or click the row label above where you want to insert the break.

▶ From the menus choose:
   Format
     Break Here

To Specify Rows or Columns to Keep Together

▶ Activate the pivot table.

▶ Select the labels of the rows or columns that you want to keep together. (Click and drag or Shift-click to select multiple row or column labels.)

▶ From the menus choose:
   Format
     Keep Together

To Rescale a Pivot Table to Fit the Page Size

▶ Activate the pivot table.

▶ From the menus choose:
   Format
     Table Properties

▶ Click the Printing tab.

▶ Select Rescale wide table to fit page.

   and/or

▶ Select Rescale long table to fit page.

How to Create a Chart from a Pivot Table

▶ Double-click the pivot table to activate it.

▶ Select the rows, columns, or cells you want to display in the chart.

▶ Right-click anywhere in the selected area.

▶ Choose Create Graph from the context menu and select a chart type.

Working with Command Syntax

SPSS provides a powerful command language that allows you to save and automate many common tasks. It also provides some functionality not found in the menus and dialog boxes.

Most commands are accessible from the menus and dialog boxes. However, some commands and options are available only by using the command language. The command language also allows you to save your jobs in a syntax file so that you can repeat your analysis at a later date or run it in an automated job with the Production Facility.

A syntax file is simply a text file that contains commands. While it is possible to open a syntax window and type in commands, it is often easier if you let the software help you build your syntax file using one of the following methods:

- Pasting command syntax from dialog boxes
- Copying syntax from the output log
- Copying syntax from the journal file

Detailed command syntax reference information is available in two forms: integrated into the overall Help system and as a separate PDF file, called the *SPSS Command Syntax Reference*, also available from the Help menu.

Context-sensitive Help for the current command in a syntax window is available by pressing the F1 key.

Syntax Rules

When you run commands from a command syntax window during an SPSS session, you are running commands in **interactive** mode.

The following rules apply to command specifications in interactive mode:

- Each command must start on a new line. Commands can begin in any column of a command line and continue for as many lines as needed. The exception is the END DATA command, which must begin in the first column of the first line after the end of data.

- Each command should end with a period as a command terminator. It is best to omit the terminator on BEGIN DATA, however, so that inline data are treated as one continuous specification.

- The command terminator must be the last nonblank character in a command.

- In the absence of a period as the command terminator, a blank line is interpreted as a command terminator.

Note: For compatibility with other modes of command execution (including command files run with INSERT or INCLUDE commands in an interactive session), each line of command syntax should not exceed 256 bytes.

- Most subcommands are separated by slashes (/). The slash before the first subcommand on a command is usually optional.

- Variable names must be spelled out fully.

- Text included within apostrophes or quotation marks must be contained on a single line.

- A period (.) must be used to indicate decimals, regardless of your regional or locale settings.

- Variable names ending in a period can cause errors in commands created by the dialog boxes. You cannot create such variable names in the dialog boxes, and you should generally avoid them.

Command syntax is case insensitive, and three- or four-letter abbreviations can be used for many command specifications. You can use as many lines as you want to specify a single command. You can add space or break lines at almost any point where a single blank is allowed, such as around slashes, parentheses, arithmetic operators, or between variable names. For example,

```
FREQUENCIES
  VARIABLES=JOBCAT GENDER
  /PERCENTILES=25 50 75
  /BARCHART.
```

and

```
freq var=jobcat gender /percent=25 50 75 /bar.
```

are both acceptable alternatives that generate the same results.

INCLUDE Files

For command files run via the INCLUDE command, batch mode syntax rules apply.

The following rules apply to command specifications in batch or production mode:

- All commands in the command file must begin in column 1. You can use plus (+) or minus (–) signs in the first column if you want to indent the command specification to make the command file more readable.
- If multiple lines are used for a command, column 1 of each continuation line must be blank.
- Command terminators are optional.
- A line cannot exceed 256 bytes; any additional characters are truncated.

Unless you have existing command files that already use the INCLUDE command, you should probably use the INSERT command instead, since it can accommodate command files that conform to either set of rules. If you generate command syntax by pasting dialog box choices into a syntax window, the format of the commands is suitable for any mode of operation. See the *SPSS Command Syntax Reference* (available in PDF format from the Help menu) for more information.

Pasting Syntax from Dialog Boxes

The easiest way to build a command syntax file is to make selections in dialog boxes and paste the syntax for the selections into a syntax window. By pasting the syntax at each step of a lengthy analysis, you can build a job file that allows you to repeat the analysis at a later date or run an automated job with the Production Facility.

In the syntax window, you can run the pasted syntax, edit it, and save it in a syntax file.

To Paste Syntax from Dialog Boxes

▶ Open the dialog box and make the selections that you want.

▶ Click Paste.

The command syntax is pasted to the designated syntax window. If you do not have an open syntax window, a new syntax window opens automatically, and the syntax is pasted there.

Figure 13-1
Command syntax pasted from a dialog box

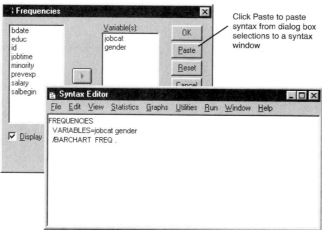

Note: If you open a dialog box from the menus in a script window, code for running syntax from a script is pasted into the script window.

Copying Syntax from the Output Log

You can build a syntax file by copying command syntax from the log that appears in the Viewer. To use this method, you must select Display commands in the log in the Viewer settings (Edit menu, Options, Viewer tab) before running the analysis. Each command will then appear in the Viewer along with the output from the analysis.

In the syntax window, you can run the pasted syntax, edit it, and save it in a syntax file.

Figure 13-2
Command syntax in the log

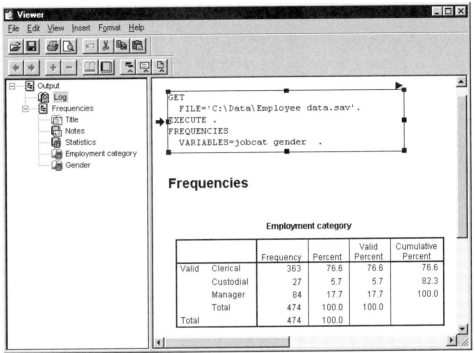

To Copy Syntax from the Output Log

▶ Before running the analysis, from the menus choose:
Edit
  Options...

▶ On the Viewer tab, select Display commands in the log.

As you run analyses, the commands for your dialog box selections are recorded in the log.

▶ Open a previously saved syntax file or create a new one. To create a new syntax file, from the menus choose:
File
  New
    Syntax

▶ In the Viewer, double-click a log item to activate it.

▶ Click and drag the mouse to highlight the syntax that you want to copy.

▶ From the Viewer menus choose:
Edit
  Copy

▶ In a syntax window, from the menus choose:
Edit
  Paste

Editing Syntax in a Journal File

By default, all commands executed during a session are recorded in a journal file named *spss.jnl* (specified using the Edit menu, Options, General tab). You can edit the journal file and save it as a syntax file that you can use to repeat a previously run analysis, or you can run it in an automated job with the Production Facility.

The journal file is a text file that can be edited like any other text file. Because error messages and warnings are also recorded in the journal file along with command syntax, you must edit out any error and warning messages that appear before saving the syntax file. Note, however, that errors must be resolved or the job will not run successfully.

Save the edited journal file with a different filename. Because the journal file is automatically appended or overwritten for each session, attempting to use the same filename for a syntax file and the journal file may yield unexpected results.

Figure 13-3
Editing the journal file

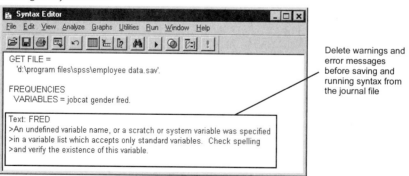

Delete warnings and error messages before saving and running syntax from the journal file

To Edit Syntax in a Journal File

▶ To open the journal file, from the menus choose:

File
  Open
    Other...

▶ Locate and open the journal file (by default, *spss.jnl* is located in the *temp* directory).

Select All files (*.*) for Files of Type or enter *.jnl in the File Name text box to display journal files in the file list. If you have difficulty locating the file, use Options on the Edit menu to see where the journal is saved in your system.

▶ Edit the file to remove any error messages or warnings, indicated by the > sign.

▶ Save the edited journal file using a different filename. (We recommend that you use a filename with the extension *.sps*, the default extension for syntax files.)

To Run Command Syntax

▶ Highlight the commands that you want to run in the syntax window.

▶ Click the Run button (the right-pointing triangle) on the Syntax Editor toolbar.

or

▶ Select one of the choices from the Run menu.

  ■ **All.** Runs all commands in the syntax window.

  ■ **Selection.** Runs the currently selected commands. This includes any commands partially highlighted.

  ■ **Current.** Runs the command where the cursor is currently located.

  ■ **To End.** Runs all commands from the current cursor location to the end of the command syntax file.

The Run button on the Syntax Editor toolbar runs the selected commands or the command where the cursor is located if there is no selection.

Figure 13-4
Syntax Editor toolbar

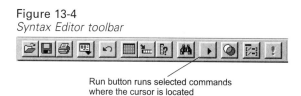

Run button runs selected commands
where the cursor is located

Multiple Execute Commands

Syntax pasted from dialog boxes or copied from the log or the journal may contain EXECUTE commands. When you run commands from a syntax window, EXECUTE commands are generally unnecessary and may slow performance, particularly with larger data files, because each EXECUTE command reads the entire data file. For more information, see the EXECUTE command in the *SPSS Command Syntax Reference* (available from the Help menu in any SPSS window).

Lag Functions

One notable exception is transformation commands that contain lag functions. In a series of transformation commands without any intervening EXECUTE commands or other commands that read the data, lag functions are calculated after all other transformations, regardless of command order. For example,

```
COMPUTE lagvar=LAG(var1).
COMPUTE var1=var1*2.
```

and

```
COMPUTE lagvar=LAG(var1).
EXECUTE.
COMPUTE var1=var1*2.
```

yield very different results for the value of *lagvar*, since the former uses the transformed value of *var1* while the latter uses the original value.

Frequencies

The Frequencies procedure provides statistics and graphical displays that are useful for describing many types of variables. The Frequencies procedure is a good place to start looking at your data.

For a frequency report and bar chart, you can arrange the distinct values in ascending or descending order, or you can order the categories by their frequencies. The frequencies report can be suppressed when a variable has many distinct values. You can label charts with frequencies (the default) or percentages.

Example. What is the distribution of a company's customers by industry type? From the output, you might learn that 37.5% of your customers are in government agencies, 24.9% are in corporations, 28.1% are in academic institutions, and 9.4% are in the healthcare industry. For continuous, quantitative data, such as sales revenue, you might learn that the average product sale is $3,576, with a standard deviation of $1,078.

Statistics and plots. Frequency counts, percentages, cumulative percentages, mean, median, mode, sum, standard deviation, variance, range, minimum and maximum values, standard error of the mean, skewness and kurtosis (both with standard errors), quartiles, user-specified percentiles, bar charts, pie charts, and histograms.

Data. Use numeric codes or short strings to code categorical variables (nominal or ordinal level measurements).

Assumptions. The tabulations and percentages provide a useful description for data from any distribution, especially for variables with ordered or unordered categories. Most of the optional summary statistics, such as the mean and standard deviation, are based on normal theory and are appropriate for quantitative variables with symmetric distributions. Robust statistics, such as the median, quartiles, and percentiles, are appropriate for quantitative variables that may or may not meet the assumption of normality.

Figure 14-1
Frequencies output

Industry

		Frequency	Percent	Valid Percent	Cumulative Percent
Valid	Government	331	37.5	37.5	37.5
	Corporate	220	24.9	24.9	62.5
	Academic	248	28.1	28.1	90.6
	Healthcare	83	9.4	9.4	100.0
	Total	882	100.0	100.0	

Statistics

	Mean	Median	Std. Deviation
Amount of Product Sale	$3,576.52	$3,417.50	$1,077.836

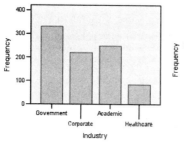
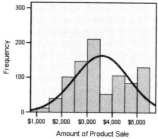

To Obtain Frequency Tables

▶ From the menus choose:

Analyze
   Descriptive Statistics
      Frequencies...

Figure 14-2
Frequencies dialog box

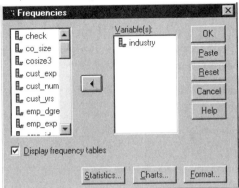

▶ Select one or more categorical or quantitative variables.

Optionally, you can:

■ Click Statistics for descriptive statistics for quantitative variables.

■ Click Charts for bar charts, pie charts, and histograms.

■ Click Format for the order in which results are displayed.

Frequencies Statistics

Figure 14-3
Frequencies Statistics dialog box

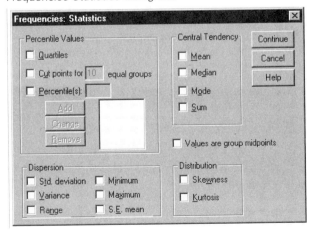

Percentile Values. Values of a quantitative variable that divide the ordered data into groups so that a certain percentage is above and another percentage is below. Quartiles (the 25th, 50th, and 75th percentiles) divide the observations into four groups of equal size. If you want an equal number of groups other than four, select Cut points for n equal groups. You can also specify individual percentiles (for example, the 95th percentile, the value below which 95% of the observations fall).

Central Tendency. Statistics that describe the location of the distribution include the mean, median, mode, and sum of all the values.

- **Mean.** A measure of central tendency. The arithmetic average, the sum divided by the number of cases.

- **Median.** The value above and below which half of the cases fall, the 50th percentile. If there is an even number of cases, the median is the average of the two middle cases when they are sorted in ascending or descending order. The median is a measure of central tendency not sensitive to outlying values (unlike the mean, which can be affected by a few extremely high or low values).

- **Mode.** The most frequently occurring value. If several values share the greatest frequency of occurrence, each of them is a mode. The Frequencies procedure reports only the smallest of such multiple modes.

- **Sum.** The sum or total of the values, across all cases with nonmissing values.

Dispersion. Statistics that measure the amount of variation or spread in the data include the standard deviation, variance, range, minimum, maximum, and standard error of the mean.

- **Std. deviation.** A measure of dispersion around the mean. In a normal distribution, 68% of cases fall within one standard deviation of the mean and 95% of cases fall within two standard deviations. For example, if the mean age is 45, with a standard deviation of 10, 95% of the cases would be between 25 and 65 in a normal distribution.

- **Variance.** A measure of dispersion around the mean, equal to the sum of squared deviations from the mean divided by one less than the number of cases. The variance is measured in units that are the square of those of the variable itself.

- **Range.** The difference between the largest and smallest values of a numeric variable, the maximum minus the minimum.

- **Minimum.** The smallest value of a numeric variable.

- **Maximum.** The largest value of a numeric variable.

- **S. E. mean.** A measure of how much the value of the mean may vary from sample to sample taken from the same distribution. It can be used to roughly compare the observed mean to a hypothesized value (that is, you can conclude the two values are different if the ratio of the difference to the standard error is less than -2 or greater than +2).

Distribution. Skewness and kurtosis are statistics that describe the shape and symmetry of the distribution. These statistics are displayed with their standard errors.

- **Skewness.** A measure of the asymmetry of a distribution. The normal distribution is symmetric and has a skewness value of 0. A distribution with a significant positive skewness has a long right tail. A distribution with a significant negative skewness has a long left tail. As a guideline, a skewness value more than twice its standard error is taken to indicate a departure from symmetry.

- **Kurtosis.** A measure of the extent to which observations cluster around a central point. For a normal distribution, the value of the kurtosis statistic is zero. Positive kurtosis indicates that the observations cluster more and have longer tails than those in the normal distribution, and negative kurtosis indicates that the observations cluster less and have shorter tails.

Values are group midpoints. If the values in your data are midpoints of groups (for example, ages of all people in their thirties are coded as 35), select this option to estimate the median and percentiles for the original, ungrouped data.

Frequencies Charts

Figure 14-4
Frequencies Charts dialog box

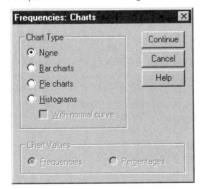

Chart Type. A pie chart displays the contribution of parts to a whole. Each slice of a pie chart corresponds to a group that is defined by a single grouping variable. A bar chart displays the count for each distinct value or category as a separate bar, allowing you to compare categories visually. A histogram also has bars, but they are plotted along an equal interval scale. The height of each bar is the count of values of a quantitative variable falling within the interval. A histogram shows the shape, center, and spread of the distribution. A normal curve superimposed on a histogram helps you judge whether the data are normally distributed.

Chart Values. For bar charts, the scale axis can be labeled by frequency counts or percentages.

Frequencies Format

Figure 14-5
Frequencies Format dialog box

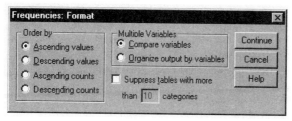

Order by. The frequency table can be arranged according to the actual values in the data or according to the count (frequency of occurrence) of those values, and the table can be arranged in either ascending or descending order. However, if you request a histogram or percentiles, Frequencies assumes that the variable is quantitative and displays its values in ascending order.

Multiple Variables. If you produce statistics tables for multiple variables, you can either display all variables in a single table (Compare variables) or display a separate statistics table for each variable (Organize output by variables).

Suppress tables with more than n categories. This option prevents the display of tables with more than the specified number of values.

Descriptives

The Descriptives procedure displays univariate summary statistics for several variables in a single table and calculates standardized values (z scores). Variables can be ordered by the size of their means (in ascending or descending order), alphabetically, or by the order in which you select the variables (the default).

When z scores are saved, they are added to the data in the Data Editor and are available for charts, data listings, and analyses. When variables are recorded in different units (for example, gross domestic product per capita and percentage literate), a z-score transformation places variables on a common scale for easier visual comparison.

Example. If each case in your data contains the daily sales totals for each member of the sales staff (for example, one entry for Bob, one entry for Kim, and one entry for Brian) collected each day for several months, the Descriptives procedure can compute the average daily sales for each staff member and can order the results from highest average sales to lowest average sales.

Statistics. Sample size, mean, minimum, maximum, standard deviation, variance, range, sum, standard error of the mean, and kurtosis and skewness with their standard errors.

Data. Use numeric variables after you have screened them graphically for recording errors, outliers, and distributional anomalies. The Descriptives procedure is very efficient for large files (thousands of cases).

Assumptions. Most of the available statistics (including z scores) are based on normal theory and are appropriate for quantitative variables (interval- or ratio-level measurements) with symmetric distributions. Avoid variables with unordered categories or skewed distributions. The distribution of z scores has the same shape as that of the original data; therefore, calculating z scores is not a remedy for problem data.

To Obtain Descriptive Statistics

▶ From the menus choose:

Analyze
  Descriptive Statistics
    Descriptives...

Figure 15-1
Descriptives dialog box

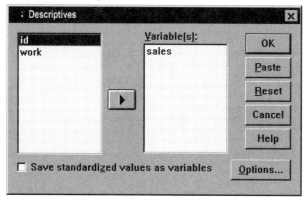

▶ Select one or more variables.

Optionally, you can:

■ Select Save standardized values as variables to save *z* scores as new variables.

■ Click Options for optional statistics and display order.

Descriptives Options

Figure 15-2
Descriptives Options dialog box

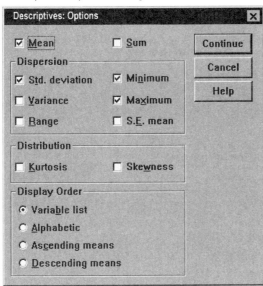

Mean and Sum. The mean, or arithmetic average, is displayed by default.

Dispersion. Statistics that measure the spread or variation in the data include the standard deviation, variance, range, minimum, maximum, and standard error of the mean.

- **Std. deviation.** A measure of dispersion around the mean. In a normal distribution, 68% of cases fall within one standard deviation of the mean and 95% of cases fall within two standard deviations. For example, if the mean age is 45, with a standard deviation of 10, 95% of the cases would be between 25 and 65 in a normal distribution.

- **Variance.** A measure of dispersion around the mean, equal to the sum of squared deviations from the mean divided by one less than the number of cases. The variance is measured in units that are the square of those of the variable itself.

- **Range.** The difference between the largest and smallest values of a numeric variable, the maximum minus the minimum.

- **Minimum.** The smallest value of a numeric variable.

- **Maximum.** The largest value of a numeric variable.

- **S.E. mean.** A measure of how much the value of the mean may vary from sample to sample taken from the same distribution. It can be used to roughly compare the observed mean to a hypothesized value (that is, you can conclude the two values are different if the ratio of the difference to the standard error is less than -2 or greater than +2).

Distribution. Kurtosis and skewness are statistics that characterize the shape and symmetry of the distribution. These statistics are displayed with their standard errors.

- **Kurtosis.** A measure of the extent to which observations cluster around a central point. For a normal distribution, the value of the kurtosis statistic is zero. Positive kurtosis indicates that the observations cluster more and have longer tails than those in the normal distribution, and negative kurtosis indicates that the observations cluster less and have shorter tails.

- **Skewness.** A measure of the asymmetry of a distribution. The normal distribution is symmetric and has a skewness value of 0. A distribution with a significant positive skewness has a long right tail. A distribution with a significant negative skewness has a long left tail. As a guideline, a skewness value more than twice its standard error is taken to indicate a departure from symmetry.

Display Order. By default, the variables are displayed in the order in which you selected them. Optionally, you can display variables alphabetically, by ascending means, or by descending means.

DESCRIPTIVES Command Additional Features

The SPSS command language also allows you to:

- Save standardized scores (z scores) for some but not all variables (with the VARIABLES subcommand).

- Specify names for new variables that contain standardized scores (with the VARIABLES subcommand).

- Exclude from the analysis cases with missing values for any variable (with the MISSING subcommand).

- Sort the variables in the display by the value of any statistic, not just the mean (with the SORT subcommand).

See the *SPSS Command Syntax Reference* for complete syntax information.

16

Explore

The Explore procedure produces summary statistics and graphical displays, either for all of your cases or separately for groups of cases. There are many reasons for using the Explore procedure—data screening, outlier identification, description, assumption checking, and characterizing differences among subpopulations (groups of cases). Data screening may show that you have unusual values, extreme values, gaps in the data, or other peculiarities. Exploring the data can help to determine whether the statistical techniques that you are considering for data analysis are appropriate. The exploration may indicate that you need to transform the data if the technique requires a normal distribution. Or, you may decide that you need nonparametric tests.

Example. Look at the distribution of maze-learning times for rats under four different reinforcement schedules. For each of the four groups, you can see if the distribution of times is approximately normal and whether the four variances are equal. You can also identify the cases with the five largest and five smallest times. The boxplots and stem-and-leaf plots graphically summarize the distribution of learning times for each of the groups.

Statistics and plots. Mean, median, 5% trimmed mean, standard error, variance, standard deviation, minimum, maximum, range, interquartile range, skewness and kurtosis and their standard errors, confidence interval for the mean (and specified confidence level), percentiles, Huber's M-estimator, Andrews' wave estimator, Hampel's redescending M-estimator, Tukey's biweight estimator, the five largest and five smallest values, the Kolmogorov-Smirnov statistic with a Lilliefors significance level for testing normality, and the Shapiro-Wilk statistic. Boxplots, stem-and-leaf plots, histograms, normality plots, and spread-versus-level plots with Levene tests and transformations.

Data. The Explore procedure can be used for quantitative variables (interval- or ratio-level measurements). A factor variable (used to break the data into groups of cases) should have a reasonable number of distinct values (categories). These values may be short string or numeric. The case label variable, used to label outliers in boxplots, can be short string, long string (first 15 characters), or numeric.

Assumptions. The distribution of your data does not have to be symmetric or normal.

Figure 16-1
Explore output

Descriptives

			Time			
			Schedule			
			1	2	3	4
Mean		Statistic	2.760	4.850	6.900	9.010
		Std. Error	.165	.422	.445	.289
95.0% Confidence Interval for Mean	Lower Bound	Statistic	2.387	3.895	5.893	8.357
	Upper Bound	Statistic	3.133	5.805	7.907	9.663
5% Trimmed Mean		Statistic	2.761	4.889	6.911	8.994
Median		Statistic	2.850	4.900	7.050	9.000
Variance		Statistic	.272	1.783	1.982	.834
Std. Deviation		Statistic	.521	1.335	1.408	.913
Minimum		Statistic	2.0	2.3	4.5	7.8
Maximum		Statistic	3.5	6.7	9.1	10.5
Range		Statistic	1.5	4.4	4.6	2.7
Interquartile Range		Statistic	.925	2.250	2.400	1.650
Skewness		Statistic	-.116	-.559	-.197	.219
		Std. Error	.687	.687	.687	.687
Kurtosis		Statistic	-1.210	-.104	-.606	-1.350
		Std. Error	1.334	1.334	1.334	1.334

Extreme Values

			Case Number	Schedule	Value
Time	Highest	1	31	4	10.5
		2	33	4	9.9
		3	39	4	9.8
		4	32	4	9.5
		5	36	4	9.3
	Lowest	1	2	1	2.0
		2	7	1	2.1
		3	1	1	2.3
		4	11	2	2.3
		5	3	1	2.5

Time Stem-and-Leaf Plot

Frequency   Stem & Leaf

```
  7.00     2 . 0133589
  6.00     3 . 014577
  3.00     4 . 568
  5.00     5 . 05779
  4.00     6 . 1379
  3.00     7 . 268
  6.00     8 . 012237
  5.00     9 . 13589
  1.00    10 . 5
```

Stem width: 1.0
Each leaf: 1 case(s)

To Explore Your Data

▶ From the menus choose:

Analyze
  Descriptive Statistics
    Explore...

Figure 16-2
Explore dialog box

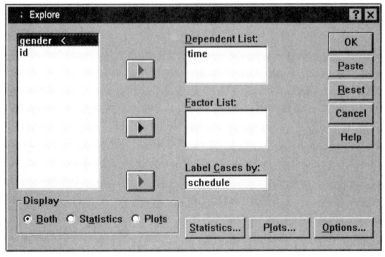

▶ Select one or more dependent variables.

Optionally, you can:

- Select one or more factor variables, whose values will define groups of cases.

- Select an identification variable to label cases.

- Click Statistics for robust estimators, outliers, percentiles, and frequency tables.

- Click Plots for histograms, normal probability plots and tests, and spread-versus-level plots with Levene's statistics.

- Click Options for the treatment of missing values.

Explore Statistics

Figure 16-3
Explore Statistics dialog box

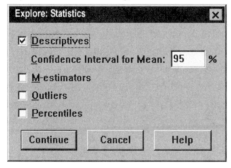

Descriptives. These measures of central tendency and dispersion are displayed by default. Measures of central tendency indicate the location of the distribution; they include the mean, median, and 5% trimmed mean. Measures of dispersion show the dissimilarity of the values; these include standard error, variance, standard deviation, minimum, maximum, range, and interquartile range. The descriptive statistics also include measures of the shape of the distribution; skewness and kurtosis are displayed with their standard errors. The 95% level confidence interval for the mean is also displayed; you can specify a different confidence level.

M-estimators. Robust alternatives to the sample mean and median for estimating the location. The estimators calculated differ in the weights they apply to cases. Huber's M-estimator, Andrews' wave estimator, Hampel's redescending M-estimator, and Tukey's biweight estimator are displayed.

Outliers. Displays the five largest and five smallest values, with case labels.

Percentiles. Displays the values for the 5th, 10th, 25th, 50th, 75th, 90th, and 95th percentiles.

Explore Plots

Figure 16-4
Explore Plots dialog box

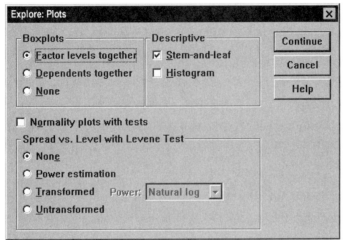

Boxplots. These alternatives control the display of boxplots when you have more than one dependent variable. Factor levels together generates a separate display for each dependent variable. Within a display, boxplots are shown for each of the groups defined by a factor variable. Dependents together generates a separate display for each group defined by a factor variable. Within a display, boxplots are shown side by side for each dependent variable. This display is particularly useful when the different variables represent a single characteristic measured at different times.

Descriptive. The Descriptive group allows you to choose stem-and-leaf plots and histograms.

Normality plots with tests. Displays normal probability and detrended normal probability plots. The Kolmogorov-Smirnov statistic, with a Lilliefors significance level for testing normality, is displayed. If non-integer weights are specified, the Shapiro-Wilk statistic is calculated when the weighted sample size lies between 3 and 50. For no weights or integer weights, the statistic is calculated when the weighted sample size lies between 3 and 5000.

Spread vs. Level with Levene Test. Controls data transformation for spread-versus-level plots. For all spread-versus-level plots, the slope of the regression line and Levene's robust tests for homogeneity of variance are displayed. If you select a transformation, Levene's tests are based on the transformed data. If no factor variable is selected, spread-versus-level plots are not produced. Power estimation produces a plot of the natural logs of the interquartile ranges against the natural logs of the medians for all cells, as well as an estimate of the power transformation for achieving equal variances in the cells. A spread-versus-level plot helps to determine the power for a transformation to stabilize (make more equal) variances across groups. Transformed allows you to select one of the power alternatives, perhaps following the recommendation from power estimation, and produces plots of transformed data. The interquartile range and median of the transformed data are plotted. Untransformed produces plots of the raw data. This is equivalent to a transformation with a power of 1.

Explore Power Transformations

These are the power transformations for spread-versus-level plots. To transform data, you must select a power for the transformation. You can choose one of the following alternatives:

- **Natural log.** Natural log transformation. This is the default.

- **1/square root.** For each data value, the reciprocal of the square root is calculated.

- **Reciprocal.** The reciprocal of each data value is calculated.

- **Square root.** The square root of each data value is calculated.

- **Square.** Each data value is squared.

- **Cube.** Each data value is cubed.

Explore Options

Figure 16-5
Explore Options dialog box

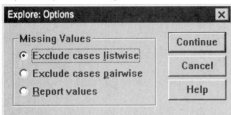

Missing Values. Controls the treatment of missing values.

■ **Exclude cases listwise.** Cases with missing values for any dependent or factor variable are excluded from all analyses. This is the default.

■ **Exclude cases pairwise.** Cases with no missing values for variables in a group (cell) are included in the analysis of that group. The case may have missing values for variables used in other groups.

■ **Report values.** Missing values for factor variables are treated as a separate category. All output is produced for this additional category. Frequency tables include categories for missing values. Missing values for a factor variable are included but labeled as missing.

EXAMINE Command Additional Features

The Explore procedure uses EXAMINE command syntax. The SPSS command language also allows you to:

■ Request total output and plots in addition to output and plots for groups defined by the factor variables (with the TOTAL subcommand).

■ Specify a common scale for a group of boxplots (with the SCALE subcommand).

■ Specify interactions of the factor variables (with the VARIABLES subcommand).

■ Specify percentiles other than the defaults (with the PERCENTILES subcommand).

■ Calculate percentiles according to any of five methods (with the PERCENTILES subcommand).

■ Specify any power transformation for spread-versus-level plots (with the PLOT subcommand).

- ■ Specify the number of extreme values to be displayed (with the STATISTICS subcommand).

- ■ Specify parameters for the M-estimators, robust estimators of location (with the MESTIMATORS subcommand).

See the *SPSS Command Syntax Reference* for complete syntax information.

Crosstabs

The Crosstabs procedure forms two-way and multiway tables and provides a variety of tests and measures of association for two-way tables. The structure of the table and whether categories are ordered determine what test or measure to use.

Crosstabs' statistics and measures of association are computed for two-way tables only. If you specify a row, a column, and a layer factor (control variable), the Crosstabs procedure forms one panel of associated statistics and measures for each value of the layer factor (or a combination of values for two or more control variables). For example, if *gender* is a layer factor for a table of *married* (yes, no) against *life* (is life exciting, routine, or dull), the results for a two-way table for the females are computed separately from those for the males and printed as panels following one another.

Example. Are customers from small companies more likely to be profitable in sales of services (for example, training and consulting) than those from larger companies? From a crosstabulation, you might learn that the majority of small companies (fewer than 500 employees) yield high service profits, while the majority of large companies (more than 2,500 employees) yield low service profits.

Statistics and measures of association. Pearson chi-square, likelihood-ratio chi-square, linear-by-linear association test, Fisher's exact test, Yates' corrected chi-square, Pearson's *r*, Spearman's rho, contingency coefficient, phi, Cramér's *V*, symmetric and asymmetric lambdas, Goodman and Kruskal's tau, uncertainty coefficient, gamma, Somers' *d*, Kendall's tau-*b*, Kendall's tau-*c*, eta coefficient, Cohen's kappa, relative risk estimate, odds ratio, McNemar test, and Cochran's and Mantel-Haenszel statistics.

Data. To define the categories of each table variable, use values of a numeric or short string (eight or fewer characters) variable. For example, for *gender*, you could code the data as 1 and 2 or as *male* and *female*.

Assumptions. Some statistics and measures assume ordered categories (ordinal data) or quantitative values (interval or ratio data), as discussed in the section on statistics. Others are valid when the table variables have unordered categories (nominal data).

For the chi-square-based statistics (phi, Cramér's V, and contingency coefficient), the data should be a random sample from a multinomial distribution.

Note: Ordinal variables can be either numeric codes that represent categories (for example, 1 = *low*, 2 = *medium*, 3 = *high*) or string values. However, the alphabetic order of string values is assumed to reflect the true order of the categories. For example, for a string variable with the values of *low*, *medium*, *high*, the order of the categories is interpreted as *high*, *low*, *medium*—which is not the correct order. In general, it is more reliable to use numeric codes to represent ordinal data.

Figure 17-1
Crosstabs output

Service Profitability * Company Size Crosstabulation

Service Profitability	Company Size			Total
	1-500	501-2,500	> 2,500	
Low	200	85	135	420
High	251	106	105	462
Total	451	191	240	882

Chi-Square Tests

	Value	df	Asymp. Sig. (2-sided)
Pearson Chi-Square	9.848	2	.007
Likelihood Ratio	9.852	2	.007
Linear-by-Linear Association	7.869	1	.005
N of Valid Cases	882		

To Obtain Crosstabulations

► From the menus choose:
Analyze
  Descriptive Statistics
    Crosstabs...

Figure 17-2
Crosstabs dialog box

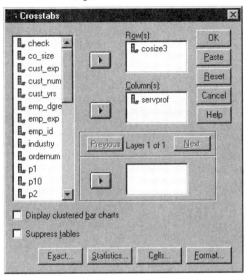

▶ Select one or more row variables and one or more column variables.

Optionally, you can:

- Select one or more control variables.

- Click Statistics for tests and measures of association for two-way tables or subtables.

- Click Cells for observed and expected values, percentages, and residuals.

- Click Format for controlling the order of categories.

Crosstabs Layers

If you select one or more layer variables, a separate crosstabulation is produced for each category of each layer variable (control variable). For example, if you have one row variable, one column variable, and one layer variable with two categories, you get a two-way table for each category of the layer variable. To make another layer of control variables, click Next. Subtables are produced for each combination of categories for each first-layer variable with each second-layer variable and so on. If statistics and measures of association are requested, they apply to two-way subtables only.

Crosstabs Clustered Bar Charts

Display clustered bar charts. A clustered bar chart helps summarize your data for groups of cases. There is one cluster of bars for each value of the variable you specified under Rows. The variable that defines the bars within each cluster is the variable you specified under Columns. There is one set of differently colored or patterned bars for each value of this variable. If you specify more than one variable under Columns or Rows, a clustered bar chart is produced for each combination of two variables.

Crosstabs Statistics

Figure 17-3
Crosstabs Statistics dialog box

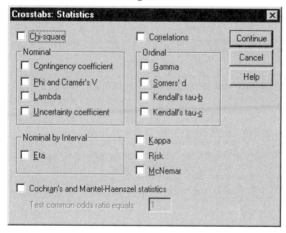

Chi-square. For tables with two rows and two columns, select Chi-square to calculate the Pearson chi-square, the likelihood-ratio chi-square, Fisher's exact test, and Yates' corrected chi-square (continuity correction). For 2 × 2 tables, Fisher's exact test is computed when a table that does not result from missing rows or columns in a larger table has a cell with an expected frequency of less than 5. Yates' corrected chi-square is computed for all other 2 × 2 tables. For tables with any number of rows and columns, select Chi-square to calculate the Pearson chi-square and the likelihood-ratio chi-square. When both table variables are quantitative, Chi-square yields the linear-by-linear association test.

Correlations. For tables in which both rows and columns contain ordered values, Correlations yields Spearman's correlation coefficient, rho (numeric data only). Spearman's rho is a measure of association between rank orders. When both table variables (factors) are quantitative, Correlations yields the Pearson correlation coefficient, r, a measure of linear association between the variables.

Nominal. For nominal data (no intrinsic order, such as Catholic, Protestant, and Jewish), you can select Phi (coefficient) and Cramér's V, Contingency coefficient, Lambda (symmetric and asymmetric lambdas and Goodman and Kruskal's tau), and Uncertainty coefficient.

- **Contingency coefficient.** A measure of association based on chi-square. The value ranges between 0 and 1, with 0 indicating no association between the row and column variables and values close to 1 indicating a high degree of association between the variables. The maximum value possible depends on the number of rows and columns in a table.

- **Phi and Cramer's V.** Phi is a chi-square-based measure of association that involves dividing the chi-square statistic by the sample size and taking the square root of the result. Cramer's V is a measure of association based on chi-square.

- **Lambda.** A measure of association that reflects the proportional reduction in error when values of the independent variable are used to predict values of the dependent variable. A value of 1 means that the independent variable perfectly predicts the dependent variable. A value of 0 means that the independent variable is no help in predicting the dependent variable.

- **Uncertainty coefficient.** A measure of association that indicates the proportional reduction in error when values of one variable are used to predict values of the other variable. For example, a value of 0.83 indicates that knowledge of one variable reduces error in predicting values of the other variable by 83%. The program calculates both symmetric and asymmetric versions of the uncertainty coefficient.

Ordinal. For tables in which both rows and columns contain ordered values, select Gamma (zero-order for 2-way tables and conditional for 3-way to 10-way tables), Kendall's tau-b, and Kendall's tau-c. For predicting column categories from row categories, select Somers' d.

- **Gamma.** A symmetric measure of association between two ordinal variables that ranges between -1 and 1. Values close to an absolute value of 1 indicate a strong relationship between the two variables. Values close to 0 indicate little or no relationship. For 2-way tables, zero-order gammas are displayed. For 3-way to n-way tables, conditional gammas are displayed.

- **Somers' d.** A measure of association between two ordinal variables that ranges from -1 to 1. Values close to an absolute value of 1 indicate a strong relationship between the two variables, and values close to 0 indicate little or no relationship between the variables. Somers' d is an asymmetric extension of gamma that differs only in the inclusion of the number of pairs not tied on the independent variable. A symmetric version of this statistic is also calculated.

- **Kendall's tau-b.** A nonparametric measure of correlation for ordinal or ranked variables that take ties into account. The sign of the coefficient indicates the direction of the relationship, and its absolute value indicates the strength, with larger absolute values indicating stronger relationships. Possible values range from -1 to 1, but a value of -1 or +1 can be obtained only from square tables.

- **Kendall's tau-c.** A nonparametric measure of association for ordinal variables that ignores ties. The sign of the coefficient indicates the direction of the relationship, and its absolute value indicates the strength, with larger absolute values indicating stronger relationships. Possible values range from -1 to 1, but a value of -1 or +1 can be obtained only from square tables.

Nominal by Interval. When one variable is categorical and the other is quantitative, select Eta. The categorical variable must be coded numerically.

- **Eta.** A measure of association that ranges from 0 to 1, with 0 indicating no association between the row and column variables and values close to 1 indicating a high degree of association. Eta is appropriate for a dependent variable measured on an interval scale (for example, income) and an independent variable with a limited number of categories (for example, gender). Two eta values are computed: one treats the row variable as the interval variable, and the other treats the column variable as the interval variable.

Kappa. Cohen's kappa measures the agreement between the evaluations of two raters when both are rating the same object. A value of 1 indicates perfect agreement. A value of 0 indicates that agreement is no better than chance. Kappa is available only for tables in which both variables use the same category values and both variables have the same number of categories.

Risk. For 2 x 2 tables, a measure of the strength of the association between the presence of a factor and the occurrence of an event. If the confidence interval for the statistic includes a value of 1, you cannot assume that the factor is associated with the event.

The odds ratio can be used as an estimate or relative risk when the occurrence of the factor is rare.

McNemar. A nonparametric test for two related dichotomous variables. Tests for changes in responses using the chi-square distribution. Useful for detecting changes in responses due to experimental intervention in "before-and-after" designs. For larger square tables, the McNemar-Bowker test of symmetry is reported.

Cochran's and Mantel-Haenszel statistics. Cochran's and Mantel-Haenszel statistics can be used to test for independence between a dichotomous factor variable and a dichotomous response variable, conditional upon covariate patterns defined by one or more layer (control) variables. Note that while other statistics are computed layer by layer, the Cochran's and Mantel-Haenszel statistics are computed once for all layers.

Crosstabs Cell Display

Figure 17-4
Crosstabs Cell Display dialog box

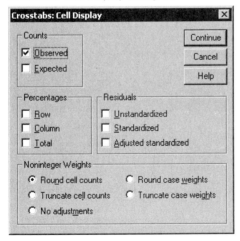

To help you uncover patterns in the data that contribute to a significant chi-square test, the Crosstabs procedure displays expected frequencies and three types of residuals (deviates) that measure the difference between observed and expected frequencies. Each cell of the table can contain any combination of counts, percentages, and residuals selected.

Counts. The number of cases actually observed and the number of cases expected if the row and column variables are independent of each other.

Percentages. The percentages can add up across the rows or down the columns. The percentages of the total number of cases represented in the table (one layer) are also available.

Residuals. Raw unstandardized residuals give the difference between the observed and expected values. Standardized and adjusted standardized residuals are also available.

- **Unstandardized.** The difference between an observed value and the expected value. The expected value is the number of cases you would expect in the cell if there were no relationship between the two variables. A positive residual indicates that there are more cases in the cell than there would be if the row and column variables were independent.

- **Standardized.** The residual divided by an estimate of its standard deviation. Standardized residuals, which are also known as Pearson residuals, have a mean of 0 and a standard deviation of 1.

- **Adjusted standardized.** The residual for a cell (observed minus expected value) divided by an estimate of its standard error. The resulting standardized residual is expressed in standard deviation units above or below the mean.

Non-integer Weights. Cell counts are normally integer values, since they represent the number of cases in each cell. But if the data file is currently weighted by a weight variable with fractional values (for example, 1.25), cell counts can also be fractional values. You can truncate or round either before or after calculating the cell counts or use fractional cell counts for both table display and statistical calculations.

- **Round cell counts.** Case weights are used as is but the accumulated weights in the cells are rounded before computing any statistics.

- **Truncate cell counts.** Case weights are used as is but the accumulated weights in the cells are truncated before computing any statistics.

- **Round case weights.** Case weights are rounded before use.

- **Truncate case weights.** Case weights are truncated before use.

- **No adjustments.** Case weights are used as is and fractional cell counts are used. However, when Exact Statistics (available only with the Exact Tests option) are requested, the accumulated weights in the cells are either truncated or rounded before computing the Exact test statistics.

Crosstabs Table Format

Figure 17-5
Crosstabs Table Format dialog box

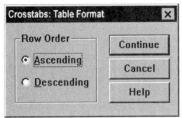

You can arrange rows in ascending or descending order of the values of the row variable.

Summarize

The Summarize procedure calculates subgroup statistics for variables within categories of one or more grouping variables. All levels of the grouping variable are crosstabulated. You can choose the order in which the statistics are displayed. Summary statistics for each variable across all categories are also displayed. Data values in each category can be listed or suppressed. With large datasets, you can choose to list only the first *n* cases.

Example. What is the average product sales amount by region and customer industry? You might discover that the average sales amount is slightly higher in the western region than in other regions, with corporate customers in the western region yielding the highest average sales amount.

Statistics. Sum, number of cases, mean, median, grouped median, standard error of the mean, minimum, maximum, range, variable value of the first category of the grouping variable, variable value of the last category of the grouping variable, standard deviation, variance, kurtosis, standard error of kurtosis, skewness, standard error of skewness, percentage of total sum, percentage of total *N*, percentage of sum in, percentage of *N* in, geometric mean, and harmonic mean.

Data. Grouping variables are categorical variables whose values can be numeric or short string. The number of categories should be reasonably small. The other variables should be able to be ranked.

Assumptions. Some of the optional subgroup statistics, such as the mean and standard deviation, are based on normal theory and are appropriate for quantitative variables with symmetric distributions. Robust statistics, such as the median and the range, are appropriate for quantitative variables that may or may not meet the assumption of normality.

Figure 18-1
Summarize output

Case Summaries
Average Product Sale by Region and Industry

Region	Industry				
	Government	Corporate	Academic	Healthcare	Total
East	$3,594.65	$3,953.76	$3,764.91	$3,722.32	$3,735.45
Central	$3,370.12	$3,268.47	$3,317.81	$3,165.11	$3,305.03
West	$3,552.50	$4,649.00	$4,276.25	$4,027.00	$4,079.46
Total	$3,503.75	$3,727.50	$3,579.76	$3,456.93	$3,576.52

To Obtain Case Summaries

▶ From the menus choose:
Analyze
  Reports
    Case Summaries...

Figure 18-2
Summarize Cases dialog box

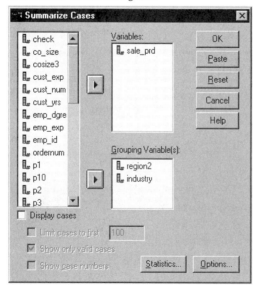

▶ Select one or more variables.

Optionally, you can:

■ Select one or more grouping variables to divide your data into subgroups.

■ Click Options to change the output title, add a caption below the output, or exclude cases with missing values.

■ Click Statistics for optional statistics.

■ Select Display cases to list the cases in each subgroup. By default, the system lists only the first 100 cases in your file. You can raise or lower the value for Limit cases to first *n* or deselect that item to list all cases.

Summarize Options

Figure 18-3
Summarize Cases Options dialog box

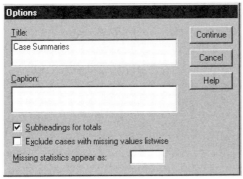

Summarize allows you to change the title of your output or add a caption that will appear below the output table. You can control line wrapping in titles and captions by typing \n wherever you want to insert a line break in the text.

You can also choose to display or suppress subheadings for totals and to include or exclude cases with missing values for any of the variables used in any of the analyses. Often it is desirable to denote missing cases in output with a period or an asterisk. Enter a character, phrase, or code that you would like to have appear when a value is missing; otherwise, no special treatment is applied to missing cases in the output.

Summarize Statistics

Figure 18-4
Summarize Cases Statistics dialog box

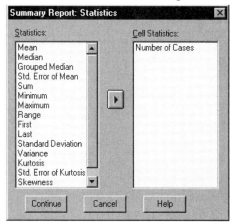

You can choose one or more of the following subgroup statistics for the variables within each category of each grouping variable: sum, number of cases, mean, median, grouped median, standard error of the mean, minimum, maximum, range, variable value of the first category of the grouping variable, variable value of the last category of the grouping variable, standard deviation, variance, kurtosis, standard error of kurtosis, skewness, standard error of skewness, percentage of total sum, percentage of total N, percentage of sum in, percentage of N in, geometric mean, harmonic mean. The order in which the statistics appear in the Cell Statistics list is the order in which they will be displayed in the output. Summary statistics are also displayed for each variable across all categories.

First. Displays the first data value encountered in the data file.

Geometric Mean. The nth root of the product of the data values, where n represents the number of cases.

Grouped Median. Median that is calculated for data that is coded into groups. For example, with age data, if each value in the 30s is coded 35, each value in the 40s is coded 45, and so on, the grouped median is the median calculated from the coded data.

Harmonic Mean. Used to estimate an average group size when the sample sizes in the groups are not equal. The harmonic mean is the total number of samples divided by the sum of the reciprocals of the sample sizes.

Kurtosis. A measure of the extent to which observations cluster around a central point. For a normal distribution, the value of the kurtosis statistic is zero. Positive kurtosis indicates that the observations cluster more and have longer tails than those in the normal distribution, and negative kurtosis indicates that the observations cluster less and have shorter tails.

Last. Displays the last data value encountered in the data file.

Maximum. The largest value of a numeric variable.

Mean. A measure of central tendency. The arithmetic average, the sum divided by the number of cases.

Median. The value above and below which half of the cases fall, the 50th percentile. If there is an even number of cases, the median is the average of the two middle cases when they are sorted in ascending or descending order. The median is a measure of central tendency not sensitive to outlying values (unlike the mean, which can be affected by a few extremely high or low values).

Minimum. The smallest value of a numeric variable.

N. The number of cases (observations or records).

Percent of Total N. Percentage of the total number of cases in each category.

Percent of Total Sum. Percentage of the total sum in each category.

Range. The difference between the largest and smallest values of a numeric variable, the maximum minus the minimum.

Skewness. A measure of the asymmetry of a distribution. The normal distribution is symmetric and has a skewness value of 0. A distribution with a significant positive skewness has a long right tail. A distribution with a significant negative skewness has a long left tail. As a guideline, a skewness value more than twice its standard error is taken to indicate a departure from symmetry.

Standard Error of Kurtosis. The ratio of kurtosis to its standard error can be used as a test of normality (that is, you can reject normality if the ratio is less than -2 or greater than +2). A large positive value for kurtosis indicates that the tails of the distribution are longer than those of a normal distribution; a negative value for kurtosis indicates shorter tails (becoming like those of a box-shaped uniform distribution).

Standard Error of Skewness. The ratio of skewness to its standard error can be used as a test of normality (that is, you can reject normality if the ratio is less than -2 or greater than +2). A large positive value for skewness indicates a long right tail; an extreme negative value indicates a long left tail.

Sum. The sum or total of the values, across all cases with nonmissing values.

Variance. A measure of dispersion around the mean, equal to the sum of squared deviations from the mean divided by one less than the number of cases. The variance is measured in units that are the square of those of the variable itself.

Means

The Means procedure calculates subgroup means and related univariate statistics for dependent variables within categories of one or more independent variables. Optionally, you can obtain a one-way analysis of variance, eta, and tests for linearity.

Example. Measure the average amount of fat absorbed by three different types of cooking oil, and perform a one-way analysis of variance to see whether the means differ.

Statistics. Sum, number of cases, mean, median, grouped median, standard error of the mean, minimum, maximum, range, variable value of the first category of the grouping variable, variable value of the last category of the grouping variable, standard deviation, variance, kurtosis, standard error of kurtosis, skewness, standard error of skewness, percentage of total sum, percentage of total N, percentage of sum in, percentage of N in, geometric mean, and harmonic mean. Options include analysis of variance, eta, eta squared, and tests for linearity R and R^2.

Data. The dependent variables are quantitative, and the independent variables are categorical. The values of categorical variables can be numeric or short string.

Assumptions. Some of the optional subgroup statistics, such as the mean and standard deviation, are based on normal theory and are appropriate for quantitative variables with symmetric distributions. Robust statistics, such as the median, are appropriate for quantitative variables that may or may not meet the assumption of normality. Analysis of variance is robust to departures from normality, but the data in each cell should be symmetric. Analysis of variance also assumes that the groups come from populations with equal variances. To test this assumption, use Levene's homogeneity-of-variance test, available in the One-Way ANOVA procedure.

Figure 19-1
Means output

Report

Absorbed Grams of Fat

Type of Oil	Peanut Oil	Mean	72.00
		N	6
		Std. Deviation	13.34
	Lard	Mean	85.00
		N	6
		Std. Deviation	7.77
	Corn Oil	Mean	62.00
		N	6
		Std. Deviation	8.22
	Total	Mean	73.00
		N	18
		Std. Deviation	13.56

ANOVA Table

			Sum of Squares	df	Mean Square	F	Significance
Absorbed Grams of Fat * Type of Oil	Between Groups	(Combined)	1596.00	2	798.000	7.824	.005
	Within Groups		1530.00	15	102.000		
	Total		3126.00	17			

To Obtain Subgroup Means

▶ From the menus choose:
Analyze
　Compare Means
　　Means...

Figure 19-2
Means dialog box

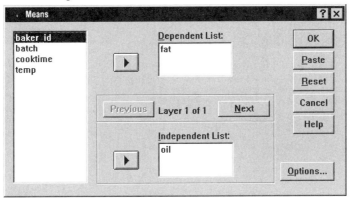

▶ Select one or more dependent variables.

▶ Use one of the following methods to select categorical independent variables:

  ■ Select one or more independent variables. Separate results are displayed for each independent variable.

  ■ Select one or more layers of independent variables. Each layer further subdivides the sample. If you have one independent variable in Layer 1 and one independent variable in Layer 2, the results are displayed in one crossed table, as opposed to separate tables for each independent variable.

▶ Optionally, click Options for optional statistics, an analysis of variance table, eta, eta squared, R, and R^2.

Means Options

Figure 19-3
Means Options dialog box

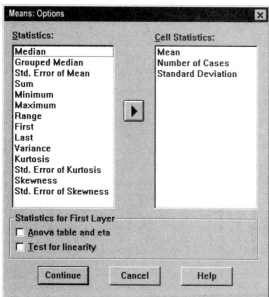

You can choose one or more of the following subgroup statistics for the variables within each category of each grouping variable: sum, number of cases, mean, median, grouped median, standard error of the mean, minimum, maximum, range, variable value of the first category of the grouping variable, variable value of the last category of the grouping variable, standard deviation, variance, kurtosis, standard error of kurtosis, skewness, standard error of skewness, percentage of total sum, percentage of total N, percentage of sum in, percentage of N in, geometric mean, and harmonic mean. You can change the order in which the subgroup statistics appear. The order in which the statistics appear in the Cell Statistics list is the order in which they are displayed in the output. Summary statistics are also displayed for each variable across all categories.

First. Displays the first data value encountered in the data file.

Geometric Mean. The nth root of the product of the data values, where n represents the number of cases.

Grouped Median. Median that is calculated for data that is coded into groups. For example, with age data, if each value in the 30s is coded 35, each value in the 40s is coded 45, and so on, the grouped median is the median calculated from the coded data.

Harmonic Mean. Used to estimate an average group size when the sample sizes in the groups are not equal. The harmonic mean is the total number of samples divided by the sum of the reciprocals of the sample sizes.

Kurtosis. A measure of the extent to which observations cluster around a central point. For a normal distribution, the value of the kurtosis statistic is zero. Positive kurtosis indicates that the observations cluster more and have longer tails than those in the normal distribution, and negative kurtosis indicates that the observations cluster less and have shorter tails.

Last. Displays the last data value encountered in the data file.

Maximum. The largest value of a numeric variable.

Mean. A measure of central tendency. The arithmetic average, the sum divided by the number of cases.

Median. The value above and below which half of the cases fall, the 50th percentile. If there is an even number of cases, the median is the average of the two middle cases when they are sorted in ascending or descending order. The median is a measure of central tendency not sensitive to outlying values (unlike the mean, which can be affected by a few extremely high or low values).

Minimum. The smallest value of a numeric variable.

N. The number of cases (observations or records).

Percent of total N. Percentage of the total number of cases in each category.

Percent of total sum. Percentage of the total sum in each category.

Range. The difference between the largest and smallest values of a numeric variable, the maximum minus the minimum.

Skewness. A measure of the asymmetry of a distribution. The normal distribution is symmetric and has a skewness value of 0. A distribution with a significant positive skewness has a long right tail. A distribution with a significant negative skewness has a long left tail. As a guideline, a skewness value more than twice its standard error is taken to indicate a departure from symmetry.

Standard Error of Kurtosis. The ratio of kurtosis to its standard error can be used as a test of normality (that is, you can reject normality if the ratio is less than -2 or greater than +2). A large positive value for kurtosis indicates that the tails of the distribution are longer than those of a normal distribution; a negative value for kurtosis indicates shorter tails (becoming like those of a box-shaped uniform distribution).

Standard Error of Skewness. The ratio of skewness to its standard error can be used as a test of normality (that is, you can reject normality if the ratio is less than -2 or greater than +2). A large positive value for skewness indicates a long right tail; an extreme negative value indicates a long left tail.

Sum. The sum or total of the values, across all cases with nonmissing values.

Variance. A measure of dispersion around the mean, equal to the sum of squared deviations from the mean divided by one less than the number of cases. The variance is measured in units that are the square of those of the variable itself.

Statistics for First Layer

Anova table and eta. Displays a one-way analysis-of-variance table and calculates eta and eta squared (measures of association) for each independent variable in the first layer.

Test for linearity. Calculates the sum of squares, degrees of freedom, and mean square associated with linear and nonlinear components, as well as the F ratio, R and R squared. Linearity is not calculated if the independent variable is a short string.

OLAP Cubes

The OLAP (Online Analytical Processing) Cubes procedure calculates totals, means, and other univariate statistics for continuous summary variables within categories of one or more categorical grouping variables. A separate layer in the table is created for each category of each grouping variable.

Example. Total and average sales for different regions and product lines within regions.

Statistics. Sum, number of cases, mean, median, grouped median, standard error of the mean, minimum, maximum, range, variable value of the first category of the grouping variable, variable value of the last category of the grouping variable, standard deviation, variance, kurtosis, standard error of kurtosis, skewness, standard error of skewness, percentage of total cases, percentage of total sum, percentage of total cases within grouping variables, percentage of total sum within grouping variables, geometric mean, and harmonic mean.

Data. The summary variables are quantitative (continuous variables measured on an interval or ratio scale), and the grouping variables are categorical. The values of categorical variables can be numeric or short string.

Assumptions. Some of the optional subgroup statistics, such as the mean and standard deviation, are based on normal theory and are appropriate for quantitative variables with symmetric distributions. Robust statistics, such as the median and range, are appropriate for quantitative variables that may or may not meet the assumption of normality.

Figure 20-1
OLAP Cubes output

**1996 Sales
by Division and Region**

Division: Total
Region: Total

Sum	$145,038,250
Mean	$371,893
Median	$307,500
Std. Deviation	$171,311

**1996 Sales
by Division and Region**

Division: Consumer Products
Region: East

Sum	$18,548,100
Mean	$289,814.06
Median	$273,600.00
Std. Deviation	$80,674.66

To Obtain OLAP Cubes

▶ From the menus choose:

Analyze
  Reports
    OLAP Cubes...

Figure 20-2
OLAP Cubes dialog box

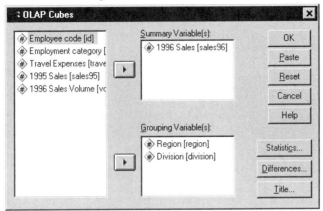

▶ Select one or more continuous summary variables.

▶ Select one or more categorical grouping variables.

▶ Optionally:

■ Select different summary statistics (click Statistics). You must select one or more grouping variables before you can select summary statistics.

■ Calculate differences between pairs of variables and pairs of groups that are defined by a grouping variable (click Differences).

■ Create custom table titles (click Title).

OLAP Cubes Statistics

Figure 20-3
OLAP Cubes Statistics dialog box

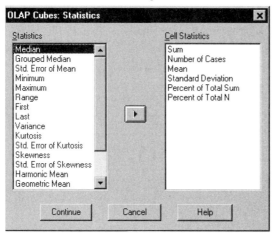

You can choose one or more of the following subgroup statistics for the summary variables within each category of each grouping variable: sum, number of cases, mean, median, grouped median, standard error of the mean, minimum, maximum, range, variable value of the first category of the grouping variable, variable value of the last category of the grouping variable, standard deviation, variance, kurtosis, standard error of kurtosis, skewness, standard error of skewness, percentage of total cases, percentage of total sum, percentage of total cases within grouping variables, percentage of total sum within grouping variables, geometric mean, and harmonic mean.

You can change the order in which the subgroup statistics appear. The order in which the statistics appear in the Cell Statistics list is the order in which they are displayed in the output. Summary statistics are also displayed for each variable across all categories.

First. Displays the first data value encountered in the data file.

Geometric Mean. The nth root of the product of the data values, where n represents the number of cases.

Grouped Median. Median that is calculated for data that is coded into groups. For example, with age data, if each value in the 30s is coded 35, each value in the 40s is coded 45, and so on, the grouped median is the median calculated from the coded data.

Harmonic Mean. Used to estimate an average group size when the sample sizes in the groups are not equal. The harmonic mean is the total number of samples divided by the sum of the reciprocals of the sample sizes.

Kurtosis. A measure of the extent to which observations cluster around a central point. For a normal distribution, the value of the kurtosis statistic is zero. Positive kurtosis indicates that the observations cluster more and have longer tails than those in the normal distribution, and negative kurtosis indicates that the observations cluster less and have shorter tails.

Last. Displays the last data value encountered in the data file.

Maximum. The largest value of a numeric variable.

Mean. A measure of central tendency. The arithmetic average, the sum divided by the number of cases.

Median. The value above and below which half of the cases fall, the 50th percentile. If there is an even number of cases, the median is the average of the two middle cases when they are sorted in ascending or descending order. The median is a measure of central tendency not sensitive to outlying values (unlike the mean, which can be affected by a few extremely high or low values).

Minimum. The smallest value of a numeric variable.

N. The number of cases (observations or records).

Percent of N in. Percentage of the number of cases for the specified grouping variable within categories of other grouping variables. If you only have one grouping variable, this value is identical to percentage of total number of cases.

Percent of Sum in. Percentage of the sum for the specified grouping variable within categories of other grouping variables. If you only have one grouping variable, this value is identical to percentage of total sum.

Percent of Total N. Percentage of the total number of cases in each category.

Percent of Total Sum. Percentage of the total sum in each category.

Range. The difference between the largest and smallest values of a numeric variable, the maximum minus the minimum.

Skewness. A measure of the asymmetry of a distribution. The normal distribution is symmetric and has a skewness value of 0. A distribution with a significant positive skewness has a long right tail. A distribution with a significant negative skewness has a long left tail. As a guideline, a skewness value more than twice its standard error is taken to indicate a departure from symmetry.

Standard Error of Kurtosis. The ratio of kurtosis to its standard error can be used as a test of normality (that is, you can reject normality if the ratio is less than -2 or greater than +2). A large positive value for kurtosis indicates that the tails of the distribution are longer than those of a normal distribution; a negative value for kurtosis indicates shorter tails (becoming like those of a box-shaped uniform distribution).

Standard Error of Skewness. The ratio of skewness to its standard error can be used as a test of normality (that is, you can reject normality if the ratio is less than -2 or greater than +2). A large positive value for skewness indicates a long right tail; an extreme negative value indicates a long left tail.

Sum. The sum or total of the values, across all cases with nonmissing values.

Variance. A measure of dispersion around the mean, equal to the sum of squared deviations from the mean divided by one less than the number of cases. The variance is measured in units that are the square of those of the variable itself.

OLAP Cubes Differences

Figure 20-4
OLAP Cubes Differences dialog box

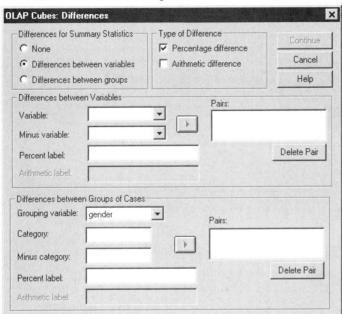

This dialog box allows you to calculate percentage and arithmetic differences between summary variables or between groups that are defined by a grouping variable. Differences are calculated for all measures that are selected in the OLAP Cubes Statistics dialog box.

Differences between variables. Calculates differences between pairs of variables. Summary statistics values for the second variable (the Minus variable) in each pair are subtracted from summary statistics values for the first variable in the pair. For percentage differences, the value of the summary variable for the Minus variable is used as the denominator. You must select at least two summary variables in the main dialog box before you can specify differences between variables.

Differences between groups. Calculates differences between pairs of groups defined by a grouping variable. Summary statistics values for the second category in each pair (the Minus category) are subtracted from summary statistics values for the first category in the pair. Percentage differences use the value of the summary statistic for the Minus category as the denominator. You must select one or more grouping variables in the main dialog box before you can specify differences between groups.

OLAP Cubes Title

Figure 20-5
OLAP Cubes Title dialog box

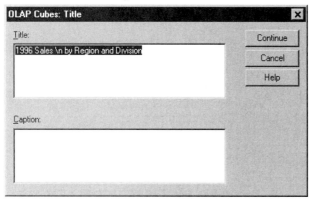

You can change the title of your output or add a caption that will appear below the output table. You can also control line wrapping of titles and captions by typing \n wherever you want to insert a line break in the text.

T Tests

Three types of *t* tests are available:

Independent-samples t test (two-sample t test). Compares the means of one variable for two groups of cases. Descriptive statistics for each group and Levene's test for equality of variances are provided, as well as both equal- and unequal-variance *t* values and a 95% confidence interval for the difference in means.

Paired-samples t test (dependent t test). Compares the means of two variables for a single group. This test is also for matched-pairs or case-control study designs. The output includes descriptive statistics for the test variables, the correlation between the variables, descriptive statistics for the paired differences, the *t* test, and a 95% confidence interval.

One-sample t test. Compares the mean of one variable with a known or hypothesized value. Descriptive statistics for the test variables are displayed along with the *t* test. A 95% confidence interval for the difference between the mean of the test variable and the hypothesized test value is part of the default output.

Independent-Samples T Test

The Independent-Samples T Test procedure compares means for two groups of cases. Ideally, for this test, the subjects should be randomly assigned to two groups, so that any difference in response is due to the treatment (or lack of treatment) and not to other factors. This is not the case if you compare average income for males and females. A person is not randomly assigned to be a male or female. In such situations, you should ensure that differences in other factors are not masking or enhancing a significant difference in means. Differences in average income may be influenced by factors such as education (and not by sex alone).

Example. Patients with high blood pressure are randomly assigned to a placebo group and a treatment group. The placebo subjects receive an inactive pill, and the treatment subjects receive a new drug that is expected to lower blood pressure. After the subjects are treated for two months, the two-sample t test is used to compare the average blood pressures for the placebo group and the treatment group. Each patient is measured once and belongs to one group.

Statistics. For each variable: sample size, mean, standard deviation, and standard error of the mean. For the difference in means: mean, standard error, and confidence interval (you can specify the confidence level). Tests: Levene's test for equality of variances, and both pooled-variances and separate-variances t tests for equality of means.

Data. The values of the quantitative variable of interest are in a single column in the data file. The procedure uses a grouping variable with two values to separate the cases into two groups. The grouping variable can be numeric (values such as 1 and 2 or 6.25 and 12.5) or short string (such as *yes* and *no*). As an alternative, you can use a quantitative variable, such as *age,* to split the cases into two groups by specifying a cut point (cut point 21 splits *age* into an under-21 group and a 21-and-over group).

Assumptions. For the equal-variance t test, the observations should be independent, random samples from normal distributions with the same population variance. For the unequal-variance t test, the observations should be independent, random samples from normal distributions. The two-sample t test is fairly robust to departures from normality. When checking distributions graphically, look to see that they are symmetric and have no outliers.

Figure 21-1
Independent-Samples T Test output

Group Statistics

			N	Mean	Std. Deviation	Std. Error Mean
Blood pressure	Treatment	placebo	10	142.50	17.04	5.39
		new_drug	10	116.40	13.62	4.31

Independent Samples Test

| | | Levene's Test for Equality of Variances | | t-test for Equality of Means | | | | | | |
| | | F | Significance | t | df | Significance (2-tailed) | Mean Difference | Std. Error Difference | 95% Confidence Interval of the Mean | |
									Lower	Upper
Blood pressure	Equal variances assumed	.134	.719	3.783	18	.001	26.10	6.90	11.61	40.59
	Equal variances not assumed			3.783	17.163	.001	26.10	6.90	11.56	40.64

To Obtain an Independent-Samples T Test

▶ From the menus choose:

Analyze
  Compare Means
    Independent-Samples T Test...

Figure 21-2
Independent-Samples T Test dialog box

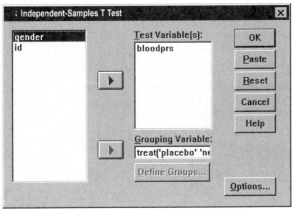

▶ Select one or more quantitative test variables. A separate *t* test is computed for each variable.

▶ Select a single grouping variable, and then click Define Groups to specify two codes for the groups that you want to compare.

▶ Optionally, click Options to control the treatment of missing data and the level of the confidence interval.

Independent-Samples T Test Define Groups

Figure 21-3
Define Groups dialog box for numeric variables

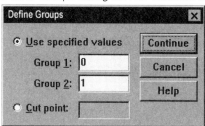

For numeric grouping variables, define the two groups for the *t* test by specifying two values or a cut point:

- **Use specified values.** Enter a value for Group 1 and another value for Group 2. Cases with any other values are excluded from the analysis. Numbers need not be integers (for example, 6.25 and 12.5 are valid).

- **Cut point.** Enter a number that splits the values of the grouping variable into two sets. All cases with values that are less than the cut point form one group, and cases with values that are greater than or equal to the cut point form the other group.

Figure 21-4
Define Groups dialog box for string variables

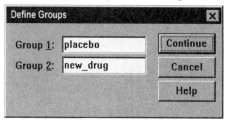

For short string grouping variables, enter a string for Group 1 and another value for Group 2, such as *yes* and *no*. Cases with other strings are excluded from the analysis.

Independent-Samples T Test Options

Figure 21-5
Independent-Samples T Test Options dialog box

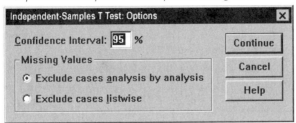

Confidence Interval. By default, a 95% confidence interval for the difference in means is displayed. Enter a value between 1 and 99 to request a different confidence level.

Missing Values. When you test several variables, and data are missing for one or more variables, you can tell the procedure which cases to include (or exclude).

- **Exclude cases analysis by analysis.** Each *t* test uses all cases that have valid data for the tested variables. Sample sizes may vary from test to test.

- **Exclude cases listwise.** Each *t* test uses only cases that have valid data for all variables that are used in the requested *t* tests. The sample size is constant across tests.

Paired-Samples T Test

The Paired-Samples T Test procedure compares the means of two variables for a single group. The procedure computes the differences between values of the two variables for each case and tests whether the average differs from 0.

Example. In a study on high blood pressure, all patients are measured at the beginning of the study, given a treatment, and measured again. Thus, each subject has two measures, often called *before* and *after* measures. An alternative design for which this test is used is a matched-pairs or case-control study, in which each record in the data file contains the response for the patient and also for his or her matched control subject. In a blood pressure study, patients and controls might be matched by age (a 75-year-old patient with a 75-year-old control group member).

Statistics. For each variable: mean, sample size, standard deviation, and standard error of the mean. For each pair of variables: correlation, average difference in means, *t* test, and confidence interval for mean difference (you can specify the confidence level). Standard deviation and standard error of the mean difference.

Data. For each paired test, specify two quantitative variables (interval level of measurement or ratio level of measurement). For a matched-pairs or case-control study, the response for each test subject and its matched control subject must be in the same case in the data file.

Assumptions. Observations for each pair should be made under the same conditions. The mean differences should be normally distributed. Variances of each variable can be equal or unequal.

Figure 21-6
Paired-Samples T Test output

Paired Samples Statistics

		Mean	N	Std. Deviation	Std. Error Mean
Pair 1	After treatment	116.40	10	13.62	4.31
	Before treatment	142.50	10	17.04	5.39

Paired Samples Test

		Paired Differences							
					95% Confidence Interval of the Difference				
		Mean	Std. Deviation	Std. Error Mean	Lower	Upper	t	df	Significance (2-tailed)
Pair 1	After treatment - Before treatment	-26.10	19.59	6.19	-40.11	-12.09	-4.214	9	.002

To Obtain a Paired-Samples T Test

▶ From the menus choose:
Analyze
  Compare Means
    Paired-Samples T Test...

Figure 21-7
Paired-Samples T Test dialog box

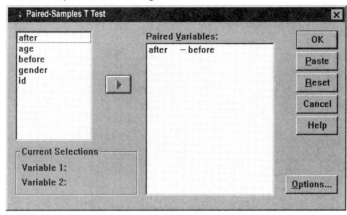

▶ Select a pair of variables, as follows:

■ Click each of two variables. The first variable appears in the Current Selections group as *Variable 1*, and the second variable appears as *Variable 2*.

■ Click the arrow button to move the pair into the Paired Variables list. You may select more pairs of variables. To remove a pair of variables from the analysis, select a pair in the Paired Variables list and click the arrow button.

▶ Optionally, click Options to control the treatment of missing data and the level of the confidence interval.

Paired-Samples T Test Options

Figure 21-8
Paired-Samples T Test Options dialog box

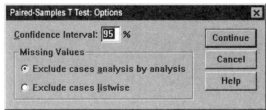

Confidence Interval. By default, a 95% confidence interval for the difference in means is displayed. Enter a value between 1 and 99 to request a different confidence level.

Missing Values. When you test several variables, and data are missing for one or more variables, you can tell the procedure which cases to include (or exclude):

- **Exclude cases analysis by analysis.** Each *t* test uses all cases that have valid data for the tested pair of variables. Sample sizes may vary from test to test.

- **Exclude cases listwise.** Each *t* test uses only cases that have valid data for all pairs of tested variables. The sample size is constant across tests.

One-Sample T Test

The One-Sample T Test procedure tests whether the mean of a single variable differs from a specified constant.

Examples. A researcher might want to test whether the average IQ score for a group of students differs from 100. Or, a cereal manufacturer can take a sample of boxes from the production line and check whether the mean weight of the samples differs from 1.3 pounds at the 95% confidence level.

Statistics. For each test variable: mean, standard deviation, and standard error of the mean. The average difference between each data value and the hypothesized test value, a *t* test that tests that this difference is 0, and a confidence interval for this difference (you can specify the confidence level).

Data. To test the values of a quantitative variable against a hypothesized test value, choose a quantitative variable and enter a hypothesized test value.

Assumptions. This test assumes that the data are normally distributed; however, this test is fairly robust to departures from normality.

Figure 21-9
One-Sample T Test output

One-Sample Statistics

	IQ
N	15
Mean	109.33
Std. Deviation	12.03
Std. Error Mean	3.11

Rows and columns have been transposed.

One-Sample Test

	Test Value = 100					
	t	df	Significance (2-tailed)	Mean Difference	95% Confidence Interval of the Difference	
					Lower	Upper
IQ	3.005	14	.009	9.33	2.67	15.99

To Obtain a One-Sample T Test

▶ From the menus choose:

Analyze
  Compare Means
    One-Sample T Test...

Figure 21-10
One-Sample T Test dialog box

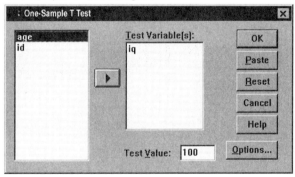

▶ Select one or more variables to be tested against the same hypothesized value.

▶ Enter a numeric test value against which each sample mean is compared.

▶ Optionally, click Options to control the treatment of missing data and the level of the confidence interval.

One-Sample T Test Options

Figure 21-11
One-Sample T Test Options dialog box

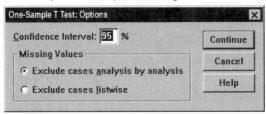

Confidence Interval. By default, a 95% confidence interval for the difference between the mean and the hypothesized test value is displayed. Enter a value between 1 and 99 to request a different confidence level.

Missing Values. When you test several variables, and data are missing for one or more variables, you can tell the procedure which cases to include (or exclude).

■ **Exclude cases analysis by analysis.** Each *t* test uses all cases that have valid data for the tested variable. Sample sizes may vary from test to test.

■ **Exclude cases listwise.** Each *t* test uses only cases that have valid data for all variables that are used in any of the requested *t* tests. The sample size is constant across tests.

T-TEST Command Additional Features

The SPSS command language also allows you to:

■ Produce both one-sample and independent-samples t tests by running a single command.

■ Test a variable against each variable on a list, in a paired t test (with the PAIRS subcommand).

See the *SPSS Command Syntax Reference* for complete syntax information.

One-Way ANOVA

The One-Way ANOVA procedure produces a one-way analysis of variance for a quantitative dependent variable by a single factor (independent) variable. Analysis of variance is used to test the hypothesis that several means are equal. This technique is an extension of the two-sample *t* test.

In addition to determining that differences exist among the means, you may want to know which means differ. There are two types of tests for comparing means: a priori contrasts and post hoc tests. Contrasts are tests set up *before* running the experiment, and post hoc tests are run *after* the experiment has been conducted. You can also test for trends across categories.

Example. Doughnuts absorb fat in various amounts when they are cooked. An experiment is set up involving three types of fat: peanut oil, corn oil, and lard. Peanut oil and corn oil are unsaturated fats, and lard is a saturated fat. Along with determining whether the amount of fat absorbed depends on the type of fat used, you could set up an a priori contrast to determine whether the amount of fat absorption differs for saturated and unsaturated fats.

Statistics. For each group: number of cases, mean, standard deviation, standard error of the mean, minimum, maximum, and 95%-confidence interval for the mean. Levene's test for homogeneity of variance, analysis-of-variance table and robust tests of the equality of means for each dependent variable, user-specified a priori contrasts, and post hoc range tests and multiple comparisons: Bonferroni, Sidak, Tukey's honestly significant difference, Hochberg's GT2, Gabriel, Dunnett, Ryan-Einot-Gabriel-Welsch *F* test (R-E-G-W *F*), Ryan-Einot-Gabriel-Welsch range test (R-E-G-W *Q*), Tamhane's T2, Dunnett's T3, Games-Howell, Dunnett's *C*, Duncan's multiple range test, Student-Newman-Keuls (S-N-K), Tukey's *b*, Waller-Duncan, Scheffé, and least-significant difference.

Data. Factor variable values should be integers, and the dependent variable should be quantitative (interval level of measurement).

Assumptions. Each group is an independent random sample from a normal population. Analysis of variance is robust to departures from normality, although the data should be symmetric. The groups should come from populations with equal variances. To test this assumption, use Levene's homogeneity-of-variance test.

Figure 22-1
One-Way ANOVA output

ANOVA

		Sum of Squares	df	Mean Square	F	Significance
Absorbed Grams of Fat	Between Groups	1596.00	2	798.00	7.824	.005
	Within Groups	1530.00	15	102.00		
	Total	3126.00	17			

Descriptives

			N	Mean	Std. Deviation	Std. Error	95% Confidence Interval for Mean		Minimum	Maximum
							Lower Bound	Upper Bound		
Absorbed Grams of Fat	Type of Oil	Peanut Oil	6	72.00	13.34	5.45	58.00	86.00	56	95
		Lard	6	85.00	7.77	3.17	76.84	93.16	77	97
		Corn Oil	6	62.00	8.22	3.36	53.37	70.63	49	70
		Total	18	73.00	13.56	3.20	66.26	79.74	49	97

Contrast Coefficients

		Type of Oil		
		Peanut Oil	Lard	Corn Oil
Contrast	1	-.5	1	-.5

Contrast Tests

				Value of Contrast	Std. Error	t	df	Significance (2-tailed)
Absorbed Grams of Fat	Assume equal variances	Contrast	1	18.00	5.05	3.565	15	.003
	Does not assume equal variances	Contrast	1	18.00	4.51	3.995	12.542	.002

Test of Homogeneity of Variances

	Levene Statistic	df1	df2	Significance
Absorbed Grams of Fat	.534	2	15	.597

To Obtain a One-Way Analysis of Variance

▶ From the menus choose:

Analyze
  Compare Means
    One-Way ANOVA...

Figure 22-2
One-Way ANOVA dialog box

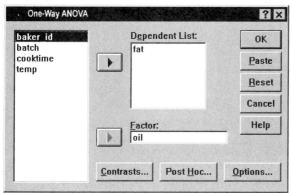

▶ Select one or more dependent variables.

▶ Select a single independent factor variable.

One-Way ANOVA Contrasts

Figure 22-3
One-Way ANOVA Contrasts dialog box

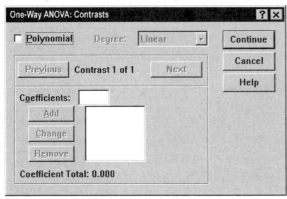

You can partition the between-groups sums of squares into trend components or specify a priori contrasts.

Polynomial. Partitions the between-groups sums of squares into trend components. You can test for a trend of the dependent variable across the ordered levels of the factor variable. For example, you could test for a linear trend (increasing or decreasing) in salary across the ordered levels of highest degree earned.

■ **Degree.** You can choose a 1st, 2nd, 3rd, 4th, or 5th degree polynomial.

Coefficients. User-specified a priori contrasts to be tested by the *t* statistic. Enter a coefficient for each group (category) of the factor variable and click Add after each entry. Each new value is added to the bottom of the coefficient list. To specify additional sets of contrasts, click Next. Use Next and Previous to move between sets of contrasts.

The order of the coefficients is important because it corresponds to the ascending order of the category values of the factor variable. The first coefficient on the list corresponds to the lowest group value of the factor variable, and the last coefficient corresponds to the highest value. For example, if there are six categories of the factor variable, the coefficients -1, 0, 0, 0, 0.5, and 0.5 contrast the first group with the fifth and sixth groups. For most applications, the coefficients should sum to 0. Sets that do not sum to 0 can also be used, but a warning message is displayed.

One-Way ANOVA Post Hoc Tests

Figure 22-4
One-Way ANOVA Post Hoc Multiple Comparisons dialog box

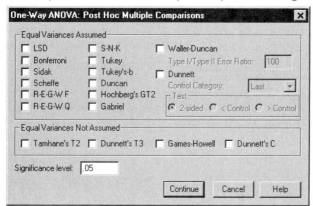

Once you have determined that differences exist among the means, post hoc range tests and pairwise multiple comparisons can determine which means differ. Range tests identify homogeneous subsets of means that are not different from each other. Pairwise multiple comparisons test the difference between each pair of means and yield a matrix where asterisks indicate significantly different group means at an alpha level of 0.05.

Equal Variances Assumed

Tukey's honestly significant difference test, Hochberg's GT2, Gabriel, and Scheffé are multiple comparison tests and range tests. Other available range tests are Tukey's *b*, S-N-K (Student-Newman-Keuls), Duncan, R-E-G-W *F* (Ryan-Einot-Gabriel-Welsch *F* test), R-E-G-W *Q* (Ryan-Einot-Gabriel-Welsch range test), and Waller-Duncan. Available multiple comparison tests are Bonferroni, Tukey's honestly significant difference test, Sidak, Gabriel, Hochberg, Dunnett, Scheffé, and LSD (least significant difference).

- **LSD.** Uses t tests to perform all pairwise comparisons between group means. No adjustment is made to the error rate for multiple comparisons.

- **Bonferroni.** Uses t tests to perform pairwise comparisons between group means, but controls overall error rate by setting the error rate for each test to the experimentwise error rate divided by the total number of tests. Hence, the observed significance level is adjusted for the fact that multiple comparisons are being made.

- **Sidak.** Pairwise multiple comparison test based on a t statistic. Sidak adjusts the significance level for multiple comparisons and provides tighter bounds than Bonferroni.

- **Scheffe.** Performs simultaneous joint pairwise comparisons for all possible pairwise combinations of means. Uses the F sampling distribution. Can be used to examine all possible linear combinations of group means, not just pairwise comparisons.

- **R-E-G-W F.** Ryan-Einot-Gabriel-Welsch multiple stepdown procedure based on an F test.

- **R-E-G-W Q.** Ryan-Einot-Gabriel-Welsch multiple stepdown procedure based on the Studentized range.

- **S-N-K.** Makes all pairwise comparisons between means using the Studentized range distribution. With equal sample sizes, it also compares pairs of means within homogeneous subsets, using a stepwise procedure. Means are ordered from highest to lowest, and extreme differences are tested first.

- **Tukey.** Uses the Studentized range statistic to make all of the pairwise comparisons between groups. Sets the experimentwise error rate at the error rate for the collection for all pairwise comparisons.

- **Tukey's b.** Uses the Studentized range distribution to make pairwise comparisons between groups. The critical value is the average of the corresponding value for the Tukey's honestly significant difference test and the Student-Newman-Keuls.

- **Duncan.** Makes pairwise comparisons using a stepwise order of comparisons identical to the order used by the Student-Newman-Keuls test, but sets a protection level for the error rate for the collection of tests, rather than an error rate for individual tests. Uses the Studentized range statistic.

- **Hochberg's GT2.** Multiple comparison and range test that uses the Studentized maximum modulus. Similar to Tukey's honestly significant difference test.

- **Gabriel.** Pairwise comparison test that used the Studentized maximum modulus and is generally more powerful than Hochberg's GT2 when the cell sizes are unequal. Gabriel's test may become liberal when the cell sizes vary greatly.

- **Waller-Duncan.** Multiple comparison test based on a t statistic; uses a Bayesian approach.

- **Dunnett.** Pairwise multiple comparison t test that compares a set of treatments against a single control mean. The last category is the default control category. Alternatively, you can choose the first category. 2-sided tests that the mean at any level (except the control category) of the factor is not equal to that of the control category. < Control tests if the mean at any level of the factor is smaller than that of the control category. > Control tests if the mean at any level of the factor is greater than that of the control category.

Equal Variances Not Assumed

Multiple comparison tests that do not assume equal variances are Tamhane's T2, Dunnett's T3, Games-Howell, and Dunnett's *C*.

- **Tamhane's T2.** Conservative pairwise comparisons test based on a t test. This test is appropriate when the variances are unequal.

- **Dunnett's T3.** Pairwise comparison test based on the Studentized maximum modulus. This test is appropriate when the variances are unequal.

- **Games-Howell.** Pairwise comparison test that is sometimes liberal. This test is appropriate when the variances are unequal.

- **Dunnett's C.** Pairwise comparison test based on the Studentized range. This test is appropriate when the variances are unequal.

Note: You may find it easier to interpret the output from post hoc tests if you deselect Hide empty rows and columns in the Table Properties dialog box (in an activated pivot table, choose Table Properties from the Format menu).

One-Way ANOVA Options

Figure 22-5
One-Way ANOVA Options dialog box

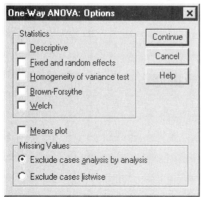

Statistics. Choose one or more of the following:

- **Descriptive.** Calculates the number of cases, mean, standard deviation, standard error of the mean, minimum, maximum, and 95%-confidence intervals for each dependent variable for each group.

- **Fixed and random effects.** Displays the standard deviation, standard error, and 95%-confidence interval for the fixed-effects model, and the standard error, 95%-confidence interval, and estimate of between-components variance for the random-effects model.

- **Homogeneity of variance test.** Calculates the Levene statistic to test for the equality of group variances. This test is not dependent on the assumption of normality.

- **Brown-Forsythe.** Calculates the Brown-Forsythe statistic to test for the equality of group means. This statistic is preferable to the F statistic when the assumption of equal variances does not hold.

- **Welch.** Calculates the Welch statistic to test for the equality of group means. This statistic is preferable to the F statistic when the assumption of equal variances does not hold.

Means plot. Displays a chart that plots the subgroup means (the means for each group defined by values of the factor variable).

Missing Values. Controls the treatment of missing values.

- **Exclude cases analysis by analysis.** A case with a missing value for either the dependent or the factor variable for a given analysis is not used in that analysis. Also, a case outside the range specified for the factor variable is not used.

- **Exclude cases listwise.** Cases with missing values for the factor variable or for any dependent variable included on the dependent list in the main dialog box are excluded from all analyses. If you have not specified multiple dependent variables, this has no effect.

ONEWAY Command Additional Features

The SPSS command language also allows you to:

- Obtain fixed- and random-effects statistics. Standard deviation, standard error of the mean, and 95% confidence intervals for the fixed-effects model. Standard error, 95% confidence intervals, and estimate of between-components variance for random-effects model (using STATISTICS=EFFECTS).

■ Specify alpha levels for the least significance difference, Bonferroni, Duncan, and Scheffé multiple comparison tests (with the RANGES subcommand).

■ Write a matrix of means, standard deviations, and frequencies, or read a matrix of means, frequencies, pooled variances, and degrees of freedom for the pooled variances. These matrices can be used in place of raw data to obtain a one-way analysis of variance (with the MATRIX subcommand).

See the *SPSS Command Syntax Reference* for complete syntax information.

GLM Univariate Analysis

The GLM Univariate procedure provides regression analysis and analysis of variance for one dependent variable by one or more factors and/or variables. The factor variables divide the population into groups. Using this General Linear Model procedure, you can test null hypotheses about the effects of other variables on the means of various groupings of a single dependent variable. You can investigate interactions between factors as well as the effects of individual factors, some of which may be random. In addition, the effects of covariates and covariate interactions with factors can be included. For regression analysis, the independent (predictor) variables are specified as covariates.

Both balanced and unbalanced models can be tested. A design is balanced if each cell in the model contains the same number of cases. In addition to testing hypotheses, GLM Univariate produces estimates of parameters.

Commonly used a priori contrasts are available to perform hypothesis testing. Additionally, after an overall F test has shown significance, you can use post hoc tests to evaluate differences among specific means. Estimated marginal means give estimates of predicted mean values for the cells in the model, and profile plots (interaction plots) of these means allow you to easily visualize some of the relationships.

Residuals, predicted values, Cook's distance, and leverage values can be saved as new variables in your data file for checking assumptions.

WLS Weight allows you to specify a variable used to give observations different weights for a weighted least-squares (WLS) analysis, perhaps to compensate for a different precision of measurement.

Example. Data are gathered for individual runners in the Chicago marathon for several years. The time in which each runner finishes is the dependent variable. Other factors include weather (cold, pleasant, or hot), number of months of training, number of previous marathons, and gender. Age is considered a covariate. You might find that gender is a significant effect and that the interaction of gender with weather is significant.

Methods. Type I, Type II, Type III, and Type IV sums of squares can be used to evaluate different hypotheses. Type III is the default.

Statistics. Post hoc range tests and multiple comparisons: least significant difference, Bonferroni, Sidak, Scheffé, Ryan-Einot-Gabriel-Welsch multiple *F*, Ryan-Einot-Gabriel-Welsch multiple range, Student-Newman-Keuls, Tukey's honestly significant difference, Tukey's *b*, Duncan, Hochberg's GT2, Gabriel, Waller-Duncan *t* test, Dunnett (one-sided and two-sided), Tamhane's T2, Dunnett's T3, Games-Howell, and Dunnett's *C*. Descriptive statistics: observed means, standard deviations, and counts for all of the dependent variables in all cells. The Levene test for homogeneity of variance.

Plots. Spread-versus-level, residual, and profile (interaction).

Data. The dependent variable is quantitative. Factors are categorical. They can have numeric values or string values of up to eight characters. Covariates are quantitative variables that are related to the dependent variable.

Assumptions. The data are a random sample from a normal population; in the population, all cell variances are the same. Analysis of variance is robust to departures from normality, although the data should be symmetric. To check assumptions, you can use homogeneity of variances tests and spread-versus-level plots. You can also examine residuals and residual plots.

Figure 23-1
GLM Univariate output

Tests of Between-Subjects Effects

Dependent Variable: SPVOL

Source	Type III Sum of Squares	df	Mean Square	F	Sig.
Corrected Model	22.520[1]	11	2.047	12.376	.000
Intercept	1016.981	1	1016.981	6147.938	.000
Flour	8.691	3	2.897	17.513	.000
Fat	10.118	2	5.059	30.583	.000
Surfactant	.997	2	.499	3.014	.082
Fat*Surfactant	5.639	4	1.410	8.522	.001
Error	2.316	14	.165		
Total	1112.960	26			
Corrected Total	24.835	25			

1. R Squared = .907 (Adjusted R Squared = .833)

fat * surfactant

Dependent Variable: SPVOL

fat	surfactant	Mean	Std. Error	95% Confidence Interval	
				Lower Bound	Upper Bound
1	1	5.536	.240	5.021	6.052
	2	5.891	.239	5.378	6.404
	3	6.123	.241	5.605	6.641
2	1	7.023	.241	6.505	7.541
	2	6.708	.301	6.064	7.353
	3	6.000	.203	5.564	6.436
3	1	6.629	.301	5.984	7.274
	2	7.200	.203	6.764	7.636
	3	8.589	.300	7.945	9.233

To Obtain GLM Univariate Tables

▶ From the menus choose:

Analyze
  General Linear Model
    Univariate...

Figure 23-2
Univariate dialog box

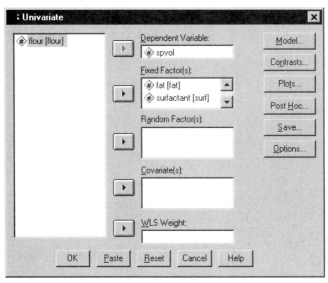

▶ Select a dependent variable.

▶ Select variables for Fixed Factor(s), Random Factor(s), and Covariate(s), as appropriate for your data.

▶ Optionally, you can use WLS Weight to specify a weight variable for weighted least-squares analysis. If the value of the weighting variable is zero, negative, or missing, the case is excluded from the analysis. A variable already used in the model cannot be used as a weighting variable.

GLM Model

Figure 23-3
Univariate Model dialog box

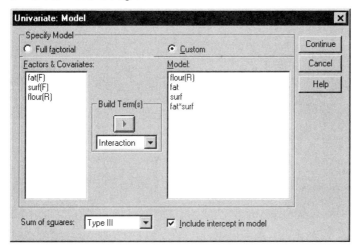

Specify Model. A full factorial model contains all factor main effects, all covariate main effects, and all factor-by-factor interactions. It does not contain covariate interactions. Select Custom to specify only a subset of interactions or to specify factor-by-covariate interactions. You must indicate all of the terms to be included in the model.

Factors and Covariates. The factors and covariates are listed with (F) for fixed factor and (C) for covariate. In a Univariate analysis, (R) indicates a random factor.

Model. The model depends on the nature of your data. After selecting Custom, you can select the main effects and interactions that are of interest in your analysis.

Sum of squares. The method of calculating the sums of squares. For balanced or unbalanced models with no missing cells, the Type III sum-of-squares method is most commonly used.

Include intercept in model. The intercept is usually included in the model. If you can assume that the data pass through the origin, you can exclude the intercept.

Build Terms

For the selected factors and covariates:

Interaction. Creates the highest-level interaction term of all selected variables. This is the default.

Main effects. Creates a main-effects term for each variable selected.

All 2-way. Creates all possible two-way interactions of the selected variables.

All 3-way. Creates all possible three-way interactions of the selected variables.

All 4-way. Creates all possible four-way interactions of the selected variables.

All 5-way. Creates all possible five-way interactions of the selected variables.

Sum of Squares

For the model, you can choose a type of sums of squares. Type III is the most commonly used and is the default.

Type I. This method is also known as the hierarchical decomposition of the sum-of-squares method. Each term is adjusted for only the term that precedes it in the model. Type I sums of squares are commonly used for:

- A balanced ANOVA model in which any main effects are specified before any first-order interaction effects, any first-order interaction effects are specified before any second-order interaction effects, and so on.

- A polynomial regression model in which any lower-order terms are specified before any higher-order terms.

- A purely nested model in which the first-specified effect is nested within the second-specified effect, the second-specified effect is nested within the third, and so on. (This form of nesting can be specified only by using syntax.)

Type II. This method calculates the sums of squares of an effect in the model adjusted for all other "appropriate" effects. An appropriate effect is one that corresponds to all effects that do not contain the effect being examined. The Type II sum-of-squares method is commonly used for:

- A balanced ANOVA model.

- Any model that has main factor effects only.

- Any regression model.

- A purely nested design. (This form of nesting can be specified by using syntax.)

Type III. The default. This method calculates the sums of squares of an effect in the design as the sums of squares adjusted for any other effects that do not contain it and orthogonal to any effects (if any) that contain it. The Type III sums of squares have one major advantage in that they are invariant with respect to the cell frequencies as long as the general form of estimability remains constant. Hence, this type of sums of squares is often considered useful for an unbalanced model with no missing cells. In a factorial design with no missing cells, this method is equivalent to the Yates' weighted-squares-of-means technique. The Type III sum-of-squares method is commonly used for:

- Any models listed in Type I and Type II.

- Any balanced or unbalanced model with no empty cells.

Type IV. This method is designed for a situation in which there are missing cells. For any effect F in the design, if F is not contained in any other effect, then Type IV = Type III = Type II. When F is contained in other effects, Type IV distributes the contrasts being made among the parameters in F to all higher-level effects equitably. The Type IV sum-of-squares method is commonly used for:

- Any models listed in Type I and Type II.

- Any balanced model or unbalanced model with empty cells.

GLM Contrasts

Figure 23-4
Univariate Contrasts dialog box

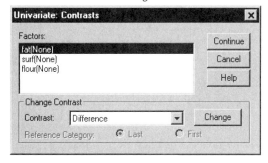

Contrasts are used to test for differences among the levels of a factor. You can specify a contrast for each factor in the model (in a repeated measures model, for each between-subjects factor). Contrasts represent linear combinations of the parameters.

Hypothesis testing is based on the null hypothesis **LB** = 0, where **L** is the contrast coefficients matrix and **B** is the parameter vector. When a contrast is specified, SPSS creates an **L** matrix in which the columns corresponding to the factor match the contrast. The remaining columns are adjusted so that the **L** matrix is estimable.

The output includes an *F* statistic for each set of contrasts. Also displayed for the contrast differences are Bonferroni-type simultaneous confidence intervals based on Student's *t* distribution.

Available Contrasts

Available contrasts are deviation, simple, difference, Helmert, repeated, and polynomial. For deviation contrasts and simple contrasts, you can choose whether the reference category is the last or first category.

Contrast Types

Deviation. Compares the mean of each level (except a reference category) to the mean of all of the levels (grand mean). The levels of the factor can be in any order.

Simple. Compares the mean of each level to the mean of a specified level. This type of contrast is useful when there is a control group. You can choose the first or last category as the reference.

Difference. Compares the mean of each level (except the first) to the mean of previous levels. (Sometimes called reverse Helmert contrasts.)

Helmert. Compares the mean of each level of the factor (except the last) to the mean of subsequent levels.

Repeated. Compares the mean of each level (except the last) to the mean of the subsequent level.

Polynomial. Compares the linear effect, quadratic effect, cubic effect, and so on. The first degree of freedom contains the linear effect across all categories; the second degree of freedom, the quadratic effect; and so on. These contrasts are often used to estimate polynomial trends.

GLM Profile Plots

Figure 23-5
Univariate Profile Plots dialog box

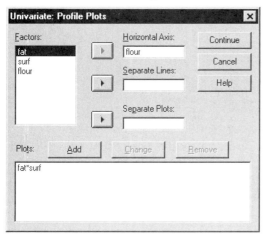

Profile plots (interaction plots) are useful for comparing marginal means in your model. A profile plot is a line plot in which each point indicates the estimated marginal mean of a dependent variable (adjusted for any covariates) at one level of a factor. The levels of a second factor can be used to make separate lines. Each level in a third factor can be used to create a separate plot. All fixed and random factors, if any, are available for plots. For multivariate analyses, profile plots are created for each dependent variable. In a repeated measures analysis, both between-subjects factors and within-subjects factors can be used in profile plots. GLM Multivariate and GLM Repeated Measures are available only if you have the Advanced Models option installed.

A profile plot of one factor shows whether the estimated marginal means are increasing or decreasing across levels. For two or more factors, parallel lines indicate that there is no interaction between factors, which means that you can investigate the levels of only one factor. Nonparallel lines indicate an interaction.

Figure 23-6
Nonparallel plot (left) and parallel plot (right)

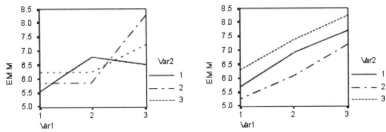

After a plot is specified by selecting factors for the horizontal axis and, optionally, factors for separate lines and separate plots, the plot must be added to the Plots list.

GLM Post Hoc Comparisons

Figure 23-7
Univariate Post Hoc Multiple Comparisons for Observed Means dialog box

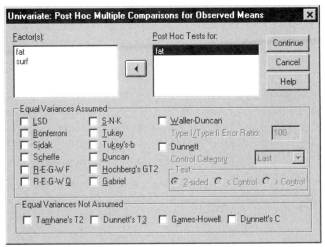

Post hoc multiple comparison tests. Once you have determined that differences exist among the means, post hoc range tests and pairwise multiple comparisons can determine which means differ. Comparisons are made on unadjusted values. These tests are used for fixed between-subjects factors only. In GLM Repeated Measures, these tests are not available if there are no between-subjects factors, and the post hoc multiple comparison tests are performed for the average across the levels of the within-subjects factors. For GLM Multivariate, the post hoc tests are performed for each dependent variable separately. GLM Multivariate and GLM Repeated Measures are available only if you have the Advanced Models option installed.

The Bonferroni and Tukey's honestly significant difference tests are commonly used multiple comparison tests. The **Bonferroni test**, based on Student's t statistic, adjusts the observed significance level for the fact that multiple comparisons are made. **Sidak's t test** also adjusts the significance level and provides tighter bounds than the Bonferroni test. **Tukey's honestly significant difference test** uses the Studentized range statistic to make all pairwise comparisons between groups and sets the experimentwise error rate to the error rate for the collection for all pairwise comparisons. When testing a large number of pairs of means, Tukey's honestly significant difference test is more powerful than the Bonferroni test. For a small number of pairs, Bonferroni is more powerful.

Hochberg's GT2 is similar to Tukey's honestly significant difference test, but the Studentized maximum modulus is used. Usually, Tukey's test is more powerful. **Gabriel's pairwise comparisons test** also uses the Studentized maximum modulus and is generally more powerful than Hochberg's GT2 when the cell sizes are unequal. Gabriel's test may become liberal when the cell sizes vary greatly.

Dunnett's pairwise multiple comparison t test compares a set of treatments against a single control mean. The last category is the default control category. Alternatively, you can choose the first category. You can also choose a two-sided or one-sided test. To test that the mean at any level (except the control category) of the factor is not equal to that of the control category, use a two-sided test. To test whether the mean at any level of the factor is smaller than that of the control category, select < Control. Likewise, to test whether the mean at any level of the factor is larger than that of the control category, select > Control.

Ryan, Einot, Gabriel, and Welsch (R-E-G-W) developed two multiple step-down range tests. Multiple step-down procedures first test whether all means are equal. If all means are not equal, subsets of means are tested for equality. **R-E-G-W F** is based on an F test and **R-E-G-W Q** is based on the Studentized range. These tests are more powerful than Duncan's multiple range test and Student-Newman-Keuls (which are also multiple step-down procedures), but they are not recommended for unequal cell sizes.

When the variances are unequal, use **Tamhane's T2** (conservative pairwise comparisons test based on a t test), **Dunnett's T3** (pairwise comparison test based on the Studentized maximum modulus), **Games-Howell pairwise comparison**

test (sometimes liberal), or **Dunnett's C** (pairwise comparison test based on the Studentized range).

Duncan's multiple range test, Student-Newman-Keuls (**S-N-K**), and **Tukey's b** are range tests that rank group means and compute a range value. These tests are not used as frequently as the tests previously discussed.

The **Waller-Duncan t test** uses a Bayesian approach. This range test uses the harmonic mean of the sample size when the sample sizes are unequal.

The significance level of the **Scheffé** test is designed to allow all possible linear combinations of group means to be tested, not just pairwise comparisons available in this feature. The result is that the Scheffé test is often more conservative than other tests, which means that a larger difference between means is required for significance.

The least significant difference (**LSD**) pairwise multiple comparison test is equivalent to multiple individual *t* tests between all pairs of groups. The disadvantage of this test is that no attempt is made to adjust the observed significance level for multiple comparisons.

Tests displayed. Pairwise comparisons are provided for LSD, Sidak, Bonferroni, Games-Howell, Tamhane's T2 and T3, Dunnett's *C*, and Dunnett's T3. Homogeneous subsets for range tests are provided for S-N-K, Tukey's *b*, Duncan, R-E-G-W *F*, R-E-G-W *Q*, and Waller. Tukey's honestly significant difference test, Hochberg's GT2, Gabriel's test, and Scheffé's test are both multiple comparison tests and range tests.

GLM Save

Figure 23-8
Save dialog box

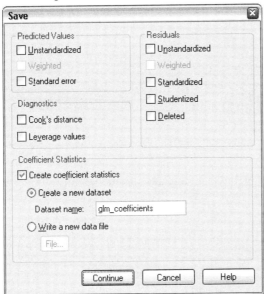

You can save values predicted by the model, residuals, and related measures as new variables in the Data Editor. Many of these variables can be used for examining assumptions about the data. To save the values for use in another SPSS session, you must save the current data file.

Predicted Values. The values that the model predicts for each case.

■ **Unstandardized.** The value the model predicts for the dependent variable.

■ **Weighted.** Weighted unstandardized predicted values. Available only if a WLS variable was previously selected.

■ **Standard error.** An estimate of the standard deviation of the average value of the dependent variable for cases that have the same values of the independent variables.

Diagnostics. Measures to identify cases with unusual combinations of values for the independent variables and cases that may have a large impact on the model.

- **Cook's distance.** A measure of how much the residuals of all cases would change if a particular case were excluded from the calculation of the regression coefficients. A large Cook's D indicates that excluding a case from computation of the regression statistics changes the coefficients substantially.

- **Leverage values.** Uncentered leverage values. The relative influence of each observation on the model's fit.

Residuals. An unstandardized residual is the actual value of the dependent variable minus the value predicted by the model. Standardized, Studentized, and deleted residuals are also available. If a WLS variable was chosen, weighted unstandardized residuals are available.

- **Unstandardized.** The difference between an observed value and the value predicted by the model.

- **Weighted.** Weighted unstandardized residuals. Available only if a WLS variable was previously selected.

- **Standardized.** The residual divided by an estimate of its standard deviation. Standardized residuals, which are also known as Pearson residuals, have a mean of 0 and a standard deviation of 1.

- **Studentized.** The residual divided by an estimate of its standard deviation that varies from case to case, depending on the distance of each case's values on the independent variables from the means of the independent variables.

- **Deleted.** The residual for a case when that case is excluded from the calculation of the regression coefficients. It is the difference between the value of the dependent variable and the adjusted predicted value.

Coefficient Statistics. Writes a variance-covariance matrix of the parameter estimates in the model to a new dataset in the current session or an external SPSS-format data file. Also, for each dependent variable, there will be a row of parameter estimates, a row of significance values for the *t* statistics corresponding to the parameter estimates, and a row of residual degrees of freedom. For a multivariate model, there are similar rows for each dependent variable. You can use this matrix file in other procedures that read an SPSS matrix file.

GLM Options

Figure 23-9
Univariate Options dialog box

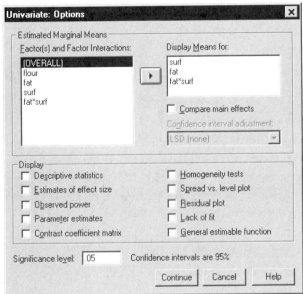

Optional statistics are available from this dialog box. Statistics are calculated using a fixed-effects model.

Estimated Marginal Means. Select the factors and interactions for which you want estimates of the population marginal means in the cells. These means are adjusted for the covariates, if any.

- **Compare main effects.** Provides uncorrected pairwise comparisons among estimated marginal means for any main effect in the model, for both between- and within-subjects factors. This item is available only if main effects are selected under the Display Means For list.

- **Confidence interval adjustment.** Select least significant difference (LSD), Bonferroni, or Sidak adjustment to the confidence intervals and significance. This item is available only if Compare main effects is selected.

Display. Select Descriptive statistics to produce observed means, standard deviations, and counts for all of the dependent variables in all cells. Estimates of effect size gives a partial eta-squared value for each effect and each parameter estimate. The eta-squared statistic describes the proportion of total variability attributable to a factor. Select Observed power to obtain the power of the test when the alternative hypothesis is set based on the observed value. Select Parameter estimates to produce the parameter estimates, standard errors, *t* tests, confidence intervals, and the observed power for each test. Select Contrast coefficient matrix to obtain the **L** matrix.

Homogeneity tests produces the Levene test of the homogeneity of variance for each dependent variable across all level combinations of the between-subjects factors, for between-subjects factors only. The spread-versus-level and residual plots options are useful for checking assumptions about the data. This item is disabled if there are no factors. Select Residual plot to produce an observed-by-predicted-by-standardized residual plot for each dependent variable. These plots are useful for investigating the assumption of equal variance. Select Lack of fit to check if the relationship between the dependent variable and the independent variables can be adequately described by the model. General estimable function allows you to construct custom hypothesis tests based on the general estimable function. Rows in any contrast coefficient matrix are linear combinations of the general estimable function.

Significance level. You might want to adjust the significance level used in post hoc tests and the confidence level used for constructing confidence intervals. The specified value is also used to calculate the observed power for the test. When you specify a significance level, the associated level of the confidence intervals is displayed in the dialog box.

UNIANOVA Command Additional Features

The SPSS command language also allows you to:

- Specify nested effects in the design (using the DESIGN subcommand).
- Specify tests of effects versus a linear combination of effects or a value (using the TEST subcommand).
- Specify multiple contrasts (using the CONTRAST subcommand).
- Include user-missing values (using the MISSING subcommand).
- Specify EPS criteria (using the CRITERIA subcommand).
- Construct a custom **L** matrix, **M** matrix, or **K** matrix (using the LMATRIX, MMATRIX, and KMATRIX subcommands).
- For deviation or simple contrasts, specify an intermediate reference category (using the CONTRAST subcommand).

■ Specify metrics for polynomial contrasts (using the CONTRAST subcommand).

■ Specify error terms for post hoc comparisons (using the POSTHOC subcommand).

■ Compute estimated marginal means for any factor or factor interaction among the factors in the factor list (using the EMMEANS subcommand).

■ Specify names for temporary variables (using the SAVE subcommand).

■ Construct a correlation matrix data file (using the OUTFILE subcommand).

■ Construct a matrix data file that contains statistics from the between-subjects ANOVA table (using the OUTFILE subcommand).

■ Save the design matrix to a new data file (using the OUTFILE subcommand).

See the *SPSS Command Syntax Reference* for complete syntax information.

Bivariate Correlations

The Bivariate Correlations procedure computes Pearson's correlation coefficient, Spearman's rho, and Kendall's tau-*b* with their significance levels. Correlations measure how variables or rank orders are related. Before calculating a correlation coefficient, screen your data for outliers (which can cause misleading results) and evidence of a linear relationship. Pearson's correlation coefficient is a measure of linear association. Two variables can be perfectly related, but if the relationship is not linear, Pearson's correlation coefficient is not an appropriate statistic for measuring their association.

Example. Is the number of games won by a basketball team correlated with the average number of points scored per game? A scatterplot indicates that there is a linear relationship. Analyzing data from the 1994–1995 NBA season yields that Pearson's correlation coefficient (0.581) is significant at the 0.01 level. You might suspect that the more games won per season, the fewer points the opponents scored. These variables are negatively correlated (–0.401), and the correlation is significant at the 0.05 level.

Statistics. For each variable: number of cases with nonmissing values, mean, and standard deviation. For each pair of variables: Pearson's correlation coefficient, Spearman's rho, Kendall's tau-*b*, cross-product of deviations, and covariance.

Data. Use symmetric quantitative variables for Pearson's correlation coefficient and quantitative variables or variables with ordered categories for Spearman's rho and Kendall's tau-*b*.

Assumptions. Pearson's correlation coefficient assumes that each pair of variables is bivariate normal.

Figure 24-1
Bivariate Correlations output

Correlations

		Number of Games Won	Scoring Points Per Game	Defense Points Per Game
Pearson Correlation	Number of Games Won	1.000	.581**	-.401*
	Scoring Points Per Game	.581**	1.000	.457*
	Defense Points Per Game	-.401*	.457*	1.000
Significance (2-tailed)	Number of Games Won	.	.001	.038
	Scoring Points Per Game	.001	.	.017
	Defense Points Per Game	.038	.017	.
N	Number of Games Won	27	27	27
	Scoring Points Per Game	27	27	27
	Defense Points Per Game	27	27	27

**. Correlation at 0.01(2-tailed):...

*. Correlation at 0.05(2-tailed):...

To Obtain Bivariate Correlations

From the menus choose:

Analyze
  Correlate
    Bivariate...

Figure 24-2
Bivariate Correlations dialog box

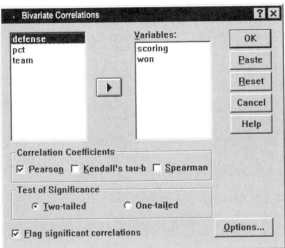

▶ Select two or more numeric variables.

The following options are also available:

- **Correlation Coefficients.** For quantitative, normally distributed variables, choose the Pearson correlation coefficient. If your data are not normally distributed or have ordered categories, choose Kendall's tau-b or Spearman, which measure the association between rank orders. Correlation coefficients range in value from –1 (a perfect negative relationship) and +1 (a perfect positive relationship). A value of 0 indicates no linear relationship. When interpreting your results, be careful not to draw any cause-and-effect conclusions due to a significant correlation.

- **Test of Significance.** You can select two-tailed or one-tailed probabilities. If the direction of association is known in advance, select One-tailed. Otherwise, select Two-tailed.

- **Flag significant correlations.** Correlation coefficients significant at the 0.05 level are identified with a single asterisk, and those significant at the 0.01 level are identified with two asterisks.

Bivariate Correlations Options

Figure 24-3
Bivariate Correlations Options dialog box

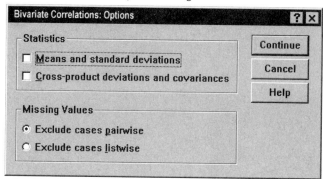

Statistics. For Pearson correlations, you can choose one or both of the following:

- **Means and standard deviations.** Displayed for each variable. The number of cases with nonmissing values is also shown. Missing values are handled on a variable-by-variable basis regardless of your missing values setting.

- **Cross-product deviations and covariances.** Displayed for each pair of variables. The cross-product of deviations is equal to the sum of the products of mean-corrected variables. This is the numerator of the Pearson correlation coefficient. The covariance is an unstandardized measure of the relationship between two variables, equal to the cross-product deviation divided by $N-1$.

Missing Values. You can choose one of the following:

- **Exclude cases pairwise.** Cases with missing values for one or both of a pair of variables for a correlation coefficient are excluded from the analysis. Since each coefficient is based on all cases that have valid codes on that particular pair of variables, the maximum information available is used in every calculation. This can result in a set of coefficients based on a varying number of cases.

- **Exclude cases listwise.** Cases with missing values for any variable are excluded from all correlations.

CORRELATIONS and NONPAR CORR Command Additional Features

The SPSS command language also allows you to:

- Write a correlation matrix for Pearson correlations that can be used in place of raw data to obtain other analyses such as factor analysis (with the MATRIX subcommand).

- Obtain correlations of each variable on a list with each variable on a second list (using the keyword WITH on the VARIABLES subcommand).

See the *SPSS Command Syntax Reference* for complete syntax information.

Partial Correlations

The Partial Correlations procedure computes partial correlation coefficients that describe the linear relationship between two variables while controlling for the effects of one or more additional variables. Correlations are measures of linear association. Two variables can be perfectly related, but if the relationship is not linear, a correlation coefficient is not an appropriate statistic for measuring their association.

Example. Is there a relationship between healthcare funding and disease rates? Although you might expect any such relationship to be a negative one, a study reports a significant *positive* correlation: as healthcare funding increases, disease rates appear to increase. Controlling for the rate of visits to healthcare providers, however, virtually eliminates the observed positive correlation. Healthcare funding and disease rates only appear to be positively related because more people have access to healthcare when funding increases, which leads to more reported diseases by doctors and hospitals.

Statistics. For each variable: number of cases with nonmissing values, mean, and standard deviation. Partial and zero-order correlation matrices, with degrees of freedom and significance levels.

Data. Use symmetric, quantitative variables.

Assumptions. The Partial Correlations procedure assumes that each pair of variables is bivariate normal.

Figure 25-1
Partial Correlations output

Correlations

Control Variables			Health care funding (amount per 100)	Reported diseases (rate per 10,000)	Visits to health care providers (rate per 10,000)
-none-[1]	Health care funding (amount per 100)	Correlation	1.000	.737	.964
		Significance (2-tailed)	.	.000	.000
		df	0	48	48
	Reported diseases (rate per 10,000)	Correlation	.737	1.000	.762
		Significance (2-tailed)	.000	.	.000
		df	48	0	48
	Visits to health care providers (rate per 10,000)	Correlation	.964	.762	1.000
		Significance (2-tailed)	.000	.000	.
		df	48	48	0
Visits to health care providers (rate per 10,000)	Health care funding (amount per 100)	Correlation	1.000	.013	
		Significance (2-tailed)	.	.928	
		df	0	47	
	Reported diseases (rate per 10,000)	Correlation	.013	1.000	
		Significance (2-tailed)	.928	.	
		df	47	0	

1. Cells contain zero-order (Pearson) correlations.

To Obtain Partial Correlations

▶ From the menus choose:

Analyze
  Correlate
    Partial...

Figure 25-2
Partial Correlations dialog box

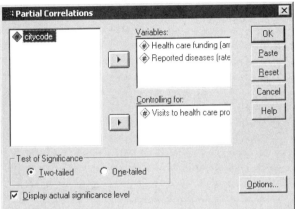

▶ Select two or more numeric variables for which partial correlations are to be computed.

▶ Select one or more numeric control variables.

The following options are also available:

■ **Test of Significance.** You can select two-tailed or one-tailed probabilities. If the direction of association is known in advance, select One-tailed. Otherwise, select Two-tailed.

■ **Display actual significance level.** By default, the probability and degrees of freedom are shown for each correlation coefficient. If you deselect this item, coefficients significant at the 0.05 level are identified with a single asterisk, coefficients significant at the 0.01 level are identified with a double asterisk, and degrees of freedom are suppressed. This setting affects both partial and zero-order correlation matrices.

Partial Correlations Options

Figure 25-3
Partial Correlations Options dialog box

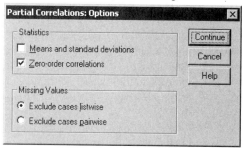

Statistics. You can choose one or both of the following:

- **Means and standard deviations.** Displayed for each variable. The number of cases with nonmissing values is also shown.

- **Zero-order correlations.** A matrix of simple correlations between all variables, including control variables, is displayed.

Missing Values. You can choose one of the following alternatives:

- **Exclude cases listwise.** Cases having missing values for any variable, including a control variable, are excluded from all computations.

- **Exclude cases pairwise.** For computation of the zero-order correlations on which the partial correlations are based, a case having missing values for both or one of a pair of variables is not used. Pairwise deletion uses as much of the data as possible. However, the number of cases may differ across coefficients. When pairwise deletion is in effect, the degrees of freedom for a particular partial coefficient are based on the smallest number of cases used in the calculation of any of the zero-order correlations.

PARTIAL CORR Command Additional Features

The SPSS command language also allows you to:

- Read a zero-order correlation matrix or write a partial correlation matrix (with the MATRIX subcommand).

- Obtain partial correlations between two lists of variables (using the keyword WITH on the VARIABLES subcommand).

- Obtain multiple analyses (with multiple VARIABLES subcommands).

- Specify order values to request (for example, both first- and second-order partial correlations) when you have two control variables (with the VARIABLES subcommand).

- Suppress redundant coefficients (with the FORMAT subcommand).

- Display a matrix of simple correlations when some coefficients cannot be computed (with the STATISTICS subcommand).

See the *SPSS Command Syntax Reference* for complete syntax information.

Distances

This procedure calculates any of a wide variety of statistics measuring either similarities or dissimilarities (distances), either between pairs of variables or between pairs of cases. These similarity or distance measures can then be used with other procedures, such as factor analysis, cluster analysis, or multidimensional scaling, to help analyze complex datasets.

Example. Is it possible to measure similarities between pairs of automobiles based on certain characteristics, such as engine size, MPG, and horsepower? By computing similarities between autos, you can gain a sense of which autos are similar to each other and which are different from each other. For a more formal analysis, you might consider applying a hierarchical cluster analysis or multidimensional scaling to the similarities to explore the underlying structure.

Statistics. Dissimilarity (distance) measures for interval data are Euclidean distance, squared Euclidean distance, Chebychev, block, Minkowski, or customized; for count data, chi-square or phi-square; for binary data, Euclidean distance, squared Euclidean distance, size difference, pattern difference, variance, shape, or Lance and Williams. Similarity measures for interval data are Pearson correlation or cosine; for binary data, Russel and Rao, simple matching, Jaccard, dice, Rogers and Tanimoto, Sokal and Sneath 1, Sokal and Sneath 2, Sokal and Sneath 3, Kulczynski 1, Kulczynski 2, Sokal and Sneath 4, Hamann, Lambda, Anderberg's *D*, Yule's *Y*, Yule's *Q*, Ochiai, Sokal and Sneath 5, phi 4-point correlation, or dispersion.

To Obtain Distance Matrices

▶ From the menus choose:

Analyze
  Correlate
    Distances...

Figure 26-1
Distances dialog box

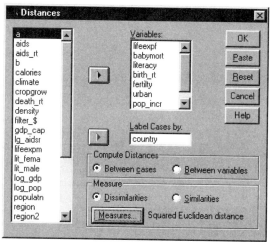

▶ Select at least one numeric variable to compute distances between cases, or select at
least two numeric variables to compute distances between variables.

▶ Select an alternative in the Compute Distances group to calculate proximities either
between cases or between variables.

Distances Dissimilarity Measures

Figure 26-2
Distances Dissimilarity Measures dialog box

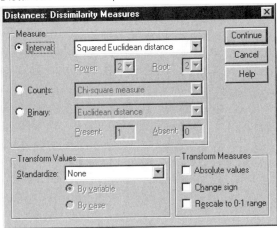

From the Measure group, select the alternative that corresponds to your type of data (interval, count, or binary); then, from the drop-down list, select one of the measures that corresponds to that type of data. Available measures, by data type, are:

- **Interval data.** Euclidean distance, squared Euclidean distance, Chebychev, block, Minkowski, or customized.

- **Count data.** Chi-square measure or phi-square measure.

- **Binary data.** Euclidean distance, squared Euclidean distance, size difference, pattern difference, variance, shape, or Lance and Williams. (Enter values for Present and Absent to specify which two values are meaningful; Distances will ignore all other values.)

The Transform Values group allows you to standardize data values for either cases or variables *before* computing proximities. These transformations are not applicable to binary data. Available standardization methods are z scores, range –1 to 1, range 0 to 1, maximum magnitude of 1, mean of 1, or standard deviation of 1.

The Transform Measures group allows you to transform the values generated by the distance measure. They are applied after the distance measure has been computed. Available options are absolute values, change sign, and rescale to 0–1 range.

Distances Similarity Measures

Figure 26-3
Distances Similarity Measures dialog box

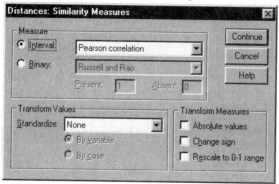

From the Measure group, select the alternative that corresponds to your type of data (interval or binary); then, from the drop-down list, select one of the measures that corresponds to that type of data. Available measures, by data type, are:

- **Interval data.** Pearson correlation or cosine.

- **Binary data.** Russell and Rao, simple matching, Jaccard, Dice, Rogers and Tanimoto, Sokal and Sneath 1, Sokal and Sneath 2, Sokal and Sneath 3, Kulczynski 1, Kulczynski 2, Sokal and Sneath 4, Hamann, Lambda, Anderberg's D, Yule's Y, Yule's Q, Ochiai, Sokal and Sneath 5, phi 4-point correlation, or dispersion. (Enter values for Present and Absent to specify which two values are meaningful; Distances will ignore all other values.)

The Transform Values group allows you to standardize data values for either cases or variables before computing proximities. These transformations are not applicable to binary data. Available standardization methods are z scores, range –1 to 1, range 0 to 1, maximum magnitude of 1, mean of 1, and standard deviation of 1.

The Transform Measures group allows you to transform the values generated by the distance measure. They are applied after the distance measure has been computed. Available options are absolute values, change sign, and rescale to 0–1 range.

PROXIMITIES Command Additional Features

The Distances procedure uses PROXIMITIES command syntax. The SPSS command language also allows you to:

- Specify any integer as the power for the Minkowski distance measure.

- Specify any integers as the power and root for a customized distance measure.

See the *SPSS Command Syntax Reference* for complete syntax information.

Linear Regression

Linear Regression estimates the coefficients of the linear equation, involving one or more independent variables, that best predict the value of the dependent variable. For example, you can try to predict a salesperson's total yearly sales (the dependent variable) from independent variables such as age, education, and years of experience.

Example. Is the number of games won by a basketball team in a season related to the average number of points the team scores per game? A scatterplot indicates that these variables are linearly related. The number of games won and the average number of points scored by the opponent are also linearly related. These variables have a negative relationship. As the number of games won increases, the average number of points scored by the opponent decreases. With linear regression, you can model the relationship of these variables. A good model can be used to predict how many games teams will win.

Statistics. For each variable: number of valid cases, mean, and standard deviation. For each model: regression coefficients, correlation matrix, part and partial correlations, multiple R, R^2, adjusted R^2, change in R^2, standard error of the estimate, analysis-of-variance table, predicted values, and residuals. Also, 95%-confidence intervals for each regression coefficient, variance-covariance matrix, variance inflation factor, tolerance, Durbin-Watson test, distance measures (Mahalanobis, Cook, and leverage values), DfBeta, DfFit, prediction intervals, and casewise diagnostics. Plots: scatterplots, partial plots, histograms, and normal probability plots.

Data. The dependent and independent variables should be quantitative. Categorical variables, such as religion, major field of study, or region of residence, need to be recoded to binary (dummy) variables or other types of contrast variables.

Assumptions. For each value of the independent variable, the distribution of the dependent variable must be normal. The variance of the distribution of the dependent variable should be constant for all values of the independent variable. The relationship between the dependent variable and each independent variable should be linear, and all observations should be independent.

Figure 27-1
Linear Regression output

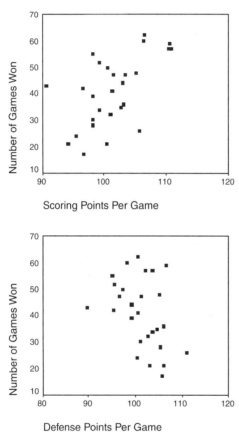

Model Summary [3,4]

		Variables		R	R Square	Adjusted R Square	Std. Error of the Estimate
		Entered	Removed				
Model	1	Defense Points Per Game, Scoring Points Per Game [1,2]	.	.947	.898	.889	4.40

[1.] Indep. vars: (constant) Defense Points Per Game, Scoring Points Per Game...

[2.] All requested variables entered.

[3.] Dependent Variable: Number of Games Won

[4.] Method: Enter

ANOVA[2]

			Sum of Squares	df	Mean Square	F	Significance
Model	1	Regression	4080.533	2	2040.266	105.198	.000[1]
		Residual	465.467	24	19.394		
		Total	4546.000	26			

[1.] Indep. vars: (constant) Defense Points Per Game, Scoring Points Per Game...

[2.] Dependent Variable: Number of Games Won

Coefficients[1]

Model		Unstandardized Coefficients		Standardized Coefficients	t	Sig.
		B	Std. Error	Beta		
1	(Constant)	28.121	21.404		1.314	.201
	Scoring Points Per Game	2.539	.193	.965	13.145	.000
	Defense Points Per Game	-2.412	.211	-.841	-11.458	.000

[1.] Dependent Variable: Number of Games Won

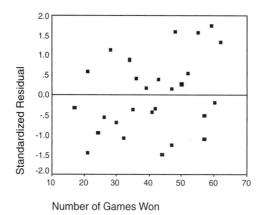

Number of Games Won

To Obtain a Linear Regression Analysis

► From the menus choose:

Analyze
  Regression
    Linear...

Figure 27-2
Linear Regression dialog box

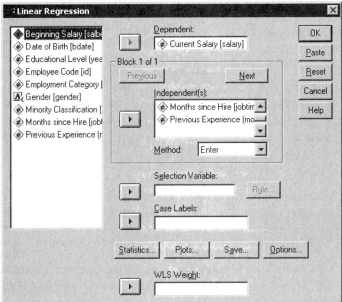

▶ In the Linear Regression dialog box, select a numeric dependent variable.

▶ Select one or more numeric independent variables.

Optionally, you can:

■ Group independent variables into blocks and specify different entry methods for different subsets of variables.

■ Choose a selection variable to limit the analysis to a subset of cases having a particular value(s) for this variable.

■ Select a case identification variable for identifying points on plots.

■ Select a numeric WLS Weight variable for a weighted least squares analysis.

WLS. Allows you to obtain a weighted least-squares model. Data points are weighted by the reciprocal of their variances. This means that observations with large variances have less impact on the analysis than observations associated with small variances. If the value of the weighting variable is zero, negative, or missing, the case is excluded from the analysis.

Linear Regression Variable Selection Methods

Method selection allows you to specify how independent variables are entered into the analysis. Using different methods, you can construct a variety of regression models from the same set of variables.

- **Enter (Regression).** A procedure for variable selection in which all variables in a block are entered in a single step.

- **Stepwise.** At each step, the independent variable not in the equation that has the smallest probability of F is entered, if that probability is sufficiently small. Variables already in the regression equation are removed if their probability of F becomes sufficiently large. The method terminates when no more variables are eligible for inclusion or removal.

- **Remove.** A procedure for variable selection in which all variables in a block are removed in a single step.

- **Backward Elimination.** A variable selection procedure in which all variables are entered into the equation and then sequentially removed. The variable with the smallest partial correlation with the dependent variable is considered first for removal. If it meets the criterion for elimination, it is removed. After the first variable is removed, the variable remaining in the equation with the smallest partial correlation is considered next. The procedure stops when there are no variables in the equation that satisfy the removal criteria.

- **Forward Selection.** A stepwise variable selection procedure in which variables are sequentially entered into the model. The first variable considered for entry into the equation is the one with the largest positive or negative correlation with the dependent variable. This variable is entered into the equation only if it satisfies the criterion for entry. If the first variable is entered, the independent variable not in the equation that has the largest partial correlation is considered next. The procedure stops when there are no variables that meet the entry criterion.

The significance values in your output are based on fitting a single model. Therefore, the significance values are generally invalid when a stepwise method (stepwise, forward, or backward) is used.

All variables must pass the tolerance criterion to be entered in the equation, regardless of the entry method specified. The default tolerance level is 0.0001. Also, a variable is not entered if it would cause the tolerance of another variable already in the model to drop below the tolerance criterion.

All independent variables selected are added to a single regression model. However, you can specify different entry methods for different subsets of variables. For example, you can enter one block of variables into the regression model using stepwise selection and a second block using forward selection. To add a second block of variables to the regression model, click Next.

Linear Regression Set Rule

Figure 27-3
Linear Regression Set Rule dialog box

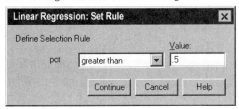

Cases defined by the selection rule are included in the analysis. For example, if you select a variable, choose equals, and type 5 for the value, then only cases for which the selected variable has a value equal to 5 are included in the analysis. A string value is also permitted.

Linear Regression Plots

Figure 27-4
Linear Regression Plots dialog box

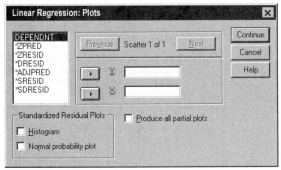

Plots can aid in the validation of the assumptions of normality, linearity, and equality of variances. Plots are also useful for detecting outliers, unusual observations, and influential cases. After saving them as new variables, predicted values, residuals, and other diagnostics are available in the Data Editor for constructing plots with the independent variables. The following plots are available:

Scatterplots. You can plot any two of the following: the dependent variable, standardized predicted values, standardized residuals, deleted residuals, adjusted predicted values, Studentized residuals, or Studentized deleted residuals. Plot the standardized residuals against the standardized predicted values to check for linearity and equality of variances.

Source variable list. Lists the dependent variable (DEPENDNT) and the following predicted and residual variables: Standardized predicted values (*ZPRED), Standardized residuals (*ZRESID), Deleted residuals (*DRESID), Adjusted predicted values (*ADJPRED), Studentized residuals (*SRESID), Studentized deleted residuals (*SDRESID).

Produce all partial plots. Displays scatterplots of residuals of each independent variable and the residuals of the dependent variable when both variables are regressed separately on the rest of the independent variables. At least two independent variables must be in the equation for a partial plot to be produced.

Standardized Residual Plots. You can obtain histograms of standardized residuals and normal probability plots comparing the distribution of standardized residuals to a normal distribution.

If any plots are requested, summary statistics are displayed for standardized predicted values and standardized residuals (*ZPRED and *ZRESID).

Linear Regression: Saving New Variables

Figure 27-5
Linear Regression Save dialog box

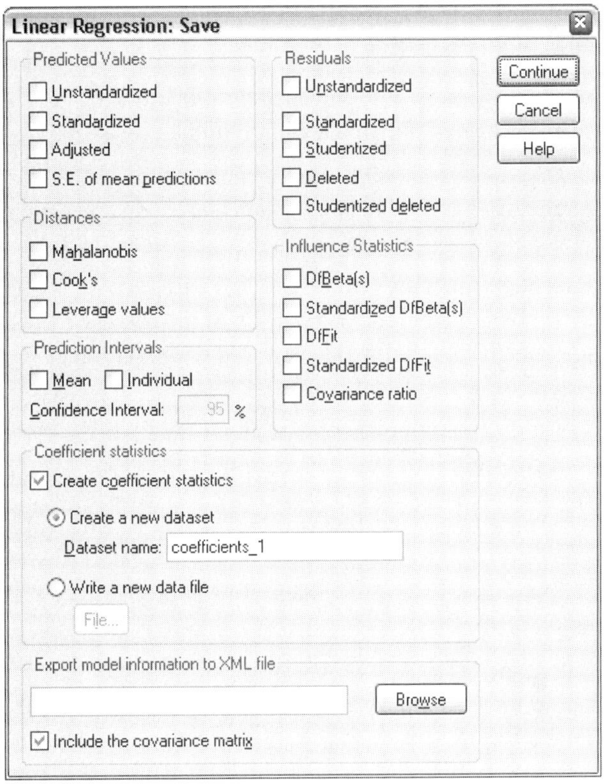

You can save predicted values, residuals, and other statistics useful for diagnostics. Each selection adds one or more new variables to your active data file.

Predicted Values. Values that the regression model predicts for each case.

- **Unstandardized.** The value the model predicts for the dependent variable.

- **Standardized.** A transformation of each predicted value into its standardized form. That is, the mean predicted value is subtracted from the predicted value, and the difference is divided by the standard deviation of the predicted values. Standardized predicted values have a mean of 0 and a standard deviation of 1.

- **Adjusted.** The predicted value for a case when that case is excluded from the calculation of the regression coefficients.

- **S.E. of mean predictions.** Standard errors of the predicted values. An estimate of the standard deviation of the average value of the dependent variable for cases that have the same values of the independent variables.

Distances. Measures to identify cases with unusual combinations of values for the independent variables and cases that may have a large impact on the regression model.

- **Mahalanobis.** A measure of how much a case's values on the independent variables differ from the average of all cases. A large Mahalanobis distance identifies a case as having extreme values on one or more of the independent variables.

- **Cook's.** A measure of how much the residuals of all cases would change if a particular case were excluded from the calculation of the regression coefficients. A large Cook's D indicates that excluding a case from computation of the regression statistics changes the coefficients substantially.

- **Leverage values.** Measures the influence of a point on the fit of the regression. The centered leverage ranges from 0 (no influence on the fit) to (N-1)/N.

Prediction Intervals. The upper and lower bounds for both mean and individual prediction intervals.

- **Mean.** Lower and upper bounds (two variables) for the prediction interval of the mean predicted response.

- **Individual.** Lower and upper bounds (two variables) for the prediction interval of the dependent variable for a single case.

- **Confidence Interval.** Enter a value between 1 and 99.99 to specify the confidence level for the two Prediction Intervals. Mean or Individual must be selected before entering this value. Typical confidence interval values are 90, 95, and 99.

Residuals. The actual value of the dependent variable minus the value predicted by the regression equation.

- **Unstandardized.** The difference between an observed value and the value predicted by the model.

- **Standardized.** The residual divided by an estimate of its standard deviation. Standardized residuals, which are also known as Pearson residuals, have a mean of 0 and a standard deviation of 1.

- **Studentized.** The residual divided by an estimate of its standard deviation that varies from case to case, depending on the distance of each case's values on the independent variables from the means of the independent variables.

- **Deleted.** The residual for a case when that case is excluded from the calculation of the regression coefficients. It is the difference between the value of the dependent variable and the adjusted predicted value.

- **Studentized deleted.** The deleted residual for a case divided by its standard error. The difference between a Studentized deleted residual and its associated Studentized residual indicates how much difference eliminating a case makes on its own prediction.

Influence Statistics. The change in the regression coefficients (DfBeta[s]) and predicted values (DfFit) that results from the exclusion of a particular case. Standardized DfBetas and DfFit values are also available along with the covariance ratio.

- **DfBeta(s).** The difference in beta value is the change in the regression coefficient that results from the exclusion of a particular case. A value is computed for each term in the model, including the constant.

- **Standardized DfBeta.** Standardized difference in beta value. The change in the regression coefficient that results from the exclusion of a particular case. You may want to examine cases with absolute values greater than 2 divided by the square root of N, where N is the number of cases. A value is computed for each term in the model, including the constant.

- **DfFit.** The difference in fit value is the change in the predicted value that results from the exclusion of a particular case.

- **Standardized DfFit.** Standardized difference in fit value. The change in the predicted value that results from the exclusion of a particular case. You may want to examine standardized values which in absolute value exceed 2 times the square root of p/N, where p is the number of parameters in the model and N is the number of cases.

- **Covariance ratio.** The ratio of the determinant of the covariance matrix with a particular case excluded from the calculation of the regression coefficients to the determinant of the covariance matrix with all cases included. If the ratio is close to 1, the case does not significantly alter the covariance matrix.

Coefficient Statistics. Saves regression coefficients to a dataset or a data file. Datasets are available for subsequent use in the same session but are not saved as files unless explicitly saved prior to the end of the session. Dataset names must conform to SPSS variable naming rules. For more information, see "Variable Names" in Chapter 5 on p. 106.

Export model information to XML file. Parameter estimates and (optionally) their covariances are exported to the specified file in XML (PMML) format. SmartScore and SPSS Server (a separate product) can use this model file to apply the model information to other data files for scoring purposes.

Linear Regression Statistics

Figure 27-6
Linear Regression Statistics dialog box

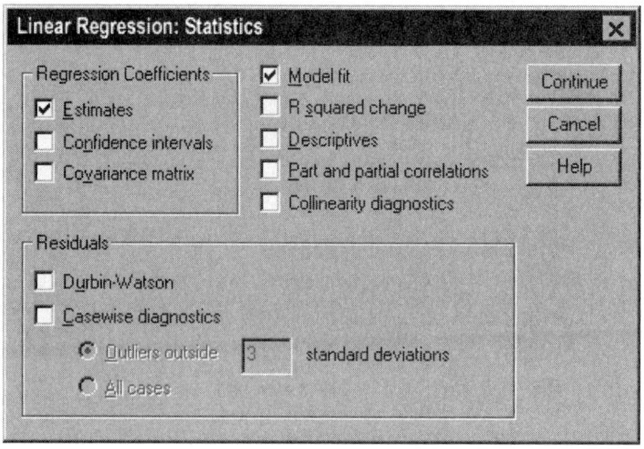

The following statistics are available:

Regression Coefficients. Estimates displays Regression coefficient B, standard error of B, standardized coefficient beta, t value for B, and two-tailed significance level of t. Confidence intervals displays 95% confidence intervals for each regression coefficient or a covariance matrix. Covariance matrix displays a variance-covariance matrix of regression coefficients with covariances off the diagonal and variances on the diagonal. A correlation matrix is also displayed.

Model fit. The variables entered and removed from the model are listed, and the following goodness-of-fit statistics are displayed: multiple R, R^2 and adjusted R^2, standard error of the estimate, and an analysis-of-variance table.

R squared change. The change in the R^2 statistic that is produced by adding or deleting an independent variable. If the R^2 change associated with a variable is large, that means that the variable is a good predictor of the dependent variable.

Descriptives. Provides the number of valid cases, the mean, and the standard deviation for each variable in the analysis. A correlation matrix with a one-tailed significance level and the number of cases for each correlation are also displayed.

Partial Correlation. The correlation that remains between two variables after removing the correlation that is due to their mutual association with the other variables. The correlation between the dependent variable and an independent variable when the linear effects of the other independent variables in the model have been removed from both.

Part Correlation. The correlation between the dependent variable and an independent variable when the linear effects of the other independent variables in the model have been removed from the independent variable. It is related to the change in R squared when a variable is added to an equation. Sometimes called the semipartial correlation.

Collinearity diagnostics. Collinearity (or multicollinearity) is the undesirable situation when one independent variable is a linear function of other independent variables. Eigenvalues of the scaled and uncentered cross-products matrix, condition indices, and variance-decomposition proportions are displayed along with variance inflation factors (VIF) and tolerances for individual variables.

Residuals. Displays the Durbin-Watson test for serial correlation of the residuals and casewise diagnostics for the cases meeting the selection criterion (outliers above n standard deviations).

Linear Regression Options

Figure 27-7
Linear Regression Options dialog box

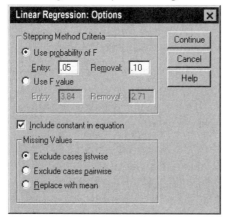

The following options are available:

Stepping Method Criteria. These options apply when either the forward, backward, or stepwise variable selection method has been specified. Variables can be entered or removed from the model depending on either the significance (probability) of the F value or the F value itself.

- **Use Probability of F.** A variable is entered into the model if the significance level of its F value is less than the Entry value and is removed if the significance level is greater than the Removal value. Entry must be less than Removal, and both values must be positive. To enter more variables into the model, increase the Entry value. To remove more variables from the model, lower the Removal value.

- **Use F Value.** A variable is entered into the model if its F value is greater than the Entry value and is removed if the F value is less than the Removal value. Entry must be greater than Removal, and both values must be positive. To enter more variables into the model, lower the Entry value. To remove more variables from the model, increase the Removal value.

Include constant in equation. By default, the regression model includes a constant term. Deselecting this option forces regression through the origin, which is rarely done. Some results of regression through the origin are not comparable to results of regression that do include a constant. For example, R^2 cannot be interpreted in the usual way.

Missing Values. You can choose one of the following:

■ **Exclude cases listwise.** Only cases with valid values for all variables are included in the analyses.

■ **Exclude cases pairwise.** Cases with complete data for the pair of variables being correlated are used to compute the correlation coefficient on which the regression analysis is based. Degrees of freedom are based on the minimum pairwise N.

■ **Replace with mean.** All cases are used for computations, with the mean of the variable substituted for missing observations.

REGRESSION Command Additional Features

The SPSS command language also allows you to:

■ Write a correlation matrix or read a matrix in place of raw data to obtain your regression analysis (with the MATRIX subcommand).

■ Specify tolerance levels (with the CRITERIA subcommand).

■ Obtain multiple models for the same or different dependent variables (with the METHOD and DEPENDENT subcommands).

■ Obtain additional statistics (with the DESCRIPTIVES and STATISTICS subcommands).

See the *SPSS Command Syntax Reference* for complete syntax information.

Ordinal Regression

Ordinal Regression allows you to model the dependence of a polytomous ordinal response on a set of predictors, which can be factors or covariates. The design of Ordinal Regression is based on the methodology of McCullagh (1980, 1998), and the procedure is referred to as PLUM in the syntax.

Standard linear regression analysis involves minimizing the sum-of-squared differences between a response (dependent) variable and a weighted combination of predictor (independent) variables. The estimated coefficients reflect how changes in the predictors affect the response. The response is assumed to be numerical, in the sense that changes in the level of the response are equivalent throughout the range of the response. For example, the difference in height between a person who is 150 cm tall and a person who is 140 cm tall is 10 cm, which has the same meaning as the difference in height between a person who is 210 cm tall and a person who is 200 cm tall. These relationships do not necessarily hold for ordinal variables, in which the choice and number of response categories can be quite arbitrary.

Example. Ordinal Regression could be used to study patient reaction to drug dosage. The possible reactions may be classified as "none," "mild," "moderate," or "severe." The difference between a mild and moderate reaction is difficult or impossible to quantify and is based on perception. Moreover, the difference between a mild and moderate response may be greater or less than the difference between a moderate and severe response.

Statistics and plots. Observed and expected frequencies and cumulative frequencies, Pearson residuals for frequencies and cumulative frequencies, observed and expected probabilities, observed and expected cumulative probabilities of each response category by covariate pattern, asymptotic correlation and covariance matrices of parameter estimates, Pearson's chi-square and likelihood-ratio chi-square, goodness-of-fit statistics, iteration history, test of parallel lines assumption, parameter estimates, standard errors, confidence intervals, and Cox and Snell's, Nagelkerke's, and McFadden's R^2 statistics.

Data. The dependent variable is assumed to be ordinal and can be numeric or string. The ordering is determined by sorting the values of the dependent variable in ascending order. The lowest value defines the first category. Factor variables are assumed to be categorical. Covariate variables must be numeric. Note that using more than one continuous covariate can easily result in the creation of a very large cell probabilities table.

Assumptions. Only one response variable is allowed, and it must be specified. Also, for each distinct pattern of values across the independent variables, the responses are assumed to be independent multinomial variables.

Related procedures. Nominal logistic regression uses similar models for nominal dependent variables.

Obtaining an Ordinal Regression

▶ From the menus choose:
Analyze
  Regression
    Ordinal...

Figure 28-1
Ordinal Regression dialog box

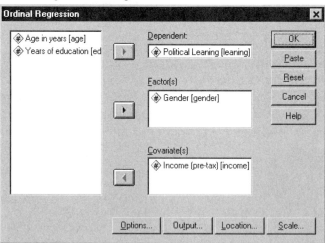

▶ Select one dependent variable.

▶ Click OK.

Ordinal Regression Options

The Options dialog box allows you to adjust parameters used in the iterative estimation algorithm, choose a level of confidence for your parameter estimates, and select a link function.

Figure 28-2
Ordinal Regression Options dialog box

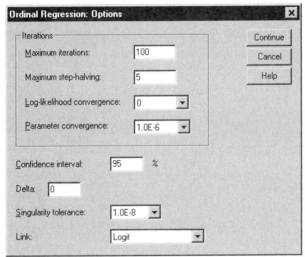

Iterations. You can customize the iterative algorithm.

■ **Maximum iterations.** Specify a non-negative integer. If 0 is specified, the procedure returns the initial estimates.

■ **Maximum step-halving.** Specify a positive integer.

■ **Log-likelihood convergence.** The algorithm stops if the absolute or relative change in the log-likelihood is less than this value. The criterion is not used if 0 is specified.

■ **Parameter convergence.** The algorithm stops if the absolute or relative change in each of the parameter estimates is less than this value. The criterion is not used if 0 is specified.

Confidence interval. Specify a value greater than or equal to 0 and less than 100.

Delta. The value added to zero cell frequencies. Specify a non-negative value less than 1.

Singularity tolerance. Used for checking for highly dependent predictors. Select a value from the list of options.

Link Function. The link function is a transformation of the cumulative probabilities that allows estimation of the model. Five link functions are available, summarized in the following table.

Function	Form	Typical application
Logit	$\log(\xi / (1-\xi))$	Evenly distributed categories
Complementary log-log	$\log(-\log(1-\xi))$	Higher categories more probable
Negative log-log	$-\log(-\log(\xi))$	Lower categories more probable
Probit	$\Phi^{-1}(\xi)$	Latent variable is normally distributed
Cauchit (inverse Cauchy)	$\tan(\pi(\xi-0.5))$	Latent variable has many extreme values

Ordinal Regression Output

The Output dialog box allows you to produce tables for display in the Viewer and save variables to the working file.

Figure 28-3
Ordinal Regression Output dialog box

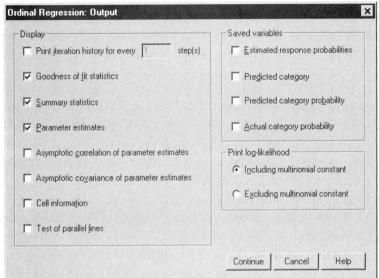

Display. Produces tables for:

- **Print iteration history.** The log-likelihood and parameter estimates are printed for the print iteration frequency specified. The first and last iterations are always printed.

- **Goodness-of-fit statistics.** The Pearson and likelihood-ratio chi-square statistics. They are computed based on the classification specified in the variable list.

- **Summary statistics.** Cox and Snell's, Nagelkerke's, and McFadden's R^2 statistics.

- **Parameter estimates.** Parameter estimates, standard errors, and confidence intervals.

- **Asymptotic correlation of parameter estimates.** Matrix of parameter estimate correlations.

- **Asymptotic covariance of parameter estimates.** Matrix of parameter estimate covariances.

- **Cell information.** Observed and expected frequencies and cumulative frequencies, Pearson residuals for frequencies and cumulative frequencies, observed and expected probabilities, and observed and expected cumulative probabilities of each response category by covariate pattern. Note that for models with many covariate patterns (for example, models with continuous covariates), this option can generate a very large, unwieldy table.

- **Test of parallel lines.** Test of the hypothesis that the location parameters are equivalent across the levels of the dependent variable. This is available only for the location-only model.

Saved variables. Saves the following variables to the working file:

- **Estimated response probabilities.** Model-estimated probabilities of classifying a factor/covariate pattern into the response categories. There are as many probabilities as the number of response categories.

- **Predicted category.** The response category that has the maximum estimated probability for a factor/covariate pattern.

- **Predicted category probability.** Estimated probability of classifying a factor/covariate pattern into the predicted category. This probability is also the maximum of the estimated probabilities of the factor/covariate pattern.

- **Actual category probability.** Estimated probability of classifying a factor/covariate pattern into the actual category.

Print log-likelihood. Controls the display of the log-likelihood. Including multinomial constant gives you the full value of the likelihood. To compare your results across products that do not include the constant, you can choose to exclude it.

Ordinal Regression Location Model

The Location dialog box allows you to specify the location model for your analysis.

Figure 28-4
Ordinal Regression Location dialog box

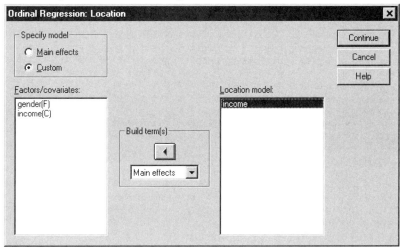

Specify model. A main-effects model contains the covariate and factor main effects but no interaction effects. You can create a custom model to specify subsets of factor interactions or covariate interactions.

Factors/covariates. The factors and covariates are listed with (F) for factor and (C) for covariate.

Location model. The model depends on the main effects and interaction effects that you select.

Build Terms

For the selected factors and covariates:

Interaction. Creates the highest-level interaction term of all selected variables. This is the default.

Main effects. Creates a main-effects term for each variable selected.

All 2-way. Creates all possible two-way interactions of the selected variables.

All 3-way. Creates all possible three-way interactions of the selected variables.

All 4-way. Creates all possible four-way interactions of the selected variables.

All 5-way. Creates all possible five-way interactions of the selected variables.

Ordinal Regression Scale Model

The Scale dialog box allows you to specify the scale model for your analysis.

Figure 28-5
Ordinal Regression Scale dialog box

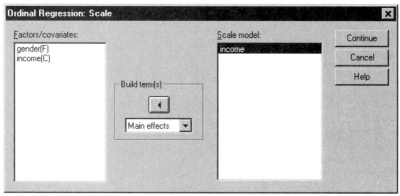

Factors/covariates. The factors and covariates are listed with (F) for factor and (C) for covariate.

Scale model. The model depends on the main and interaction effects that you select.

Build Terms

For the selected factors and covariates:

Interaction. Creates the highest-level interaction term of all selected variables. This is the default.

Main effects. Creates a main-effects term for each variable selected.

All 2-way. Creates all possible two-way interactions of the selected variables.

All 3-way. Creates all possible three-way interactions of the selected variables.

All 4-way. Creates all possible four-way interactions of the selected variables.

All 5-way. Creates all possible five-way interactions of the selected variables.

PLUM Command Additional Features

You can customize your Ordinal Regression if you paste your selections into a syntax window and edit the resulting `PLUM` command syntax. The SPSS command language also allows you to:

- Create customized hypothesis tests by specifying null hypotheses as linear combinations of parameters.

See the *SPSS Command Syntax Reference* for complete syntax information.

Curve Estimation

The Curve Estimation procedure produces curve estimation regression statistics and related plots for 11 different curve estimation regression models. A separate model is produced for each dependent variable. You can also save predicted values, residuals, and prediction intervals as new variables.

Example. An Internet service provider tracks the percentage of virus-infected e-mail traffic on its networks over time. A scatterplot reveals that the relationship is nonlinear. You might fit a quadratic or cubic model to the data and check the validity of assumptions and the goodness of fit of the model.

Statistics. For each model: regression coefficients, multiple R, R^2, adjusted R^2, standard error of the estimate, analysis-of-variance table, predicted values, residuals, and prediction intervals. Models: linear, logarithmic, inverse, quadratic, cubic, power, compound, S-curve, logistic, growth, and exponential.

Data. The dependent and independent variables should be quantitative. If you select Time from the active dataset as the independent variable (instead of selecting a variable), the Curve Estimation procedure generates a time variable where the length of time between cases is uniform. If Time is selected, the dependent variable should be a time-series measure. Time-series analysis requires a data file structure in which each case (row) represents a set of observations at a different time and the length of time between cases is uniform.

Assumptions. Screen your data graphically to determine how the independent and dependent variables are related (linearly, exponentially, etc.). The residuals of a good model should be randomly distributed and normal. If a linear model is used, the following assumptions should be met: For each value of the independent variable, the distribution of the dependent variable must be normal. The variance of the distribution of the dependent variable should be constant for all values of the independent variable. The relationship between the dependent variable and the independent variable should be linear, and all observations should be independent.

Figure 29-1
Curve Estimation summary table

R	R Square	Adjusted R Square	Std. Error of the Estimate
.885	.783	.766	.083

The independent variable is Hours since detection.

Figure 29-2
Curve Estimation ANOVA

	Sum of Squares	df	Mean Square	F	Sig.
Regression	.949	3	.316	45.736	.000
Residual	.263	38	.007		
Total	1.212	41			

The independent variable is Hours since detection.

Figure 29-3
Curve Estimation coefficients

	Unstandardized Coefficients		Standardized Coefficients		
	B	Std. Error	Beta	t	Sig.
Hours since detection	.088	.011	6.305	7.888	.000
Hours since detection ** 2	-.004	.001	-12.111	-6.363	.000
Hours since detection ** 3	4.399E-05	.000	5.638	4.780	.000
(Constant)	-.123	.056		-2.187	.035

Figure 29-4
Curve Estimation chart

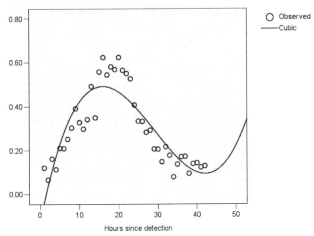

To Obtain a Curve Estimation

▶ From the menus choose:

Analyze
  Regression
    Curve Estimation...

Figure 29-5
Curve Estimation dialog box

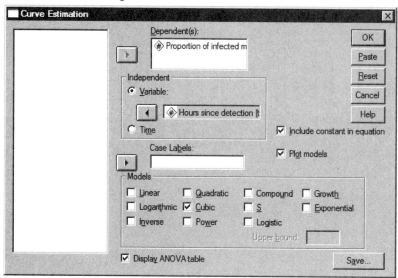

▶ Select one or more dependent variables. A separate model is produced for each dependent variable.

▶ Select an independent variable (either select a variable in the active dataset or select Time).

▶ Optionally:

■ Select a variable for labeling cases in scatterplots. For each point in the scatterplot, you can use the Point Selection tool to display the value of the Case Label variable.

■ Click Save to save predicted values, residuals, and prediction intervals as new variables.

The following options are also available:

■ **Include constant in equation.** Estimates a constant term in the regression equation. The constant is included by default.

■ **Plot models.** Plots the values of the dependent variable and each selected model against the independent variable. A separate chart is produced for each dependent variable.

■ **Display ANOVA table.** Displays a summary analysis-of-variance table for each selected model.

Curve Estimation Models

You can choose one or more curve estimation regression models. To determine which model to use, plot your data. If your variables appear to be related linearly, use a simple linear regression model. When your variables are not linearly related, try transforming your data. When a transformation does not help, you may need a more complicated model. View a scatterplot of your data; if the plot resembles a mathematical function you recognize, fit your data to that type of model. For example, if your data resemble an exponential function, use an exponential model.

Linear. Model whose equation is $Y = b0 + (b1 * t)$. The series values are modeled as a linear function of time.

Logarithmic. Model whose equation is $Y = b0 + (b1 * \ln(t))$.

Inverse. Model whose equation is $Y = b0 + (b1 / t)$.

Quadratic. Model whose equation is $Y = b0 + (b1 * t) + (b2 * t**2)$. The quadratic model can be used to model a series that "takes off" or a series that dampens.

Cubic. Model that is defined by the equation $Y = b0 + (b1 * t) + (b2 * t**2) + (b3 * t**3)$.

Power. Model whose equation is $Y = b0 * (t**b1)$ or $\ln(Y) = \ln(b0) + (b1 * \ln(t))$.

Compound. Model whose equation is $Y = b0 * (b1**t)$ or $\ln(Y) = \ln(b0) + (\ln(b1) * t)$.

S-curve. Model whose equation is $Y = e**(b0 + (b1/t))$ or $\ln(Y) = b0 + (b1/t)$.

Logistic. Model whose equation is $Y = 1 / (1/u + (b0 * (b1**t)))$ or $\ln(1/y-1/u) = \ln(b0) + (\ln(b1) * t)$ where u is the upper boundary value. After selecting Logistic, specify the upper boundary value to use in the regression equation. The value must be a positive number that is greater than the largest dependent variable value.

Growth. Model whose equation is $Y = e**(b0 + (b1 * t))$ or $\ln(Y) = b0 + (b1 * t)$.

Exponential. Model whose equation is $Y = b0 * (e**(b1 * t))$ or $\ln(Y) = \ln(b0) + (b1 * t)$.

Curve Estimation Save

Figure 29-6
Curve Estimation Save dialog box

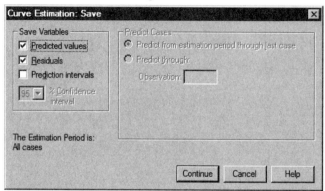

Save Variables. For each selected model, you can save predicted values, residuals (observed value of the dependent variable minus the model predicted value), and prediction intervals (upper and lower bounds). The new variable names and descriptive labels are displayed in a table in the output window.

Predict Cases. In the active dataset, if you select Time instead of a variable as the independent variable, you can specify a forecast period beyond the end of the time series. You can choose one of the following alternatives:

- **Predict from estimation period through last case.** Predicts values for all cases in the file, based on the cases in the estimation period. The estimation period, displayed at the bottom of the dialog box, is defined with the Range subdialog box of the Select Cases option on the Data menu. If no estimation period has been defined, all cases are used to predict values.

- **Predict through.** Predicts values through the specified date, time, or observation number, based on the cases in the estimation period. This feature can be used to forecast values beyond the last case in the time series. The currently defined date variables determine what text boxes are available for specifying the end of the prediction period. If there are no defined date variables, you can specify the ending observation (case) number.

Use the Define Dates option on the Data menu to create date variables.

Discriminant Analysis

Discriminant analysis is useful for building a predictive model of group membership based on observed characteristics of each case. The procedure generates a discriminant function (or, for more than two groups, a set of discriminant functions) based on linear combinations of the predictor variables that provide the best discrimination between the groups. The functions are generated from a sample of cases for which group membership is known; the functions can then be applied to new cases that have measurements for the predictor variables but have unknown group membership.

Note: The grouping variable can have more than two values. The codes for the grouping variable must be integers, however, and you need to specify their minimum and maximum values. Cases with values outside of these bounds are excluded from the analysis.

Example. On average, people in temperate zone countries consume more calories per day than people in the tropics, and a greater proportion of the people in the temperate zones are city dwellers. A researcher wants to combine this information into a function to determine how well an individual can discriminate between the two groups of countries. The researcher thinks that population size and economic information may also be important. Discriminant analysis allows you to estimate coefficients of the linear discriminant function, which looks like the right side of a multiple linear regression equation. That is, using coefficients a, b, c, and d, the function is:

D = a * climate + b * urban + c * population + d * gross domestic product per capita

If these variables are useful for discriminating between the two climate zones, the values of D will differ for the temperate and tropic countries. If you use a stepwise variable selection method, you may find that you do not need to include all four variables in the function.

Statistics. For each variable: means, standard deviations, univariate ANOVA. For each analysis: Box's *M*, within-groups correlation matrix, within-groups covariance matrix, separate-groups covariance matrix, total covariance matrix. For each canonical discriminant function: eigenvalue, percentage of variance, canonical correlation, Wilks' lambda, chi-square. For each step: prior probabilities, Fisher's function coefficients, unstandardized function coefficients, Wilks' lambda for each canonical function.

Data. The grouping variable must have a limited number of distinct categories, coded as integers. Independent variables that are nominal must be recoded to dummy or contrast variables.

Assumptions. Cases should be independent. Predictor variables should have a multivariate normal distribution, and within-group variance-covariance matrices should be equal across groups. Group membership is assumed to be mutually exclusive (that is, no case belongs to more than one group) and collectively exhaustive (that is, all cases are members of a group). The procedure is most effective when group membership is a truly categorical variable; if group membership is based on values of a continuous variable (for example, high IQ versus low IQ), consider using linear regression to take advantage of the richer information that is offered by the continuous variable itself.

Figure 30-1
Discriminant analysis output

Eigenvalues

Function	Eigenvalue	% of Variance	Cumulative %	Canonical Correlation
1	1.002	100.0	100.0	.707

Wilks' Lambda

Test of Function(s)	Wilks' Lambda	Chi-square	df	Sig.
1	.499	31.934	4	.000

Structure Matrix

	Function
	1
CALORIES	.986
LOG_GDP	.790
URBAN	.488
LOG_POP	.082

Functions at Group Centroids

	Function
CLIMATE	1
tropical	-.869
temperate	1.107

To Obtain a Discriminant Analysis

▶ From the menus choose:

Analyze
  Classify
    Discriminant...

Figure 30-2
Discriminant Analysis dialog box

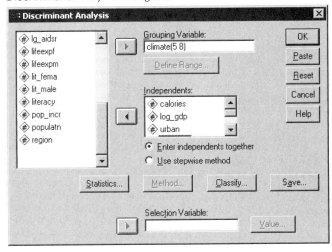

▶ Select an integer-valued grouping variable and click Define Range to specify the categories of interest.

▶ Select the independent, or predictor, variables. (If your grouping variable does not have integer values, Automatic Recode on the Transform menu will create a variable that does.)

▶ Select the method for entering the independent variables.

■ **Enter independents together.** Simultaneously enters all independent variables that satisfy tolerance criteria.

■ **Use stepwise method.** Uses stepwise analysis to control variable entry and removal.

▶ Optionally, select cases with a selection variable.

Discriminant Analysis Define Range

Figure 30-3
Discriminant Analysis Define Range dialog box

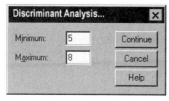

Specify the minimum and maximum value of the grouping variable for the analysis. Cases with values outside of this range are not used in the discriminant analysis but are classified into one of the existing groups based on the results of the analysis. The minimum and maximum values must be integers.

Discriminant Analysis Select Cases

Figure 30-4
Discriminant Analysis Set Value dialog box

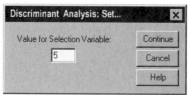

To select cases for your analysis:

▶ In the Discriminant Analysis dialog box, choose a selection variable.

▶ Click Value to enter an integer as the selection value.

Only cases with the specified value for the selection variable are used to derive the discriminant functions. Statistics and classification results are generated for both selected and unselected cases. This process provides a mechanism for classifying new cases based on previously existing data or for partitioning your data into training and testing subsets to perform validation on the model generated.

Discriminant Analysis Statistics

Figure 30-5
Discriminant Analysis Statistics dialog box

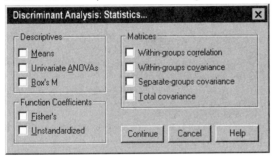

Descriptives. Available options are means (including standard deviations), univariate ANOVAs, and Box's *M* test.

- **Means.** Displays total and group means, as well as standard deviations for the independent variables.

- **Univariate ANOVAs.** Performs a one-way analysis-of-variance test for equality of group means for each independent variable.

- **Box's M.** A test for the equality of the group covariance matrices. For sufficiently large samples, a nonsignificant p value means there is insufficient evidence that the matrices differ. The test is sensitive to departures from multivariate normality.

Function Coefficients. Available options are Fisher's classification coefficients and unstandardized coefficients.

■ **Fisher's.** Displays Fisher's classification function coefficients that can be used directly for classification. A set of coefficients is obtained for each group, and a case is assigned to the group for which it has the largest discriminant score.

■ **Unstandardized.** Displays the unstandardized discriminant function coefficients.

Matrices. Available matrices of coefficients for independent variables are within-groups correlation matrix, within-groups covariance matrix, separate-groups covariance matrix, and total covariance matrix.

■ **Within-groups correlation.** Displays a pooled within-groups correlation matrix that is obtained by averaging the separate covariance matrices for all groups before computing the correlations.

■ **Within-groups covariance.** Displays a pooled within-groups covariance matrix, which may differ from the total covariance matrix. The matrix is obtained by averaging the separate covariance matrices for all groups.

■ **Separate-groups covariance.** Displays separate covariance matrices for each group.

■ **Total covariance.** Displays a covariance matrix from all cases as if they were from a single sample.

Discriminant Analysis Stepwise Method

Figure 30-6
Discriminant Analysis Stepwise Method dialog box

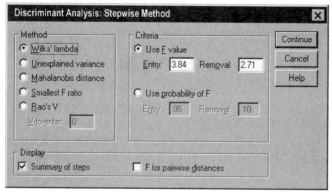

Method. Select the statistic to be used for entering or removing new variables. Available alternatives are Wilks' lambda, unexplained variance, Mahalanobis distance, smallest F ratio, and Rao's V. With Rao's V, you can specify the minimum increase in V for a variable to enter.

■ **Wilks' lambda.** A variable selection method for stepwise discriminant analysis that chooses variables for entry into the equation on the basis of how much they lower Wilks' lambda. At each step, the variable that minimizes the overall Wilks' lambda is entered.

■ **Unexplained variance.** At each step, the variable that minimizes the sum of the unexplained variation between groups is entered.

■ **Mahalanobis distance.** A measure of how much a case's values on the independent variables differ from the average of all cases. A large Mahalanobis distance identifies a case as having extreme values on one or more of the independent variables.

■ **Smallest F ratio.** A method of variable selection in stepwise analysis based on maximizing an F ratio computed from the Mahalanobis distance between groups.

■ **Rao's V.** A measure of the differences between group means. Also called the Lawley-Hotelling trace. At each step, the variable that maximizes the increase in Rao's V is entered. After selecting this option, enter the minimum value a variable must have to enter the analysis.

Criteria. Available alternatives are Use F value and Use probability of F. Enter values for entering and removing variables.

■ **Use F value.** A variable is entered into the model if its F value is greater than the Entry value and is removed if the F value is less than the Removal value. Entry must be greater than Removal, and both values must be positive. To enter more variables into the model, lower the Entry value. To remove more variables from the model, increase the Removal value.

■ **Use probability of F.** A variable is entered into the model if the significance level of its F value is less than the Entry value and is removed if the significance level is greater than the Removal value. Entry must be less than Removal, and both values must be positive. To enter more variables into the model, increase the Entry value. To remove more variables from the model, lower the Removal value.

Display. Summary of steps displays statistics for all variables after each step; F for pairwise distances displays a matrix of pairwise F ratios for each pair of groups.

Discriminant Analysis Classification

Figure 30-7
Discriminant Analysis Classification dialog box

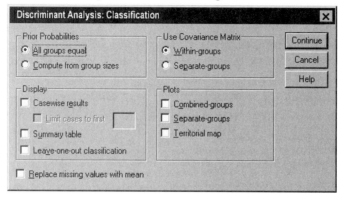

Prior Probabilities. These values are used in classification. You can specify equal prior probabilities for all groups (All groups equal), or you can let the observed group sizes in your sample determine the probabilities of group membership (Compute from group sizes).

Display. Available display options are casewise results, summary table, and leave-one-out classification.

■ **Casewise results.** Codes for actual group, predicted group, posterior probabilities, and discriminant scores are displayed for each case.

■ **Summary table.** The number of cases correctly and incorrectly assigned to each of the groups based on the discriminant analysis. Sometimes called the "Confusion Matrix."

■ **Leave-one-out classification.** Each case in the analysis is classified by the functions derived from all cases other than that case. It is also known as the "U-method."

Replace missing values with mean. Select this option to substitute the mean of an independent variable for a missing value during the classification phase only.

Use Covariance Matrix. You can choose to classify cases using a within-groups covariance matrix or a separate-groups covariance matrix.

- **Within-groups.** The pooled within-groups covariance matrix is used to classify cases.

- **Separate-groups.** Separate-groups covariance matrices are used for classification. Because classification is based on the discriminant functions (not based on the original variables), this option is not always equivalent to quadratic discrimination.

Plots. Available plot options are combined-groups, separate-groups, and territorial map.

- **Combined-groups.** Creates an all-groups scatterplot of the first two discriminant function values. If there is only one function, a histogram is displayed instead.

- **Separate-groups.** Creates separate-group scatterplots of the first two discriminant function values. If there is only one function, histograms are displayed instead.

- **Territorial map.** A plot of the boundaries used to classify cases into groups based on function values. The numbers correspond to groups into which cases are classified. The mean for each group is indicated by an asterisk within its boundaries. The map is not displayed if there is only one discriminant function.

Discriminant Analysis Save

Figure 30-8
Discriminant Analysis Save dialog box

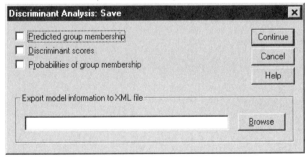

You can add new variables to your active data file. Available options are predicted group membership (a single variable), discriminant scores (one variable for each discriminant function in the solution), and probabilities of group membership given the discriminant scores (one variable for each group).

You can also export model information to the specified file in XML (PMML) format. SmartScore and SPSS Server (a separate product) can use this model file to apply the model information to other data files for scoring purposes.

DISCRIMINANT Command Additional Features

The SPSS command language also allows you to:

- Perform multiple discriminant analyses (with one command) and control the order in which variables are entered (with the ANALYSIS subcommand).

- Specify prior probabilities for classification (with the PRIORS subcommand).

- Display rotated pattern and structure matrices (with the ROTATE subcommand).

- Limit the number of extracted discriminant functions (with the FUNCTIONS subcommand).

- Restrict classification to the cases that are selected (or unselected) for the analysis (with the SELECT subcommand).

- Read and analyze a correlation matrix (with the MATRIX subcommand).

- Write a correlation matrix for later analysis (with the MATRIX subcommand).

See the *SPSS Command Syntax Reference* for complete syntax information.

Factor Analysis

Factor analysis attempts to identify underlying variables, or **factors**, that explain the pattern of correlations within a set of observed variables. Factor analysis is often used in data reduction to identify a small number of factors that explain most of the variance that is observed in a much larger number of manifest variables. Factor analysis can also be used to generate hypotheses regarding causal mechanisms or to screen variables for subsequent analysis (for example, to identify collinearity prior to performing a linear regression analysis).

The factor analysis procedure offers a high degree of flexibility:

■ Seven methods of factor extraction are available.

■ Five methods of rotation are available, including direct oblimin and promax for nonorthogonal rotations.

■ Three methods of computing factor scores are available, and scores can be saved as variables for further analysis.

Example. What underlying attitudes lead people to respond to the questions on a political survey as they do? Examining the correlations among the survey items reveals that there is significant overlap among various subgroups of items—questions about taxes tend to correlate with each other, questions about military issues correlate with each other, and so on. With factor analysis, you can investigate the number of underlying factors and, in many cases, you can identify what the factors represent conceptually. Additionally, you can compute factor scores for each respondent, which can then be used in subsequent analyses. For example, you might build a logistic regression model to predict voting behavior based on factor scores.

Statistics. For each variable: number of valid cases, mean, and standard deviation. For each factor analysis: correlation matrix of variables, including significance levels, determinant, and inverse; reproduced correlation matrix, including anti-image; initial solution (communalities, eigenvalues, and percentage of variance explained); Kaiser-Meyer-Olkin measure of sampling adequacy and Bartlett's test of sphericity;

unrotated solution, including factor loadings, communalities, and eigenvalues; rotated solution, including rotated pattern matrix and transformation matrix. For oblique rotations: rotated pattern and structure matrices; factor score coefficient matrix and factor covariance matrix. Plots: scree plot of eigenvalues and loading plot of first two or three factors.

Data. The variables should be quantitative at the **interval** or **ratio** level. Categorical data (such as religion or country of origin) are not suitable for factor analysis. Data for which Pearson correlation coefficients can sensibly be calculated should be suitable for factor analysis.

Assumptions. The data should have a bivariate normal distribution for each pair of variables, and observations should be independent. The factor analysis model specifies that variables are determined by common factors (the factors estimated by the model) and unique factors (which do not overlap between observed variables); the computed estimates are based on the assumption that all unique factors are uncorrelated with each other and with the common factors.

Figure 31-1
Factor analysis output

Descriptive Statistics

	Mean	Std. Deviation	Analysis N
Average female life expectancy	72.833	8.272	72
Infant mortality (deaths per 1000 live births)	35.132	32.222	72
People who read (%)	82.472	18.625	72
Birth rate per 1000 people	24.375	10.552	72
Fertility: average number of kids	3.205	1.593	72
People living in cities (%)	62.583	22.835	72
Log (base 10) of GDP_CAP	3.504	.608	72
Population increase (% per year)	1.697	1.156	72
Birth to death ratio	3.577	2.313	72
Death rate per 1000 people	8.038	3.174	72
Log (base 10) of Population	4.153	.686	72

Communalities

	Initial	Extraction
LIFEEXPF	1.000	.953
BABYMORT	1.000	.949
LITERACY	1.000	.825
BIRTH_RT	1.000	.943
FERTILTY	1.000	.875
URBAN	1.000	.604
LOG_GDP	1.000	.738
POP_INCR	1.000	.945
B_TO_D	1.000	.925
DEATH_RT	1.000	.689
LOG_POP	1.000	.292

Extraction Method: Principal
Component Analysis.

Total Variance Explained

		Initial Eigenvalues			Extraction Sums of Squared Loadings			Rotation Sums of Squared Loadings		
		Total	% of Variance	Cumulative %	Total	% of Variance	Cumulative %	Total	% of Variance	Cumulative %
Component	1	6.242	56.750	56.750	6.242	56.750	56.750	6.108	55.525	55.525
	2	2.495	22.685	79.435	2.495	22.685	79.435	2.630	23.910	79.435
	3	.988	8.986	88.421						
	4	.591	5.372	93.793						
	5	.236	2.142	95.935						
	6	.172	1.561	97.496						
	7	.124	1.126	98.622						
	8	7.0E-02	.633	99.254						
	9	4.5E-02	.405	99.660						
	10	2.4E-02	.222	99.882						
	11	1.3E-02	.118	100.000						

Extraction Method: Principal Component Analysis.

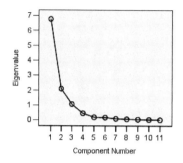

Rotated Component Matrix

	Component	
	1	2
BIRTH_RT	.969	
FERTILTY	.931	
LITERACY	-.880	.226
LIFEEXPF	-.856	.469
BABYMORT	.853	-.469
POP_INCR	.847	.476
LOG_GDP	-.794	.327
URBAN	-.561	.539
DEATH_RT		-.827
B_TO_D	.614	.741
LOG_POP		-.520

Extraction Method: Principal
Component Analysis.
Rotation Method: Varimax
with Kaiser Normalization.

Component Transformation Matrix

		1	2
Component	1	.982	-.190
	2	.190	.982

Extraction Method: Principal Component
Analysis.
Rotation Method: Varimax with Kaiser
Normalization.

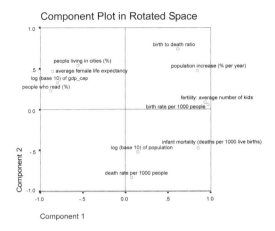

Component Plot in Rotated Space

To Obtain a Factor Analysis

▶ From the menus choose:

Analyze
  Data Reduction
    Factor...

▶ Select the variables for the factor analysis.

Figure 31-2
Factor Analysis dialog box

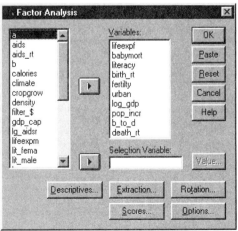

Factor Analysis Select Cases

Figure 31-3
Factor Analysis Set Value dialog box

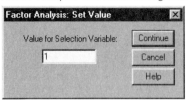

To select cases for your analysis:

▶ Choose a selection variable.

▶ Click Value to enter an integer as the selection value.

Only cases with that value for the selection variable are used in the factor analysis.

Factor Analysis Descriptives

Figure 31-4
Factor Analysis Descriptives dialog box

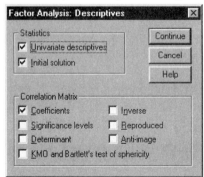

Statistics. Univariate descriptives includes the mean, standard deviation, and number of valid cases for each variable. Initial solution displays initial communalities, eigenvalues, and the percentage of variance explained.

Correlation Matrix. The available options are coefficients, significance levels, determinant, KMO and Bartlett's test of sphericity, inverse, reproduced, and anti-image.

- **KMO and Bartlett's Test of Sphericity.** The Kaiser-Meyer-Olkin measure of sampling adequacy tests whether the partial correlations among variables are small. Bartlett's test of sphericity tests whether the correlation matrix is an identity matrix, which would indicate that the factor model is inappropriate.

- **Reproduced.** The estimated correlation matrix from the factor solution. Residuals (difference between estimated and observed correlations) are also displayed.

- **Anti-image.** The anti-image correlation matrix contains the negatives of the partial correlation coefficients, and the anti-image covariance matrix contains the negatives of the partial covariances. In a good factor model, most of the off-diagonal elements will be small. The measure of sampling adequacy for a variable is displayed on the diagonal of the anti-image correlation matrix.

Factor Analysis Extraction

Figure 31-5
Factor Analysis Extraction dialog box

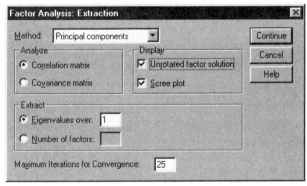

Method. Allows you to specify the method of factor extraction. Available methods are principal components, unweighted least squares, generalized least squares, maximum likelihood, principal axis factoring, alpha factoring, and image factoring.

- **Principal Components Analysis.** A factor extraction method used to form uncorrelated linear combinations of the observed variables. The first component has maximum variance. Successive components explain progressively smaller portions of the variance and are all uncorrelated with each other. Principal components analysis is used to obtain the initial factor solution. It can be used when a correlation matrix is singular.

- **Unweighted Least-Squares Method.** A factor extraction method that minimizes the sum of the squared differences between the observed and reproduced correlation matrices (ignoring the diagonals).

- **Generalized Least-Squares Method.** A factor extraction method that minimizes the sum of the squared differences between the observed and reproduced correlation matrices. Correlations are weighted by the inverse of their uniqueness, so that variables with high uniqueness are given less weight than those with low uniqueness.

- **Maximum-Likelihood Method.** A factor extraction method that produces parameter estimates that are most likely to have produced the observed correlation matrix if the sample is from a multivariate normal distribution. The correlations are weighted by the inverse of the uniqueness of the variables, and an iterative algorithm is employed.

- **Principal Axis Factoring.** A method of extracting factors from the original correlation matrix, with squared multiple correlation coefficients placed in the diagonal as initial estimates of the communalities. These factor loadings are used to estimate new communalities that replace the old communality estimates in the diagonal. Iterations continue until the changes in the communalities from one iteration to the next satisfy the convergence criterion for extraction.

- **Alpha.** A factor extraction method that considers the variables in the analysis to be a sample from the universe of potential variables. This method maximizes the alpha reliability of the factors.

- **Image Factoring.** A factor extraction method developed by Guttman and based on image theory. The common part of the variable, called the partial image, is defined as its linear regression on remaining variables, rather than a function of hypothetical factors.

Analyze. Allows you to specify either a correlation matrix or a covariance matrix.

- **Correlation matrix.** Useful if variables in your analysis are measured on different scales.

- **Covariance matrix.** Useful when you want to apply your factor analysis to multiple groups with different variances for each variable.

Extract. You can either retain all factors whose eigenvalues exceed a specified value, or you can retain a specific number of factors.

Display. Allows you to request the unrotated factor solution and a scree plot of the eigenvalues.

- **Unrotated Factor Solution.** Displays unrotated factor loadings (factor pattern matrix), communalities, and eigenvalues for the factor solution.

- **Scree plot.** A plot of the variance that is associated with each factor. This plot is used to determine how many factors should be kept. Typically the plot shows a distinct break between the steep slope of the large factors and the gradual trailing of the rest (the scree).

Maximum Iterations for Convergence. Allows you to specify the maximum number of steps that the algorithm can take to estimate the solution.

Factor Analysis Rotation

Figure 31-6
Factor Analysis Rotation dialog box

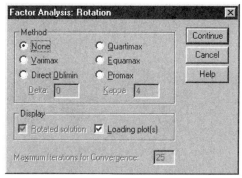

Method. Allows you to select the method of factor rotation. Available methods are varimax, direct oblimin, quartimax, equamax, or promax.

- **Varimax Method.** An orthogonal rotation method that minimizes the number of variables that have high loadings on each factor. This method simplifies the interpretation of the factors.

- **Direct Oblimin Method.** A method for oblique (nonorthogonal) rotation. When delta equals 0 (the default), solutions are most oblique. As delta becomes more negative, the factors become less oblique. To override the default delta of 0, enter a number less than or equal to 0.8.

- **Quartimax Method.** A rotation method that minimizes the number of factors needed to explain each variable. This method simplifies the interpretation of the observed variables.

- **Equamax Method.** A rotation method that is a combination of the varimax method, which simplifies the factors, and the quartimax method, which simplifies the variables. The number of variables that load highly on a factor and the number of factors needed to explain a variable are minimized.

- **Promax Rotation.** An oblique rotation, which allows factors to be correlated. This rotation can be calculated more quickly than a direct oblimin rotation, so it is useful for large datasets.

Display. Allows you to include output on the rotated solution, as well as loading plots for the first two or three factors.

- **Rotated Solution.** A rotation method must be selected to obtain a rotated solution. For orthogonal rotations, the rotated pattern matrix and factor transformation matrix are displayed. For oblique rotations, the pattern, structure, and factor correlation matrices are displayed.

- **Factor Loading Plot.** Three-dimensional factor loading plot of the first three factors. For a two-factor solution, a two-dimensional plot is shown. The plot is not displayed if only one factor is extracted. Plots display rotated solutions if rotation is requested.

Maximum Iterations for Convergence. Allows you to specify the maximum number of steps that the algorithm can take to perform the rotation.

Factor Analysis Scores

Figure 31-7
Factor Analysis Factor Scores dialog box

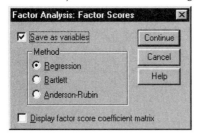

Save as variables. Creates one new variable for each factor in the final solution. Select one of the following alternative methods for calculating the factor scores: regression, Bartlett, or Anderson-Rubin.

- **Regression Method.** A method for estimating factor score coefficients. The scores that are produced have a mean of 0 and a variance equal to the squared multiple correlation between the estimated factor scores and the true factor values. The scores may be correlated even when factors are orthogonal.

- **Bartlett Scores.** A method of estimating factor score coefficients. The scores that are produced have a mean of 0. The sum of squares of the unique factors over the range of variables is minimized.

- **Anderson-Rubin Method.** A method of estimating factor score coefficients; a modification of the Bartlett method which ensures orthogonality of the estimated factors. The scores that are produced have a mean of 0, have a standard deviation of 1, and are uncorrelated.

Display factor score coefficient matrix. Shows the coefficients by which variables are multiplied to obtain factor scores. Also shows the correlations between factor scores.

Factor Analysis Options

Figure 31-8
Factor Analysis Options dialog box

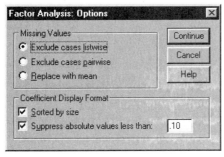

Missing Values. Allows you to specify how missing values are handled. The available choices are to exclude cases **listwise**, exclude cases **pairwise**, or replace with mean.

Coefficient Display Format. Allows you to control aspects of the output matrices. You sort coefficients by size and suppress coefficients with absolute values that are less than the specified value.

FACTOR Command Additional Features

The SPSS command language also allows you to:

- Specify convergence criteria for iteration during extraction and rotation.
- Specify individual rotated-factor plots.
- Specify how many factor scores to save.
- Specify diagonal values for the principal axis factoring method.
- Write correlation matrices or factor-loading matrices to disk for later analysis.
- Read and analyze correlation matrices or factor-loading matrices.

See the *SPSS Command Syntax Reference* for complete syntax information.

Choosing a Procedure for Clustering

Cluster analyses can be performed using the TwoStep, Hierarchical, or K-Means Cluster Analysis procedure. Each procedure employs a different algorithm for creating clusters, and each has options not available in the others.

TwoStep Cluster Analysis. For many applications, the TwoStep Cluster Analysis procedure will be the method of choice. It provides the following unique features:

- Automatic selection of the best number of clusters, in addition to measures for choosing between cluster models.

- Ability to create cluster models simultaneously based on categorical and continuous variables.

- Ability to save the cluster model to an external XML file and then read that file and update the cluster model using newer data.

Additionally, the TwoStep Cluster Analysis procedure can analyze large data files.

Hierarchical Cluster Analysis. The Hierarchical Cluster Analysis procedure is limited to smaller data files (hundreds of objects to be clustered) but has the following unique features:

- Ability to cluster cases or variables.

- Ability to compute a range of possible solutions and save cluster memberships for each of those solutions.

- Several methods for cluster formation, variable transformation, and measuring the dissimilarity between clusters.

As long as all the variables are of the same type, the Hierarchical Cluster Analysis procedure can analyze interval (continuous), count, or binary variables.

K-Means Cluster Analysis. The K-Means Cluster Analysis procedure is limited to continuous data and requires you to specify the number of clusters in advance, but it has the following unique features:

- Ability to save distances from cluster centers for each object.

- Ability to read initial cluster centers from and save final cluster centers to an external SPSS file.

Additionally, the K-Means Cluster Analysis procedure can analyze large data files.

TwoStep Cluster Analysis

The TwoStep Cluster Analysis procedure is an exploratory tool designed to reveal natural groupings (or clusters) within a dataset that would otherwise not be apparent. The algorithm employed by this procedure has several desirable features that differentiate it from traditional clustering techniques:

- **Handling of categorical and continuous variables.** By assuming variables to be independent, a joint multinomial-normal distribution can be placed on categorical and continuous variables.

- **Automatic selection of number of clusters.** By comparing the values of a model-choice criterion across different clustering solutions, the procedure can automatically determine the optimal number of clusters.

- **Scalability.** By constructing a cluster features (CF) tree that summarizes the records, the TwoStep algorithm allows you to analyze large data files.

Example. Retail and consumer product companies regularly apply clustering techniques to data that describe their customers' buying habits, gender, age, income level, etc. These companies tailor their marketing and product development strategies to each consumer group to increase sales and build brand loyalty.

Statistics. The procedure produces information criteria (AIC or BIC) by number of clusters in the solution, cluster frequencies for the final clustering, and descriptive statistics by cluster for the final clustering.

Plots. The procedure produces bar charts of cluster frequencies, pie charts of cluster frequencies, and variable importance charts.

Figure 33-1
TwoStep Cluster Analysis dialog box

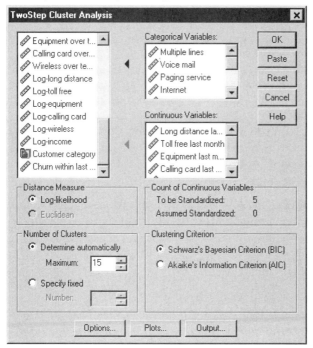

Distance Measure. This selection determines how the similarity between two clusters is computed.

■ **Log-likelihood.** The likelihood measure places a probability distribution on the variables. Continuous variables are assumed to be normally distributed, while categorical variables are assumed to be multinomial. All variables are assumed to be independent.

■ **Euclidean.** The Euclidean measure is the "straight line" distance between two clusters. It can be used only when all of the variables are continuous.

Number of Clusters. This selection allows you to specify how the number of clusters is to be determined.

■ **Determine automatically.** The procedure will automatically determine the "best" number of clusters, using the criterion specified in the Clustering Criterion group. Optionally, enter a positive integer specifying the maximum number of clusters that the procedure should consider.

■ **Specify fixed.** Allows you to fix the number of clusters in the solution. Enter a positive integer.

Count of Continuous Variables. This group provides a summary of the continuous variable standardization specifications made in the Options dialog box. For more information, see "TwoStep Cluster Analysis Options" on p. 517.

Clustering Criterion. This selection determines how the automatic clustering algorithm determines the number of clusters. Either the Bayesian Information Criterion (BIC) or the Akaike Information Criterion (AIC) can be specified.

Data. This procedure works with both continuous and categorical variables. Cases represent objects to be clustered, and the variables represent attributes upon which the clustering is based.

Case Order. Note that the cluster features tree and the final solution may depend on the order of cases. To minimize order effects, randomly order the cases. You may want to obtain several different solutions with cases sorted in different random orders to verify the stability of a given solution. In situations where this is difficult due to extremely large file sizes, multiple runs with a sample of cases sorted in different random orders might be substituted.

Assumptions. The likelihood distance measure assumes that variables in the cluster model are independent. Further, each continuous variable is assumed to have a normal (Gaussian) distribution, and each categorical variable is assumed to have a multinomial distribution. Empirical internal testing indicates that the procedure is fairly robust to violations of both the assumption of independence and the distributional assumptions, but you should try to be aware of how well these assumptions are met.

Use the Bivariate Correlations procedure to test the independence of two continuous variables. Use the Crosstabs procedure to test the independence of two categorical variables. Use the Means procedure to test the independence between a continuous variable and categorical variable. Use the Explore procedure to test the normality of a continuous variable. Use the Chi-Square Test procedure to test whether a categorical variable has a specified multinomial distribution.

To Obtain a TwoStep Cluster Analysis

▶ From the menus choose:

Analyze
　Classify
　　TwoStep Cluster...

▶ Select one or more categorical or continuous variables.

Optionally, you can:

- Adjust the criteria by which clusters are constructed.
- Select settings for noise handling, memory allocation, variable standardization, and cluster model input.
- Request optional tables and plots.
- Save model results to the working file or to an external XML file.

TwoStep Cluster Analysis Options

Figure 33-2
TwoStep Cluster Analysis Options dialog box

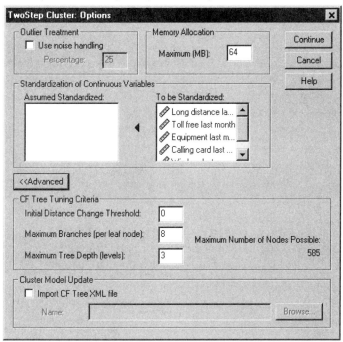

Outlier Treatment. This group allows you to treat outliers specially during clustering if the cluster features (CF) tree fills. The CF tree is full if it cannot accept any more cases in a leaf node, and no leaf node can be split.

- If you select noise handling and the CF tree fills, it will be regrown after placing cases in sparse leaves into a "noise" leaf. A leaf is considered sparse if it contains fewer than the specified percentage of cases of the maximum leaf size. After the tree is regrown, the outliers will be placed in the CF tree if possible. If not, the outliers are discarded.

- If you do not select noise handling and the CF tree fills, it will be regrown using a larger distance change threshold. After final clustering, values that cannot be assigned to a cluster are labeled outliers. The outlier cluster is given an identification number of –1 and is not included in the count of the number of clusters.

Memory Allocation. This group allows you to specify the maximum amount of memory in megabytes (MB) that the cluster algorithm should use. If the procedure exceeds this maximum, it will use the disk to store information that will not fit in memory. Specify a number greater than or equal to 4.

■ Consult your system administrator for the largest value that you can specify on your system.

■ The algorithm may fail to find the correct or desired number of clusters if this value is too low.

Variable standardization. The clustering algorithm works with standardized continuous variables. Any continuous variables that are not standardized should be left as variables in the To be Standardized list. To save some time and computational effort, you can select any continuous variables that you have already standardized as variables in the Assumed Standardized list.

Advanced Options

CF Tree Tuning Criteria. The following clustering algorithm settings apply specifically to the cluster features (CF) tree and should be changed with care:

■ **Initial Distance Change Threshold.** This is the initial threshold used to grow the CF tree. If inserting a given case into a leaf of the CF tree would yield tightness less than the threshold, the leaf is not split. If the tightness exceeds the threshold, the leaf is split.

■ **Maximum Branches (per leaf node).** The maximum number of child nodes that a leaf node can have.

■ **Maximum Tree Depth.** The maximum number of levels that the CF tree can have.

■ **Maximum Number of Nodes Possible.** This indicates the maximum number of CF tree nodes that could potentially be generated by the procedure, based on the function $(b^{d+1} - 1) / (b - 1)$, where b is the maximum branches and d is the maximum tree depth. Be aware that an overly large CF tree can be a drain on system resources and can adversely affect the performance of the procedure. At a minimum, each node requires 16 bytes.

Cluster Model Update. This group allows you to import and update a cluster model generated in a prior analysis. The input file contains the CF tree in XML format. The model will then be updated with the data in the active file. You must select the variable names in the main dialog box in the same order in which they were specified in the prior analysis. The XML file remains unaltered, unless you specifically write the new model information to the same filename. For more information, see "TwoStep Cluster Analysis Output" on p. 521.

If a cluster model update is specified, the options pertaining to generation of the CF tree that were specified for the original model are used. More specifically, the distance measure, noise handling, memory allocation, or CF tree tuning criteria settings for the saved model are used, and any settings for these options in the dialog boxes are ignored.

Note: When performing a cluster model update, the procedure assumes that none of the selected cases in the active dataset were used to create the original cluster model. The procedure also assumes that the cases used in the model update come from the same population as the cases used to create the original model; that is, the means and variances of continuous variables and levels of categorical variables are assumed to be the same across both sets of cases. If your "new" and "old" sets of cases come from heterogeneous populations, you should run the TwoStep Cluster Analysis procedure on the combined sets of cases for the best results.

TwoStep Cluster Analysis Plots

Figure 33-3
TwoStep Cluster Analysis Plots dialog box

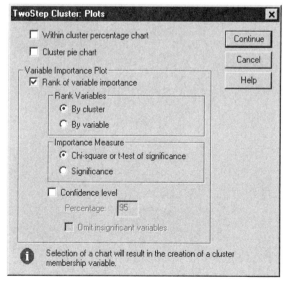

Within cluster percentage chart. Displays charts showing the within-cluster variation of each variable. For each categorical variable, a clustered bar chart is produced, showing the category frequency by cluster ID. For each continuous variable, an error bar chart is produced, showing error bars by cluster ID.

Cluster pie chart. Displays a pie chart showing the percentage and counts of observations within each cluster.

Variable Importance Plot. Displays several different charts showing the importance of each variable within each cluster. The output is sorted by the importance rank of each variable.

- **Rank Variables.** This option determines whether plots will be created for each cluster (By variable) or for each variable (By cluster).

- **Importance Measure.** This option allows you to select which measure of variable importance to plot. Chi-square or t-test of significance reports a Pearson chi-square statistic as the importance of a categorical variable and a *t* statistic as the importance of a continuous variable. Significance reports one minus the *p* value for the test of equality of means for a continuous variable and the expected frequency with the overall dataset for a categorical variable.

- **Confidence level.** This option allows you to set the confidence level for the test of equality of a variable's distribution within a cluster versus the variable's overall distribution. Specify a number less than 100 and greater than or equal to 50. The value of the confidence level is shown as a vertical line in the variable importance plots, if the plots are created by variable or if the significance measure is plotted.

- **Omit insignificant variables.** Variables that are not significant at the specified confidence level are not displayed in the variable importance plots.

TwoStep Cluster Analysis Output

Figure 33-4
TwoStep Cluster Analysis Output dialog box

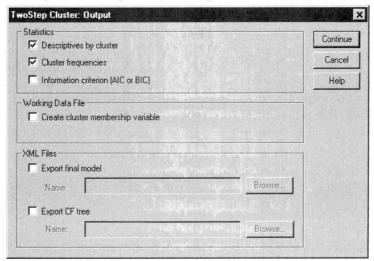

Statistics. This group provides options for displaying tables of the clustering results. The descriptive statistics and cluster frequencies are produced for the final cluster model, while the information criterion table displays results for a range of cluster solutions.

■ **Descriptives by cluster.** Displays two tables that describe the variables in each cluster. In one table, means and standard deviations are reported for continuous variables by cluster. The other table reports frequencies of categorical variables by cluster.

■ **Cluster frequencies.** Displays a table that reports the number of observations in each cluster.

■ **Information criterion (AIC or BIC).** Displays a table containing the values of the AIC or BIC, depending on the criterion chosen in the main dialog box, for different numbers of clusters. This table is provided only when the number of clusters is being determined automatically. If the number of clusters is fixed, this setting is ignored, and the table is not provided.

Active dataset. This group allows you to save variables to the active dataset.

■ **Create cluster membership variable.** This variable contains a cluster identification number for each case. The name of this variable is *tsc_n*, where *n* is a positive integer indicating the ordinal of the active dataset save operation completed by this procedure in a given session.

XML Files. The final cluster model and CF tree are two types of output files that can be exported in XML format.

■ **Export final model.** The final cluster model is exported to the specified file in XML (PMML) format. SmartScore and SPSS Server (a separate product) can use this model file to apply the model information to other data files for scoring purposes.

■ **Export CF tree.** This option allows you to save the current state of the cluster tree and update it later using newer data.

Hierarchical Cluster Analysis

This procedure attempts to identify relatively homogeneous groups of cases (or variables) based on selected characteristics, using an algorithm that starts with each case (or variable) in a separate cluster and combines clusters until only one is left. You can analyze raw variables or you can choose from a variety of standardizing transformations. Distance or similarity measures are generated by the Proximities procedure. Statistics are displayed at each stage to help you select the best solution.

Example. Are there identifiable groups of television shows that attract similar audiences within each group? With hierarchical cluster analysis, you could cluster television shows (cases) into homogeneous groups based on viewer characteristics. This can be used to identify segments for marketing. Or you can cluster cities (cases) into homogeneous groups so that comparable cities can be selected to test various marketing strategies.

Statistics. Agglomeration schedule, distance (or similarity) matrix, and cluster membership for a single solution or a range of solutions. Plots: dendrograms and icicle plots.

Data. The variables can be quantitative, binary, or count data. Scaling of variables is an important issue—differences in scaling may affect your cluster solution(s). If your variables have large differences in scaling (for example, one variable is measured in dollars and the other is measured in years), you should consider standardizing them (this can be done automatically by the Hierarchical Cluster Analysis procedure).

Case order. If tied distances or similarities exist in the input data or occur among updated clusters during joining, the resulting cluster solution may depend on the order of cases in the file. You may want to obtain several different solutions with cases sorted in different random orders to verify the stability of a given solution.

Assumptions. The distance or similarity measures used should be appropriate for the data analyzed (see the Proximities procedure for more information on choices of distance and similarity measures). Also, you should include all relevant variables in your analysis. Omission of influential variables can result in a misleading solution. Because hierarchical cluster analysis is an exploratory method, results should be treated as tentative until they are confirmed with an independent sample.

Figure 34-1
Hierarchical cluster analysis output

Agglomeration Schedule

		Cluster Combined			Stage Cluster First Appears		
		Cluster 1	Cluster 2	Coefficients	Cluster 1	Cluster 2	Next Stage
Stage	1	11	12	.112	0	0	2
	2	6	11	.132	0	1	4
	3	7	9	.185	0	0	5
	4	6	8	.227	2	0	7
	5	7	10	.274	3	0	7
	6	1	3	.423	0	0	10
	7	6	7	.438	4	5	14
	8	13	14	.484	0	0	15
	9	2	5	.547	0	0	11
	10	1	4	.691	6	0	11
	11	1	2	1.023	10	9	13
	12	15	16	1.370	0	0	13
	13	1	15	1.716	11	12	14
	14	1	6	2.642	13	7	15
	15	1	13	4.772	14	8	0

Cluster Membership

		Label	4 Clusters	3 Clusters	2 Clusters
Case	1	Argentina	1	1	1
	2	Brazil	1	1	1
	3	Chile	1	1	1
	4	Domincan R.	1	1	1
	5	Indonesia	1	1	1
	6	Austria	2	2	1
	7	Canada	2	2	1
	8	Denmark	2	2	1
	9	Italy	2	2	1
	10	Japan	2	2	1
	11	Norway	2	2	1
	12	Switzerland	2	2	1
	13	Bangladesh	3	3	2
	14	India	3	3	2
	15	Bolivia	4	1	1
	16	Paraguay	4	1	1

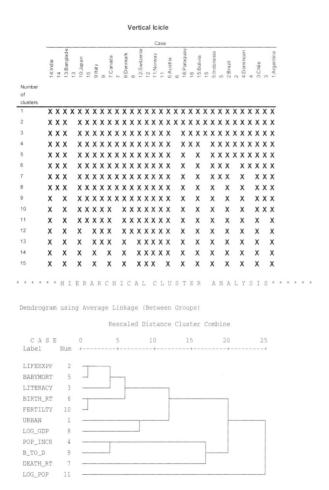

Vertical Icicle

To Obtain a Hierarchical Cluster Analysis

▶ From the menus choose:

Analyze
  Classify
    Hierarchical Cluster...

Figure 34-2
Hierarchical Cluster Analysis dialog box

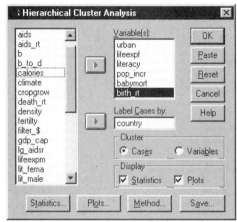

▶ If you are clustering cases, select at least one numeric variable. If you are clustering variables, select at least three numeric variables.

Optionally, you can select an identification variable to label cases.

Hierarchical Cluster Analysis Method

Figure 34-3
Hierarchical Cluster Analysis Method dialog box

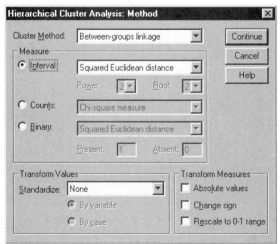

Cluster Method. Available alternatives are between-groups linkage, within-groups linkage, nearest neighbor, furthest neighbor, centroid clustering, median clustering, and Ward's method.

Measure. Allows you to specify the distance or similarity measure to be used in clustering. Select the type of data and the appropriate distance or similarity measure:

- **Interval.** Available alternatives are Euclidean distance, squared Euclidean distance, cosine, Pearson correlation, Chebychev, block, Minkowski, and customized.

- **Counts.** Available alternatives are chi-square measure and phi-square measure.

- **Binary.** Available alternatives are Euclidean distance, squared Euclidean distance, size difference, pattern difference, variance, dispersion, shape, simple matching, phi 4-point correlation, lambda, Anderberg's D, dice, Hamann, Jaccard, Kulczynski 1, Kulczynski 2, Lance and Williams, Ochiai, Rogers and Tanimoto, Russel and Rao, Sokal and Sneath 1, Sokal and Sneath 2, Sokal and Sneath 3, Sokal and Sneath 4, Sokal and Sneath 5, Yule's Y, and Yule's Q.

Transform Values. Allows you to standardize data values for either cases or values before computing proximities (not available for binary data). Available standardization methods are z scores, range -1 to 1, range 0 to 1, maximum magnitude of 1, mean of 1, and standard deviation of 1.

Transform Measures. Allows you to transform the values generated by the distance measure. They are applied after the distance measure has been computed. Available alternatives are absolute values, change sign, and rescale to 0–1 range.

Hierarchical Cluster Analysis Statistics

Figure 34-4
Hierarchical Cluster Analysis Statistics dialog box

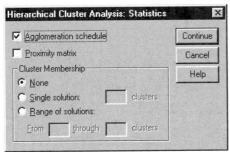

Agglomeration schedule. Displays the cases or clusters combined at each stage, the distances between the cases or clusters being combined, and the last cluster level at which a case (or variable) joined the cluster.

Proximity matrix. Gives the distances or similarities between items.

Cluster Membership. Displays the cluster to which each case is assigned at one or more stages in the combination of clusters. Available options are single solution and range of solutions.

Hierarchical Cluster Analysis Plots

Figure 34-5
Hierarchical Cluster Analysis Plots dialog box

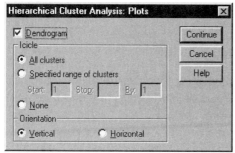

Dendrogram. Displays a **dendrogram**. Dendrograms can be used to assess the cohesiveness of the clusters formed and can provide information about the appropriate number of clusters to keep.

Icicle. Displays an **icicle plot**, including all clusters or a specified range of clusters. Icicle plots display information about how cases are combined into clusters at each iteration of the analysis. Orientation allows you to select a vertical or horizontal plot.

Hierarchical Cluster Analysis Save New Variables

Figure 34-6
Hierarchical Cluster Analysis Save dialog box

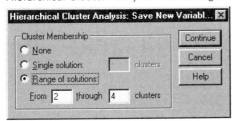

Cluster Membership. Allows you to save cluster memberships for a single solution or a range of solutions. Saved variables can then be used in subsequent analyses to explore other differences between groups.

CLUSTER Command Syntax Additional Features

The Hierarchical Cluster procedure uses CLUSTER command syntax. The SPSS command language also allows you to:

■ Use several clustering methods in a single analysis.

■ Read and analyze a proximity matrix.

■ Write a proximity matrix to disk for later analysis.

■ Specify any values for power and root in the customized (Power) distance measure.

■ Specify names for saved variables.

See the *SPSS Command Syntax Reference* for complete syntax information.

K-Means Cluster Analysis

This procedure attempts to identify relatively homogeneous groups of cases based on selected characteristics, using an algorithm that can handle large numbers of cases. However, the algorithm requires you to specify the number of clusters. You can specify initial cluster centers if you know this information. You can select one of two methods for classifying cases, either updating cluster centers iteratively or classifying only. You can save cluster membership, distance information, and final cluster centers. Optionally, you can specify a variable whose values are used to label casewise output. You can also request analysis of variance F statistics. While these statistics are opportunistic (the procedure tries to form groups that do differ), the relative size of the statistics provides information about each variable's contribution to the separation of the groups.

Example. What are some identifiable groups of television shows that attract similar audiences within each group? With *k*-means cluster analysis, you could cluster television shows (cases) into *k* homogeneous groups based on viewer characteristics. This process can be used to identify segments for marketing. Or you can cluster cities (cases) into homogeneous groups so that comparable cities can be selected to test various marketing strategies.

Statistics. Complete solution: initial cluster centers, ANOVA table. Each case: cluster information, distance from cluster center.

Data. Variables should be quantitative at the interval or ratio level. If your variables are binary or counts, use the Hierarchical Cluster Analysis procedure.

Case and initial cluster center order. The default algorithm for choosing initial cluster centers is not invariant to case ordering. The Use running means option on the Iterate dialog box makes the resulting solution potentially dependent on case order, regardless of how initial cluster centers are chosen. If you are using either of these methods, you may want to obtain several different solutions with cases sorted in different random orders to verify the stability of a given solution. Specifying initial cluster centers and not using the Use running means option will avoid issues related to case order. However, ordering of the initial cluster centers may affect the solution, if there are tied distances from cases to cluster centers. To assess the stability of a given solution, you can compare results from analyses with different permutations of the initial center values.

Assumptions. Distances are computed using simple Euclidean distance. If you want to use another distance or similarity measure, use the Hierarchical Cluster Analysis procedure. Scaling of variables is an important consideration. If your variables are measured on different scales (for example, one variable is expressed in dollars and another variable is expressed in years), your results may be misleading. In such cases, you should consider standardizing your variables before you perform the k-means cluster analysis (this task can be done in the Descriptives procedure). The procedure assumes that you have selected the appropriate number of clusters and that you have included all relevant variables. If you have chosen an inappropriate number of clusters or omitted important variables, your results may be misleading.

Figure 35-1
K-means cluster analysis output

Initial Cluster Centers

	Cluster			
	1	2	3	4
ZURBAN	-1.88606	-1.54314	1.45741	.55724
ZLIFEEXP	-3.52581	-1.69358	.62725	.99370
ZLITERAC	-2.89320	-1.65146	-.51770	.88601
ZPOP_INC	.93737	.16291	3.03701	-1.12785
ZBABYMOR	4.16813	1.38422	-.69589	-.88983
ZBIRTH_R	2.68796	.42699	.33278	-1.08033
ZDEATH_R	4.41517	.63185	-1.89037	.63185
ZLOG_GDP	-1.99641	-1.78455	.53091	1.22118
ZB_TO_D	-.52182	-.31333	4.40082	-.99285
ZFERTILT	2.24070	.75481	.46008	-.76793
ZLOG_POP	.24626	2.65246	-1.29624	-.74406

Iteration History

		Change in Cluster Centers			
		1	2	3	4
Iteration	1	1.932	2.724	3.343	1.596
	2	.000	.471	.466	.314
	3	.861	.414	.172	.195
	4	.604	.337	.000	.150
	5	.000	.253	.237	.167
	6	.000	.199	.287	.071
	7	.623	.160	.000	.000
	8	.000	.084	.000	.074
	9	.000	.080	.000	.077
	10	.000	.097	.185	.000

Final Cluster Centers

	Cluster			
	1	2	3	4
ZURBAN	-1.70745	-.30863	.16816	.62767
ZLIFEEXP	-2.52826	-.15939	-.28417	.80611
ZLITERAC	-2.30833	.13880	-.81671	.73368
ZPOP_INC	.59747	.13400	1.45301	-.95175
ZBABYMOR	2.43210	.22286	.25622	-.80817
ZBIRTH_R	1.52607	.12929	1.13716	-.99285
ZDEATH_R	2.10314	-.44640	-.71414	.31319
ZLOG_GDP	-1.77704	-.58745	-.16871	.94249
ZB_TO_D	-.29856	.19154	1.45251	-.84758
ZFERTILT	1.51003	-.12150	1.27010	-.87669
ZLOG_POP	.83475	.34577	-.49499	-.22199

Distances between Final Cluster Centers

		1	2	3	4
Cluster	1		5.627	5.640	7.924
	2	5.627		2.897	3.249
	3	5.640	2.897		5.246
	4	7.924	3.249	5.246	

ANOVA

	Cluster		Error			
	Mean Square	df	Mean Square	df	F	Sig.
ZURBAN	10.409	3	.541	68	19.234	.000
ZLIFEEXP	19.410	3	.210	68	92.614	.000
ZLITERAC	18.731	3	.229	68	81.655	.000
ZPOP_INC	18.464	3	.219	68	84.428	.000
ZBABYMOR	18.621	3	.239	68	77.859	.000
ZBIRTH_R	19.599	3	.167	68	117.339	.000
ZDEATH_R	13.628	3	.444	68	30.676	.000
ZLOG_GDP	17.599	3	.287	68	61.313	.000
ZB_TO_D	16.316	3	.288	68	56.682	.000
ZFERTILT	18.829	3	.168	68	112.273	.000
ZLOG_POP	3.907	3	.877	68	4.457	.006

The F tests should be used only for descriptive purposes because the clusters have been chosen to maximize the differences among cases in different clusters. The observed significance levels are not corrected for this and thus cannot be interpreted as tests of the hypothesis that the cluster means are equal.

To Obtain a K-Means Cluster Analysis

▶ From the menus choose:

Analyze
  Classify
    K-Means Cluster...

Figure 35-2
K-Means Cluster Analysis dialog box

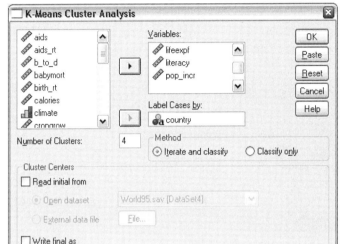

▶ Select the variables to be used in the cluster analysis.

▶ Specify the number of clusters. (The number of clusters must be at least 2 and must not be greater than the number of cases in the data file.)

▶ Select either Iterate and classify or Classify only.

▶ Optionally, select an identification variable to label cases.

K-Means Cluster Analysis Efficiency

The *k*-means cluster analysis command is efficient primarily because it does not compute the distances between all pairs of cases, as do many clustering algorithms, including the algorithm that is used by the hierarchical clustering command.

For maximum efficiency, take a sample of cases and select the Iterate and classify method to determine cluster centers. Select Write final as. Then restore the entire data file and select Classify only as the method and select Read initial from to classify the entire file using the centers that are estimated from the sample. You can write to and read from a file or a dataset. Datasets are available for subsequent use in the same session but are not saved as files unless explicitly saved prior to the end of the session. Dataset names must conform to SPSS variable-naming rules. For more information, see "Variable Names" in Chapter 5 on p. 106.

K-Means Cluster Analysis Iterate

Figure 35-3
K-Means Cluster Analysis Iterate dialog box

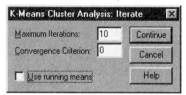

Note: These options are available only if you select the Iterate and classify method from the K-Means Cluster Analysis dialog box.

Maximum Iterations. Limits the number of iterations in the *k*-means algorithm. Iteration stops after this many iterations even if the convergence criterion is not satisfied. This number must be between 1 and 999.

To reproduce the algorithm used by the Quick Cluster command prior to version 5.0, set Maximum Iterations to 1.

Convergence Criterion. Determines when iteration ceases. It represents a proportion of the minimum distance between initial cluster centers, so it must be greater than 0 but not greater than 1. If the criterion equals 0.02, for example, iteration ceases when a complete iteration does not move any of the cluster centers by a distance of more than 2% of the smallest distance between any initial cluster centers.

Use running means. Allows you to request that cluster centers be updated after each case is assigned. If you do not select this option, new cluster centers are calculated after all cases have been assigned.

K-Means Cluster Analysis Save

Figure 35-4
K-Means Cluster Analysis Save New Variables dialog box

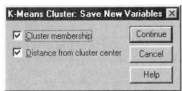

You can save information about the solution as new variables to be used in subsequent analyses:

Cluster membership. Creates a new variable indicating the final cluster membership of each case. Values of the new variable range from 1 to the number of clusters.

Distance from cluster center. Creates a new variable indicating the Euclidean distance between each case and its classification center.

K-Means Cluster Analysis Options

Figure 35-5
K-Means Cluster Analysis Options dialog box

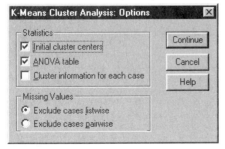

Statistics. You can select the following statistics: initial cluster centers, ANOVA table, and cluster information for each case.

■ **Initial cluster centers.** First estimate of the variable means for each of the clusters. By default, a number of well-spaced cases equal to the number of clusters is selected from the data. Initial cluster centers are used for a first round of classification and are then updated.

- **ANOVA table.** Displays an analysis-of-variance table which includes univariate F tests for each clustering variable. The F tests are only descriptive and the resulting probabilities should not be interpreted. The ANOVA table is not displayed if all cases are assigned to a single cluster.

- **Cluster information for each case.** Displays for each case the final cluster assignment and the Euclidean distance between the case and the cluster center used to classify the case. Also displays Euclidean distance between final cluster centers.

Missing Values. Available options are Exclude cases listwise or Exclude cases pairwise.

- **Exclude cases listwise.** Excludes cases with missing values for any clustering variable from the analysis.

- **Exclude cases pairwise.** Assigns cases to clusters based on distances that are computed from all variables with nonmissing values.

QUICK CLUSTER Command Additional Features

The K-Means Cluster procedure uses QUICK CLUSTER command syntax. The SPSS command language also allows you to:

- Accept the first k cases as initial cluster centers, thereby avoiding the data pass that is normally used to estimate them.

- Specify initial cluster centers directly as a part of the command syntax.

- Specify names for saved variables.

See the *SPSS Command Syntax Reference* for complete syntax information.

Nonparametric Tests

The Nonparametric Tests procedure provides several tests that do not require assumptions about the shape of the underlying distribution.

Chi-Square Test. Tabulates a variable into categories and computes a chi-square statistic based on the differences between observed and expected frequencies.

Binomial Test. Compares the observed frequency in each category of a dichotomous variable with expected frequencies from the binomial distribution.

Runs Test. Tests whether the order of occurrence of two values of a variable is random.

One-Sample Kolmogorov-Smirnov Test. Compares the observed cumulative distribution function for a variable with a specified theoretical distribution, which may be normal, uniform, exponential, or Poisson.

Two-Independent-Samples Tests. Compares two groups of cases on one variable. The Mann-Whitney U test, two-sample Kolmogorov-Smirnov test, Moses test of extreme reactions, and Wald-Wolfowitz runs test are available.

Two-Related-Samples Tests. Compares the distributions of two variables. The Wilcoxon signed-rank test, the sign test, and the McNemar test are available.

Tests for Several Independent Samples. Compares two or more groups of cases on one variable. The Kruskal-Wallis test, the Median test, and the Jonckheere-Terpstra test are available.

Tests for Several Related Samples. Compares the distributions of two or more variables. Friedman's test, Kendall's W, and Cochran's Q are available.

Quartiles and the mean, standard deviation, minimum, maximum, and number of nonmissing cases are available for all of the above tests.

Chi-Square Test

The Chi-Square Test procedure tabulates a variable into categories and computes a chi-square statistic. This goodness-of-fit test compares the observed and expected frequencies in each category to test that all categories contain the same proportion of values or test that each category contains a user-specified proportion of values.

Example. The chi-square test could be used to determine whether a bag of jelly beans contains equal proportions of blue, brown, green, orange, red, and yellow candies. You could also test to see whether a bag of jelly beans contains 5% blue, 30% brown, 10% green, 20% orange, 15% red, and 15% yellow candies.

Statistics. Mean, standard deviation, minimum, maximum, and quartiles. The number and the percentage of nonmissing and missing cases; the number of cases observed and expected for each category; residuals; and the chi-square statistic.

Data. Use ordered or unordered numeric categorical variables (ordinal or nominal levels of measurement). To convert string variables to numeric variables, use the Automatic Recode procedure, which is available on the Transform menu.

Assumptions. Nonparametric tests do not require assumptions about the shape of the underlying distribution. The data are assumed to be a random sample. The expected frequencies for each category should be at least 1. No more than 20% of the categories should have expected frequencies of less than 5.

Figure 36-1
Chi-Square Test output

Color of Jelly Bean

	Observed N	Expected N	Residual
Blue	6	18.8	-12.8
Brown	33	18.8	14.2
Green	9	18.8	-9.8
Yellow	17	18.8	-1.8
Orange	22	18.8	3.2
Red	26	18.8	7.2
Total	113		

Test Statistics

	Color of Jelly Bean
Chi-Square [1]	27.973
df	5
Asymptotic Significance	.000

[1]. 0 Cells .0% low freqs 18.8 expected low...

Color of Jelly Bean

	Observed N	Expected N	Residual
Blue	6	5.7	.3
Brown	33	33.9	-.9
Green	9	11.3	-2.3
Yellow	17	17.0	.0
Orange	22	22.6	-.6
Red	26	22.6	3.4
Total	113		

Test Statistics

	Color of Jelly Bean
Chi-Square [1]	1.041
df	5
Asymptotic Significance	.959

[1]. 0 Cells .0% low freqs 5.7 expected low...

To Obtain a Chi-Square Test

▶ From the menus choose:

Analyze
  Nonparametric Tests
    Chi-Square...

Figure 36-2
Chi-Square Test dialog box

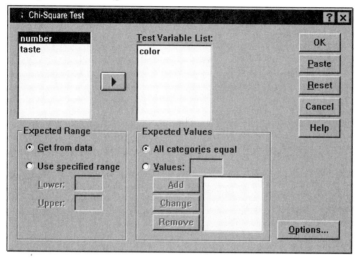

▶ Select one or more test variables. Each variable produces a separate test.

▶ Optionally, click Options for descriptive statistics, quartiles, and control of the treatment of missing data.

Chi-Square Test Expected Range and Expected Values

Expected Range. By default, each distinct value of the variable is defined as a category. To establish categories within a specific range, select Use specified range and enter integer values for lower and upper bounds. Categories are established for each integer value within the inclusive range, and cases with values outside of the bounds are excluded. For example, if you specify a value of 1 for Lower and a value of 4 for Upper, only the integer values of 1 through 4 are used for the chi-square test.

Expected Values. By default, all categories have equal expected values. Categories can have user-specified expected proportions. Select Values, enter a value that is greater than 0 for each category of the test variable, and then click Add. Each time you add a value, it appears at the bottom of the value list. The order of the values is important; it corresponds to the ascending order of the category values of the test variable. The first value of the list corresponds to the lowest group value of the test variable, and the last value corresponds to the highest value. Elements of the value list are summed, and then each value is divided by this sum to calculate the proportion of cases expected in the corresponding category. For example, a value list of 3, 4, 5, 4 specifies expected proportions of 3/16, 4/16, 5/16, and 4/16.

Chi-Square Test Options

Figure 36-3
Chi-Square Test Options dialog box

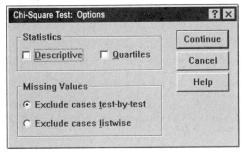

Statistics. You can choose one or both summary statistics.

- **Descriptive.** Displays the mean, standard deviation, minimum, maximum, and number of nonmissing cases.

- **Quartiles.** Displays values corresponding to the 25th, 50th, and 75th percentiles.

Missing Values. Controls the treatment of missing values.

- **Exclude cases test-by-test.** When several tests are specified, each test is evaluated separately for missing values.

- **Exclude cases listwise.** Cases with missing values for any variable are excluded from all analyses.

NPAR TESTS Command Additional Features (Chi-Square Test)

The SPSS command language also allows you to:

- Specify different minimum and maximum values or expected frequencies for different variables (with the CHISQUARE subcommand).

- Test the same variable against different expected frequencies or use different ranges (with the EXPECTED subcommand).

See the *SPSS Command Syntax Reference* for complete syntax information.

Binomial Test

The Binomial Test procedure compares the observed frequencies of the two categories of a dichotomous variable to the frequencies that are expected under a binomial distribution with a specified probability parameter. By default, the probability parameter for both groups is 0.5. To change the probabilities, you can enter a test proportion for the first group. The probability for the second group will be 1 minus the specified probability for the first group.

Example. When you toss a dime, the probability of a head equals 1/2. Based on this hypothesis, a dime is tossed 40 times, and the outcomes are recorded (heads or tails). From the binomial test, you might find that 3/4 of the tosses were heads and that the observed significance level is small (0.0027). These results indicate that it is not likely that the probability of a head equals 1/2; the coin is probably biased.

Statistics. Mean, standard deviation, minimum, maximum, number of nonmissing cases, and quartiles.

Data. The variables that are tested should be numeric and dichotomous. To convert string variables to numeric variables, use the Automatic Recode procedure, which is available on the Transform menu. A **dichotomous variable** is a variable that can take only two possible values: *yes* or *no*, *true* or *false*, 0 or 1, and so on. If the variables are not dichotomous, you must specify a cut point. The cut point assigns cases with values that are greater than the cut point to one group and assigns the rest of the cases to another group.

Assumptions. Nonparametric tests do not require assumptions about the shape of the underlying distribution. The data are assumed to be a random sample.

Figure 36-4
Binomial Test output

Binomial Test

		Category	N	Observed Proportion	Test Proportion	Asymptotic Significance (2-tailed)
Coin	Group 1	Head	30	.75	.50	.003[1]
	Group 2	Tail	10	.25		
	Total		40	1.00		

[1] Based on Z Approximation

To Obtain a Binomial Test

▶ From the menus choose:

Analyze
  Nonparametric Tests
    Binomial...

Figure 36-5
Binomial Test dialog box

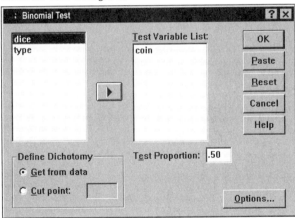

▶ Select one or more numeric test variables.

▶ Optionally, click Options for descriptive statistics, quartiles, and control of the treatment of missing data.

Binomial Test Options

Figure 36-6
Binomial Test Options dialog box

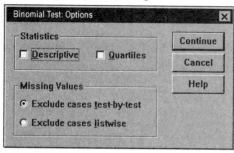

Statistics. You can choose one or both summary statistics.

- **Descriptive.** Displays the mean, standard deviation, minimum, maximum, and number of nonmissing cases.

- **Quartiles.** Displays values corresponding to the 25th, 50th, and 75th percentiles.

Missing Values. Controls the treatment of missing values.

- **Exclude cases test-by-test.** When several tests are specified, each test is evaluated separately for missing values.

- **Exclude cases listwise.** Cases with missing values for any variable that is tested are excluded from all analyses.

NPAR TESTS Command Additional Features (Binomial Test)

The SPSS command language also allows you to:

- Select specific groups (and exclude other groups) when a variable has more than two categories (with the BINOMIAL subcommand).

- Specify different cut points or probabilities for different variables (with the BINOMIAL subcommand).

- Test the same variable against different cut points or probabilities (with the EXPECTED subcommand).

See the *SPSS Command Syntax Reference* for complete syntax information.

Runs Test

The Runs Test procedure tests whether the order of occurrence of two values of a variable is random. A run is a sequence of like observations. A sample with too many or too few runs suggests that the sample is not random.

Example. Suppose that 20 people are polled to find out whether they would purchase a product. The assumed randomness of the sample would be seriously questioned if all 20 people were of the same gender. The runs test can be used to determine whether the sample was drawn at random.

Statistics. Mean, standard deviation, minimum, maximum, number of nonmissing cases, and quartiles.

Data. The variables must be numeric. To convert string variables to numeric variables, use the Automatic Recode procedure, which is available on the Transform menu.

Assumptions. Nonparametric tests do not require assumptions about the shape of the underlying distribution. Use samples from continuous probability distributions.

Figure 36-7
Runs Test output

Runs Test

	Gender
Test Value [1]	1.00
Cases < Test Value	7
Cases >= Test Value	13
Total Cases	20
Number of Runs	15
Z	2.234
Asymptotic Significance (2-tailed)	.025

[1]. Median

To Obtain a Runs Test

▶ From the menus choose:

Analyze
    Nonparametric Tests
        Runs...

Figure 36-8
Runs Test dialog box

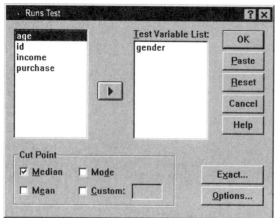

▶ Select one or more numeric test variables.

▶ Optionally, click Options for descriptive statistics, quartiles, and control of the treatment of missing data.

Runs Test Cut Point

Cut Point. Specifies a cut point to dichotomize the variables that you have chosen. You can use the observed mean, median, or mode, or you can use a specified value as a cut point. Cases with values that are less than the cut point are assigned to one group, and cases with values that are greater than or equal to the cut point are assigned to another group. One test is performed for each chosen cut point.

Runs Test Options

Figure 36-9
Runs Test Options dialog box

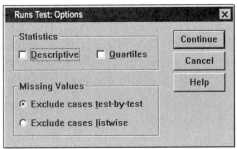

Statistics. You can choose one or both summary statistics.

- **Descriptive.** Displays the mean, standard deviation, minimum, maximum, and number of nonmissing cases.

- **Quartiles.** Displays values corresponding to the 25th, 50th, and 75th percentiles.

Missing Values. Controls the treatment of missing values.

- **Exclude cases test-by-test.** When several tests are specified, each test is evaluated separately for missing values.

- **Exclude cases listwise.** Cases with missing values for any variable are excluded from all analyses.

NPAR TESTS Command Additional Features (Runs Test)

The SPSS command language also allows you to:

- Specify different cut points for different variables (with the RUNS subcommand).

- Test the same variable against different custom cut points (with the RUNS subcommand).

See the *SPSS Command Syntax Reference* for complete syntax information.

One-Sample Kolmogorov-Smirnov Test

The One-Sample Kolmogorov-Smirnov Test procedure compares the observed cumulative distribution function for a variable with a specified theoretical distribution, which may be normal, uniform, Poisson, or exponential. The Kolmogorov-Smirnov Z is computed from the largest difference (in absolute value) between the observed and theoretical cumulative distribution functions. This goodness-of-fit test tests whether the observations could reasonably have come from the specified distribution.

Example. Many parametric tests require normally distributed variables. The one-sample Kolmogorov-Smirnov test can be used to test that a variable (for example, *income*) is normally distributed.

Statistics. Mean, standard deviation, minimum, maximum, number of nonmissing cases, and quartiles.

Data. Use quantitative variables (interval or ratio level of measurement).

Assumptions. The Kolmogorov-Smirnov test assumes that the parameters of the test distribution are specified in advance. This procedure estimates the parameters from the sample. The sample mean and sample standard deviation are the parameters for a normal distribution, the sample minimum and maximum values define the range of the uniform distribution, the sample mean is the parameter for the Poisson distribution, and the sample mean is the parameter for the exponential distribution. The power of the test to detect departures from the hypothesized distribution may be seriously diminished. For testing against a normal distribution with estimated parameters, consider the adjusted K-S Lilliefors test (available in the Explore procedure).

Figure 36-10
One-Sample Kolmogorov-Smirnov Test output

One-Sample Kolmogorov-Smirnov Test

		Income
N		20
Normal Parameters [1,2]	Mean	56250.00
	Std. Deviation	45146.40
Most Extreme Differences	Absolute	.170
	Positive	.170
	Negative	-.164
Kolmogorov-Smirnov Z		.760
Asymptotic Significance (2-tailed)		.611

[1] Test Distribution is Normal

[2] Calculated from data

To Obtain a One-Sample Kolmogorov-Smirnov Test

▶ From the menus choose:

Analyze
  Nonparametric Tests
    1-Sample K-S...

Figure 36-11
One-Sample Kolmogorov-Smirnov Test dialog box

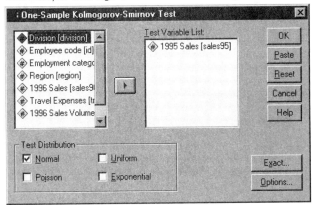

▶ Select one or more numeric test variables. Each variable produces a separate test.

▶ Optionally, click Options for descriptive statistics, quartiles, and control of the treatment of missing data.

One-Sample Kolmogorov-Smirnov Test Options

Figure 36-12
One-Sample K-S Options dialog box

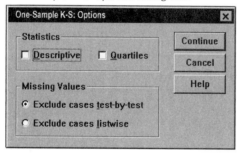

Statistics. You can choose one or both summary statistics.

■ **Descriptive.** Displays the mean, standard deviation, minimum, maximum, and number of nonmissing cases.

■ **Quartiles.** Displays values corresponding to the 25th, 50th, and 75th percentiles.

Missing Values. Controls the treatment of missing values.

■ **Exclude cases test-by-test.** When several tests are specified, each test is evaluated separately for missing values.

■ **Exclude cases listwise.** Cases with missing values for any variable are excluded from all analyses.

NPAR TESTS Command Additional Features (One-Sample Kolmogorov-Smirnov Test)

The SPSS command language also allows you to specify the parameters of the test distribution (with the K-S subcommand).

See the *SPSS Command Syntax Reference* for complete syntax information.

Two-Independent-Samples Tests

The Two-Independent-Samples Tests procedure compares two groups of cases on one variable.

Example. New dental braces have been developed that are intended to be more comfortable, to look better, and to provide more rapid progress in realigning teeth. To find out whether the new braces have to be worn as long as the old braces, 10 children are randomly chosen to wear the old braces, and another 10 children are chosen to wear the new braces. From the Mann-Whitney U test, you might find that, on average, children with the new braces did not have to wear the braces as long as children with the old braces.

Statistics. Mean, standard deviation, minimum, maximum, number of nonmissing cases, and quartiles. Tests: Mann-Whitney U, Moses extreme reactions, Kolmogorov-Smirnov Z, Wald-Wolfowitz runs.

Data. Use numeric variables that can be ordered.

Assumptions. Use independent, random samples. The Mann-Whitney U test requires that the two tested samples be similar in shape.

Figure 36-13
Two-Independent-Samples output

Ranks

			N	Mean Rank	Sum of Ranks
Time Worn in Days	Type of Braces	Old Braces	10	14.10	141.00
		New Braces	10	6.90	69.00
		Total	20		

Test Statistics [2]

	Time Worn in Days
Mann-Whitney U	14.000
Wilcoxon W	69.000
Z	-2.721
Asymptotic Significance (2-tailed)	.007
Exact Significance [2*(1-tailed Sig.)]	.005 [1]

[1.] Not corrected for ties.

[2.] Grouping Variable: Type of Braces

To Obtain Two-Independent-Samples Tests

► From the menus choose:

Analyze
    Nonparametric Tests
        2 Independent Samples...

Figure 36-14
Two-Independent-Samples Tests dialog box

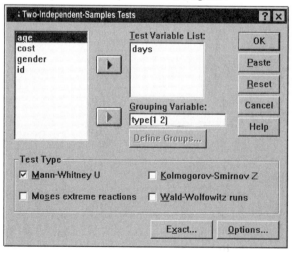

▶ Select one or more numeric variables.

▶ Select a grouping variable and click Define Groups to split the file into two groups or samples.

Two-Independent-Samples Test Types

Test Type. Four tests are available to test whether two independent samples (groups) come from the same population.

The **Mann-Whitney U test** is the most popular of the two-independent-samples tests. It is equivalent to the Wilcoxon rank sum test and the Kruskal-Wallis test for two groups. Mann-Whitney tests that two sampled populations are equivalent in location. The observations from both groups are combined and ranked, with the average rank assigned in the case of ties. The number of ties should be small relative to the total number of observations. If the populations are identical in location, the ranks should be randomly mixed between the two samples. The test calculates the number of times that a score from group 1 precedes a score from group 2 and the number of times that a score from group 2 precedes a score from group 1. The Mann-Whitney U statistic is the smaller of these two numbers. The Wilcoxon rank sum W statistic, also displayed, is the rank sum of the smaller sample. If both samples have the same number of observations, W is the rank sum of the group that is named first in the Two-Independent-Samples Define Groups dialog box.

The **Kolmogorov-Smirnov Z test** and the **Wald-Wolfowitz runs test** are more general tests that detect differences in both the locations and shapes of the distributions. The Kolmogorov-Smirnov test is based on the maximum absolute difference between the observed cumulative distribution functions for both samples. When this difference is significantly large, the two distributions are considered different. The Wald-Wolfowitz runs test combines and ranks the observations from both groups. If the two samples are from the same population, the two groups should be randomly scattered throughout the ranking.

The **Moses extreme reactions test** assumes that the experimental variable will affect some subjects in one direction and other subjects in the opposite direction. The test tests for extreme responses compared to a control group. This test focuses on the span of the control group and is a measure of how much extreme values in the experimental group influence the span when combined with the control group. The control group is defined by the group 1 value in the Two-Independent-Samples Define Groups dialog box. Observations from both groups are combined and ranked. The span of the control group is computed as the difference between the ranks of the largest and smallest values in the control group plus 1. Because chance outliers can easily distort the range of the span, 5% of the control cases are trimmed automatically from each end.

Two-Independent-Samples Tests Define Groups

Figure 36-15
Two-Independent-Samples Define Groups dialog box

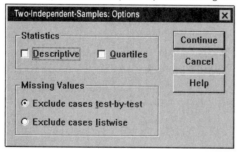

To split the file into two groups or samples, enter an integer value for Group 1 and another value for Group 2. Cases with other values are excluded from the analysis.

Two-Independent-Samples Tests Options

Figure 36-16
Two-Independent-Samples Options dialog box

Statistics. You can choose one or both summary statistics.

■ **Descriptive.** Displays the mean, standard deviation, minimum, maximum, and the number of nonmissing cases.

■ **Quartiles.** Displays values corresponding to the 25th, 50th, and 75th percentiles.

Missing Values. Controls the treatment of missing values.

- **Exclude cases test-by-test.** When several tests are specified, each test is evaluated separately for missing values.

- **Exclude cases listwise.** Cases with missing values for any variable are excluded from all analyses.

NPAR TESTS Command Additional Features (Two-Independent-Samples Tests)

The SPSS command language also allows you to specify the number of cases to be trimmed for the Moses test (with the `MOSES` subcommand).

See the *SPSS Command Syntax Reference* for complete syntax information.

Two-Related-Samples Tests

The Two-Related-Samples Tests procedure compares the distributions of two variables.

Example. In general, do families receive the asking price when they sell their homes? By applying the Wilcoxon signed-rank test to data for 10 homes, you might learn that seven families receive less than the asking price, one family receives more than the asking price, and two families receive the asking price.

Statistics. Mean, standard deviation, minimum, maximum, number of nonmissing cases, and quartiles. Tests: Wilcoxon signed-rank, sign, McNemar.

Data. Use numeric variables that can be ordered.

Assumptions. Although no particular distributions are assumed for the two variables, the population distribution of the paired differences is assumed to be symmetric.

Figure 36-17
Two-Related-Samples output

Ranks

		N	Mean Rank	Sum of Ranks
Asking Price - Sale Price	Negative Ranks	7[1]	4.93	34.50
	Positive Ranks	1[2]	1.50	1.50
	Ties	2[3]		
	Total	10		

[1]. Asking Price < Sale Price
[2]. Asking Price > Sale Price
[3]. Asking Price = Sale Price

Test Statistics [2]

	Asking Price - Sale Price
Z	-2.313[1]
Asymptotic Significance (2-tailed)	.021

[1]. Based on positive ranks
[2]. Wilcoxon Signed Ranks Test

To Obtain Two-Related-Samples Tests

▶ From the menus choose:

Analyze
  Nonparametric Tests
    2 Related Samples...

Figure 36-18
Two-Related-Samples Tests dialog box

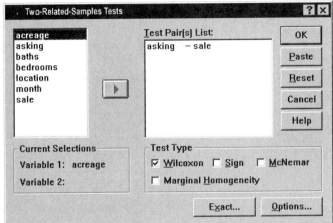

▶ Select one or more pairs of variables, as follows:

■ Click each of two variables. The first variable appears in the Current Selections group as *Variable 1*, and the second variable appears as *Variable 2*.

■ Click the arrow button to move the pair into the Test Pair(s) List. You may select more pairs of variables. To remove a pair of variables from the analysis, select a pair in the Test Pair(s) List and click the arrow button.

Two-Related-Samples Test Types

The tests in this section compare the distributions of two related variables. The appropriate test to use depends on the type of data.

If your data are continuous, use the sign test or the Wilcoxon signed-rank test. The **sign test** computes the differences between the two variables for all cases and classifies the differences as positive, negative, or tied. If the two variables are similarly distributed, the number of positive and negative differences will not differ significantly. The **Wilcoxon signed-rank test** considers information about both the sign of the differences and the magnitude of the differences between pairs. Because the Wilcoxon signed-rank test incorporates more information about the data, it is more powerful than the sign test.

If your data are binary, use the **McNemar test**. This test is typically used in a repeated measures situation, in which each subject's response is elicited twice, once before and once after a specified event occurs. The McNemar test determines whether the initial response rate (before the event) equals the final response rate (after the event). This test is useful for detecting changes in responses due to experimental intervention in before-and-after designs.

If your data are categorical, use the **marginal homogeneity test**. This test is an extension of the McNemar test from binary response to multinomial response. It tests for changes in response (using the chi-square distribution) and is useful for detecting response changes due to experimental intervention in before-and-after designs. The marginal homogeneity test is available only if you have installed Exact Tests.

Two-Related-Samples Tests Options

Figure 36-19
Two-Related-Samples Options dialog box

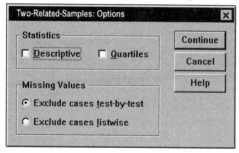

Statistics. You can choose one or both summary statistics.

- **Descriptive.** Displays the mean, standard deviation, minimum, maximum, and the number of nonmissing cases.

- **Quartiles.** Displays values corresponding to the 25th, 50th, and 75th percentiles.

Missing Values. Controls the treatment of missing values.

- **Exclude cases test-by-test.** When several tests are specified, each test is evaluated separately for missing values.

- **Exclude cases listwise.** Cases with missing values for any variable are excluded from all analyses.

NPAR TESTS Command Additional Features (Two Related Samples)

The SPSS command language also allows you to test a variable with each variable on a list.

See the *SPSS Command Syntax Reference* for complete syntax information.

Tests for Several Independent Samples

The Tests for Several Independent Samples procedure compares two or more groups of cases on one variable.

Example. Do three brands of 100-watt lightbulbs differ in the average time that the bulbs will burn? From the Kruskal-Wallis one-way analysis of variance, you might learn that the three brands do differ in average lifetime.

Statistics. Mean, standard deviation, minimum, maximum, number of nonmissing cases, and quartiles. Tests: Kruskal-Wallis H, median.

Data. Use numeric variables that can be ordered.

Assumptions. Use independent, random samples. The Kruskal-Wallis H test requires that the tested samples be similar in shape.

Figure 36-20
Tests for Several Independent Samples output

Ranks

			N	Mean Rank
Hours	Brand	Brand A	10	15.20
		Brand B	10	25.50
		Brand C	10	5.80
		Total	30	

Test Statistics [1,2]

	Hours
Chi-Square	25.061
df	2
Asymptotic Significance	.000

[1.] Kruskal Wallis Test

[2.] Grouping Variable: Brand

To Obtain Tests for Several Independent Samples

▶ From the menus choose:

Analyze
  Nonparametric Tests
    K Independent Samples...

Figure 36-21
Tests for Several Independent Samples dialog box

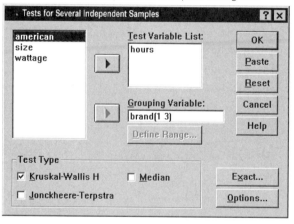

▶ Select one or more numeric variables.

▶ Select a grouping variable and click Define Range to specify minimum and maximum integer values for the grouping variable.

Tests for Several Independent Samples Test Types

Three tests are available to determine if several independent samples come from the same population. The Kruskal-Wallis *H* test, the median test, and the Jonckheere-Terpstra test all test whether several independent samples are from the same population.

The **Kruskal-Wallis H test**, an extension of the Mann-Whitney *U* test, is the nonparametric analog of one-way analysis of variance and detects differences in distribution location. The **median test**, which is a more general test (but not as powerful), detects distributional differences in location and shape. The Kruskal-Wallis *H* test and the median test assume that there is no *a priori* ordering of the *k* populations from which the samples are drawn.

When there *is* a natural *a priori* ordering (ascending or descending) of the *k* populations, the **Jonckheere-Terpstra test** is more powerful. For example, the *k* populations might represent *k* increasing temperatures. The hypothesis that different temperatures produce the same response distribution is tested against the alternative that as the temperature increases, the magnitude of the response increases. Here, the alternative hypothesis is ordered; therefore, Jonckheere-Terpstra is the most appropriate test to use. The Jonckheere-Terpstra test is available only if you have installed SPSS Exact Tests.

Tests for Several Independent Samples Define Range

Figure 36-22
Several Independent Samples Define Range dialog box

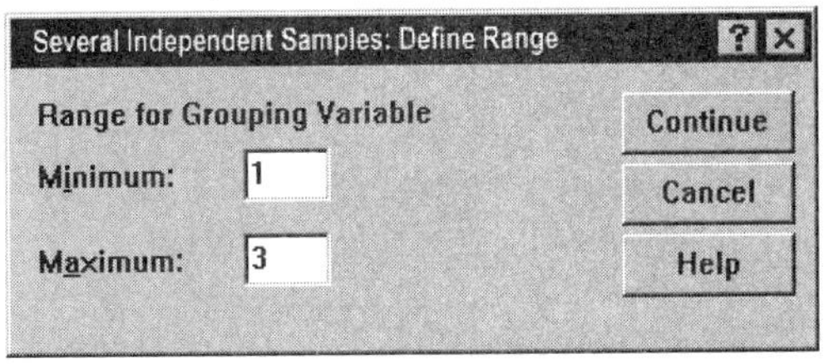

To define the range, enter integer values for Minimum and Maximum that correspond to the lowest and highest categories of the grouping variable. Cases with values outside of the bounds are excluded. For example, if you specify a minimum value of 1 and a maximum value of 3, only the integer values of 1 through 3 are used. The minimum value must be less than the maximum value, and both values must be specified.

Tests for Several Independent Samples Options

Figure 36-23
Several Independent Samples Options dialog box

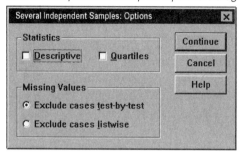

Statistics. You can choose one or both summary statistics.

■ **Descriptive.** Displays the mean, standard deviation, minimum, maximum, and the number of nonmissing cases.

■ **Quartiles.** Displays values corresponding to the 25th, 50th, and 75th percentiles.

Missing Values. Controls the treatment of missing values.

■ **Exclude cases test-by-test.** When several tests are specified, each test is evaluated separately for missing values.

■ **Exclude cases listwise.** Cases with missing values for any variable are excluded from all analyses.

NPAR TESTS Command Additional Features (K Independent Samples)

The SPSS command language also allows you to specify a value other than the observed median for the median test (with the MEDIAN subcommand).

See the *SPSS Command Syntax Reference* for complete syntax information.

Tests for Several Related Samples

The Tests for Several Related Samples procedure compares the distributions of two or more variables.

Example. Does the public associate different amounts of prestige with a doctor, a lawyer, a police officer, and a teacher? Ten people are asked to rank these four occupations in order of prestige. Friedman's test indicates that the public does associate different amounts of prestige with these four professions.

Statistics. Mean, standard deviation, minimum, maximum, number of nonmissing cases, and quartiles. Tests: Friedman, Kendall's *W*, and Cochran's *Q*.

Data. Use numeric variables that can be ordered.

Assumptions. Nonparametric tests do not require assumptions about the shape of the underlying distribution. Use dependent, random samples.

Figure 36-24
Tests for Several Related Samples output

Ranks

	Mean Rank
Doctor	1.50
Lawyer	2.50
Police	3.40
Teacher	2.60

Test Statistics [1]

N	10
Chi-Square	10.920
df	3
Asymptotic Significance	.012

[1] Friedman Test

To Obtain Tests for Several Related Samples

▶ From the menus choose:
Analyze
  Nonparametric Tests
    K Related Samples...

Figure 36-25
Tests for Several Related Samples dialog box

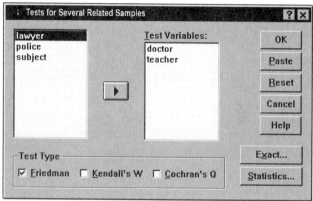

▶ Select two or more numeric test variables.

Tests for Several Related Samples Test Types

Three tests are available to compare the distributions of several related variables.

The **Friedman test** is the nonparametric equivalent of a one-sample repeated measures design or a two-way analysis of variance with one observation per cell. Friedman tests the null hypothesis that k related variables come from the same population. For each case, the k variables are ranked from 1 to k. The test statistic is based on these ranks.

Kendall's W is a normalization of the Friedman statistic. Kendall's W is interpretable as the coefficient of concordance, which is a measure of agreement among raters. Each case is a judge or rater, and each variable is an item or person being judged. For each variable, the sum of ranks is computed. Kendall's W ranges between 0 (no agreement) and 1 (complete agreement).

Cochran's Q is identical to the Friedman test but is applicable when all responses are binary. This test is an extension of the McNemar test to the k-sample situation. Cochran's Q tests the hypothesis that several related dichotomous variables have the same mean. The variables are measured on the same individual or on matched individuals.

Tests for Several Related Samples Statistics

Figure 36-26
Several Related Samples Statistics dialog box

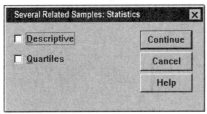

You can choose statistics.

■ **Descriptive.** Displays the mean, standard deviation, minimum, maximum, and the number of nonmissing cases.

■ **Quartiles.** Displays values corresponding to the 25th, 50th, and 75th percentiles.

NPAR TESTS Command Additional Features (K Related Samples)

See the *SPSS Command Syntax Reference* for complete syntax information.

Multiple Response Analysis

Two procedures are available for analyzing multiple dichotomy and multiple category sets. The Multiple Response Frequencies procedure displays frequency tables. The Multiple Response Crosstabs procedure displays two- and three-dimensional crosstabulations. Before using either procedure, you must define multiple response sets.

Example. This example illustrates the use of multiple response items in a market research survey. The data are fictitious and should not be interpreted as real. An airline might survey passengers flying a particular route to evaluate competing carriers. In this example, American Airlines wants to know about its passengers' use of other airlines on the Chicago–New York route and the relative importance of schedule and service in selecting an airline. The flight attendant hands each passenger a brief questionnaire upon boarding. The first question reads: Circle all airlines you have flown at least once in the last six months on this route—American, United, TWA, USAir, Other. This is a multiple response question, since the passenger can circle more than one response. However, this question cannot be coded directly because a variable can have only one value for each case. You must use several variables to map responses to each question. There are two ways to do this. One is to define a variable corresponding to each of the choices (for example, American, United, TWA, USAir, and Other). If the passenger circles United, the variable *united* is assigned a code of 1, otherwise 0. This is a **multiple dichotomy method** of mapping variables. The other way to map responses is the **multiple category method**, in which you estimate the maximum number of possible responses to the question and set up the same number of variables, with codes used to specify the airline flown. By perusing a sample of questionnaires, you might discover that no user has flown more than three different airlines on this route in the last six months. Further, you find that due to the deregulation of airlines, 10 other airlines are named in the Other category. Using the multiple response method, you would define three variables, each coded as 1 = *american*, 2 = *united*, 3 = *twa*, 4 = *usair*, 5 = *delta*, and so on. If a given passenger circles American and TWA, the first variable has a code of 1, the second has a code of 3, and the third has a missing-value code. Another passenger might have circled American and entered Delta. Thus, the first variable has a code of 1, the second has a code of 5, and the third a missing-value code. If you use the multiple dichotomy method, on the other hand, you end up with 14 separate variables. Although either method of mapping is feasible for this survey, the method you choose depends on the distribution of responses.

Multiple Response Define Sets

The Define Multiple Response Sets procedure groups elementary variables into multiple dichotomy and multiple category sets, for which you can obtain frequency tables and crosstabulations. You can define up to 20 multiple response sets. Each set must have a unique name. To remove a set, highlight it on the list of multiple response sets and click Remove. To change a set, highlight it on the list, modify any set definition characteristics, and click Change.

You can code your elementary variables as dichotomies or categories. To use dichotomous variables, select Dichotomies to create a multiple dichotomy set. Enter an integer value for Counted value. Each variable having at least one occurrence of the counted value becomes a category of the multiple dichotomy set. Select Categories to create a multiple category set having the same range of values as the component variables. Enter integer values for the minimum and maximum values of the range for categories of the multiple category set. The procedure totals each distinct integer value in the inclusive range across all component variables. Empty categories are not tabulated.

Each multiple response set must be assigned a unique name of up to seven characters. The procedure prefixes a dollar sign ($) to the name you assign. You cannot use the following reserved names: *casenum, sysmis, jdate, date, time, length*, and *width*. The name of the multiple response set exists only for use in multiple response procedures. You cannot refer to multiple response set names in other procedures. Optionally, you can enter a descriptive variable label for the multiple response set. The label can be up to 40 characters long.

To Define Multiple Response Sets

▶ From the menus choose:

Analyze
  Multiple Response
    Define Sets...

Figure 37-1
Define Multiple Response Sets dialog box

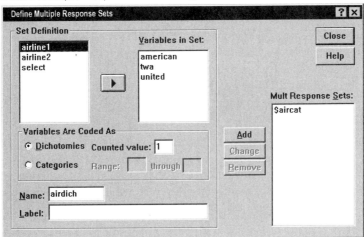

▶ Select two or more variables.

▶ If your variables are coded as dichotomies, indicate which value you want to have counted. If your variables are coded as categories, define the range of the categories.

▶ Enter a unique name for each multiple response set.

▶ Click Add to add the multiple response set to the list of defined sets.

Multiple Response Frequencies

The Multiple Response Frequencies procedure produces frequency tables for multiple response sets. You must first define one or more multiple response sets (see "Multiple Response Define Sets").

For multiple dichotomy sets, category names shown in the output come from variable labels defined for elementary variables in the group. If the variable labels are not defined, variable names are used as labels. For multiple category sets, category labels come from the value labels of the first variable in the group. If categories missing for the first variable are present for other variables in the group, define a value label for the missing categories.

Missing Values. Cases with missing values are excluded on a table-by-table basis. Alternatively, you can choose one or both of the following:

■ **Exclude cases listwise within dichotomies.** Excludes cases with missing values for any variable from the tabulation of the multiple dichotomy set. This applies only to multiple response sets defined as dichotomy sets. By default, a case is considered missing for a multiple dichotomy set if none of its component variables contains the counted value. Cases with missing values for some (but not all variables) are included in the tabulations of the group if at least one variable contains the counted value.

■ **Exclude cases listwise within categories.** Excludes cases with missing values for any variable from tabulation of the multiple category set. This applies only to multiple response sets defined as category sets. By default, a case is considered missing for a multiple category set only if none of its components has valid values within the defined range.

Example. Each variable created from a survey question is an elementary variable. To analyze a multiple response item, you must combine the variables into one of two types of multiple response sets: a multiple dichotomy set or a multiple category set. For example, if an airline survey asked which of three airlines (American, United, TWA) you have flown in the last six months and you used dichotomous variables and defined a **multiple dichotomy set**, each of the three variables in the set would become a category of the group variable. The counts and percentages for the three airlines are displayed in one frequency table. If you discover that no respondent mentioned more than two airlines, you could create two variables, each having three codes, one for each airline. If you define a **multiple category set**, the values are tabulated by adding the same codes in the elementary variables together. The resulting set of values is the same as those for each of the elementary variables. For example, 30 responses for United are the sum of the five United responses for airline 1 and the 25 United responses for airline 2. The counts and percentages for the three airlines are displayed in one frequency table.

Statistics. Frequency tables displaying counts, percentages of responses, percentages of cases, number of valid cases, and number of missing cases.

Data. Use multiple response sets.

Assumptions. The counts and percentages provide a useful description for data from any distribution.

Related procedures. The Multiple Response Define Sets procedure allows you to define multiple response sets.

Figure 37-2
Multiple Response Frequencies output

		Responses		Percent of Cases
		N	Percent	
Name[a]	American	75	67.6%	92.6%
	TWA	6	5.4%	7.4%
	United	30	27.0%	37.0%
Total		111	100.0%	137.0%

a. Dichotomy group tabulated at value 1.

To Obtain Multiple Response Frequencies

▶ From the menus choose:

Analyze
  Multiple Response
    Frequencies...

Figure 37-3
Multiple Response Frequencies dialog box

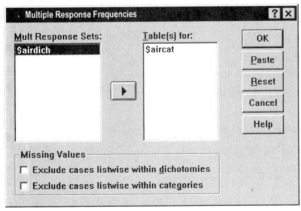

▶ Select one or more multiple response sets.

Multiple Response Crosstabs

The Multiple Response Crosstabs procedure crosstabulates defined multiple response sets, elementary variables, or a combination. You can also obtain cell percentages based on cases or responses, modify the handling of missing values, or get paired crosstabulations. You must first define one or more multiple response sets (see "To Define Multiple Response Sets").

For multiple dichotomy sets, category names shown in the output come from variable labels defined for elementary variables in the group. If the variable labels are not defined, variable names are used as labels. For multiple category sets, category labels come from the value labels of the first variable in the group. If categories missing for the first variable are present for other variables in the group, define a value label for the missing categories. The procedure displays category labels for columns on three lines, with up to eight characters per line. To avoid splitting words, you can reverse row and column items or redefine labels.

Example. Both multiple dichotomy and multiple category sets can be crosstabulated with other variables in this procedure. An airline passenger survey asks passengers for the following information: Circle all of the following airlines you have flown at least once in the last six months (American, United, TWA). Which is more important in selecting a flight—schedule or service? Select only one. After entering the data as dichotomies or multiple categories and combining them into a set, you can crosstabulate the airline choices with the question involving service or schedule.

Statistics. Crosstabulation with cell, row, column, and total counts, and cell, row, column, and total percentages. The cell percentages can be based on cases or responses.

Data. Use multiple response sets or numeric categorical variables.

Assumptions. The counts and percentages provide a useful description of data from any distribution.

Related procedures. The Multiple Response Define Sets procedure allows you to define multiple response sets.

Figure 37-4
Multiple Response Crosstabs output

			Select airline because of		Total
			Schedule	Service	
Name	American	Count	41	34	75
	TWA	Count	3	3	6
	United	Count	27	3	30
Total		Count	44	37	81

Percentages and totals are based on respondents.

a. Dichotomy group tabulated at value 1.

To Obtain Multiple Response Crosstabs

▶ From the menus choose:

Analyze
  Multiple Response
    Crosstabs...

Figure 37-5
Multiple Response Crosstabs dialog box

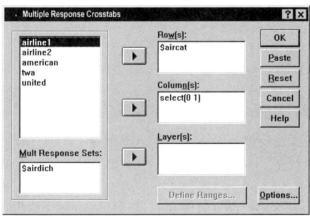

▶ Select one or more numeric variables or multiple response sets for each dimension of the crosstabulation.

▶ Define the range of each elementary variable.

Optionally, you can obtain a two-way crosstabulation for each category of a control variable or multiple response set. Select one or more items for the Layer(s) list.

Multiple Response Crosstabs Define Ranges

Figure 37-6
Multiple Response Crosstabs Define Variable Range dialog box

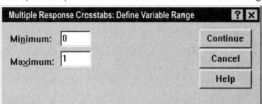

Value ranges must be defined for any elementary variable in the crosstabulation. Enter the integer minimum and maximum category values that you want to tabulate. Categories outside the range are excluded from analysis. Values within the inclusive range are assumed to be integers (non-integers are truncated).

Multiple Response Crosstabs Options

Figure 37-7
Multiple Response Crosstabs Options dialog box

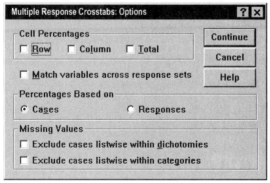

Cell Percentages. Cell counts are always displayed. You can choose to display row percentages, column percentages, and two-way table (total) percentages.

Percentages Based on. You can base cell percentages on cases (or respondents). This is not available if you select matching of variables across multiple category sets. You can also base cell percentages on responses. For multiple dichotomy sets, the number of responses is equal to the number of counted values across cases. For multiple category sets, the number of responses is the number of values in the defined range.

Missing Values. You can choose one or both of the following:

■ **Exclude cases listwise within dichotomies.** Excludes cases with missing values for any variable from the tabulation of the multiple dichotomy set. This applies only to multiple response sets defined as dichotomy sets. By default, a case is considered missing for a multiple dichotomy set if none of its component variables contains the counted value. Cases with missing values for some, but not all, variables are included in the tabulations of the group if at least one variable contains the counted value.

■ **Exclude cases listwise within categories.** Excludes cases with missing values for any variable from tabulation of the multiple category set. This applies only to multiple response sets defined as category sets. By default, a case is considered missing for a multiple category set only if none of its components has valid values within the defined range.

By default, when crosstabulating two multiple category sets, the procedure tabulates each variable in the first group with each variable in the second group and sums the counts for each cell; therefore, some responses can appear more than once in a table. You can choose the following option:

Match variables across response sets. Pairs the first variable in the first group with the first variable in the second group, and so on. If you select this option, the procedure bases cell percentages on responses rather than respondents. Pairing is not available for multiple dichotomy sets or elementary variables.

MULT RESPONSE Command Additional Features

The SPSS command language also allows you to:

■ Obtain crosstabulation tables with up to five dimensions (with the BY subcommand).

■ Change output formatting options, including suppression of value labels (with the FORMAT subcommand).

See the *SPSS Command Syntax Reference* for complete syntax information.

Reporting Results

Case listings and descriptive statistics are basic tools for studying and presenting data. You can obtain case listings with the Data Editor or the Summarize procedure, frequency counts and descriptive statistics with the Frequencies procedure, and subpopulation statistics with the Means procedure. Each of these uses a format designed to make information clear. If you want to display the information in a different format, Report Summaries in Rows and Report Summaries in Columns give you the control you need over data presentation.

Report Summaries in Rows

Report Summaries in Rows produces reports in which different summary statistics are laid out in rows. Case listings are also available, with or without summary statistics.

Example. A company with a chain of retail stores keeps records of employee information, including salary, job tenure, and the store and division in which each employee works. You could generate a report that provides individual employee information (listing) broken down by store and division (break variables), with summary statistics (for example, mean salary) for each store, division, and division within each store.

Data Columns. Lists the report variables for which you want case listings or summary statistics and controls the display format of data columns.

Break Columns. Lists optional break variables that divide the report into groups and controls the summary statistics and display formats of break columns. For multiple break variables, there will be a separate group for each category of each break variable within categories of the preceding break variable in the list. Break variables should be discrete categorical variables that divide cases into a limited number of meaningful categories. Individual values of each break variable appear, sorted, in a separate column to the left of all data columns.

Report. Controls overall report characteristics, including overall summary statistics, display of missing values, page numbering, and titles.

Display cases. Displays the actual values (or value labels) of the data-column variables for every case. This produces a listing report, which can be much longer than a summary report.

Preview. Displays only the first page of the report. This option is useful for previewing the format of your report without processing the whole report.

Data are already sorted. For reports with break variables, the data file must be sorted by break variable values before generating the report. If your data file is already sorted by values of the break variables, you can save processing time by selecting this option. This option is particularly useful after running a preview report.

Sample Output

Figure 38-1
Combined report with case listings and summary statistics

Division	Age	Tenure in Company	Grade	Salary--Annual
Carpeting	27.00	3.67	2.17	$9,200
	22.00	3.92	3.08	$10,900
	23.00	3.92	3.08	$10,900
	24.00	4.00	3.25	$10,000
	30.00	4.08	3.08	$10,000
	27.00	4.33	3.17	$10,000
	33.00	2.67	2.67	$9,335
	33.00	3.75	3.25	$10,000
	44.00	4.83	4.33	$15,690
	36.00	3.83	3.25	$10,000
	35.00	3.50	3.00	$15,520
	35.00	6.00	5.33	$19,500
Mean	30.75	4.04	3.31	$11,754
Appliances	21.00	2.67	2.67	$8,700
	26.00	2.92	2.08	$8,000
	32.00	2.92	2.92	$8,900
	33.00	3.42	2.92	$8,900
	34.00	5.08	4.50	$15,300
	24.00	3.17	3.17	$8,975
	42.00	6.50	6.50	$18,000
	30.00	2.67	2.67	$7,500
	38.00	5.00	4.42	$28,300
Mean	31.11	3.81	3.54	$12,508

To Obtain a Summary Report: Summaries in Rows

▶ From the menus choose:

Analyze
 Reports
  Report Summaries in Rows...

▶ Select one or more variables for Data Columns. One column in the report is generated for each variable selected.

▶ For reports sorted and displayed by subgroups, select one or more variables for Break Columns.

▶ For reports with summary statistics for subgroups defined by break variables, select the break variable in the Break Columns list and click Summary in the Break Columns group to specify the summary measure(s).

▶ For reports with overall summary statistics, click Summary in the Report group to specify the summary measure(s).

Figure 38-2
Report Summaries in Rows dialog box

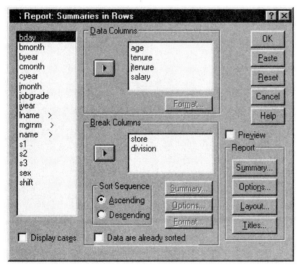

Report Data Column/Break Format

The Format dialog boxes control column titles, column width, text alignment, and the display of data values or value labels. Data Column Format controls the format of data columns on the right side of the report page. Break Format controls the format of break columns on the left side.

Figure 38-3
Report Data Column Format dialog box

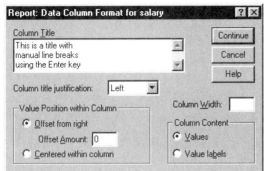

Column Title. For the selected variable, controls the column title. Long titles are automatically wrapped within the column. Use the Enter key to manually insert line breaks where you want titles to wrap.

Value Position within Column. For the selected variable, controls the alignment of data values or value labels within the column. Alignment of values or labels does not affect alignment of column headings. You can either indent the column contents by a specified number of characters or center the contents.

Column Content. For the selected variable, controls the display of either data values or defined value labels. Data values are always displayed for any values that do not have defined value labels. (Not available for data columns in column summary reports.)

Report Summary Lines for/Final Summary Lines

The two Summary Lines dialog boxes control the display of summary statistics for break groups and for the entire report. Summary Lines controls subgroup statistics for each category defined by the break variable(s). Final Summary Lines controls overall statistics, displayed at the end of the report.

Figure 38-4
Report Summary Lines dialog box

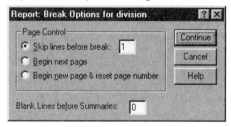

Available summary statistics are sum, mean, minimum, maximum, number of cases, percentage of cases above or below a specified value, percentage of cases within a specified range of values, standard deviation, kurtosis, variance, and skewness.

Report Break Options

Break Options controls spacing and pagination of break category information.

Figure 38-5
Report Break Options dialog box

Page Control. Controls spacing and pagination for categories of the selected break variable. You can specify a number of blank lines between break categories or start each break category on a new page.

Blank Lines before Summaries. Controls the number of blank lines between break category labels or data and summary statistics. This is particularly useful for combined reports that include both individual case listings and summary statistics for break categories; in these reports, you can insert space between the case listings and the summary statistics.

Report Options

Report Options controls the treatment and display of missing values and report page numbering.

Figure 38-6
Report Options dialog box

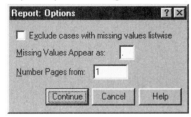

Exclude cases with missing values listwise. Eliminates (from the report) any case with missing values for any of the report variables.

Missing Values Appear as. Allows you to specify the symbol that represents missing values in the data file. The symbol can be only one character and is used to represent both **system-missing** and **user-missing** values.

Number Pages from. Allows you to specify a page number for the first page of the report.

Report Layout

Report Layout controls the width and length of each report page, placement of the report on the page, and the insertion of blank lines and labels.

Figure 38-7
Report Layout dialog box

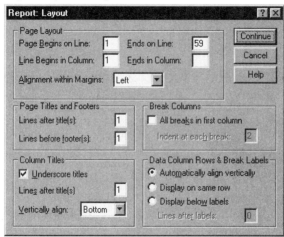

Page Layout. Controls the page margins expressed in lines (top and bottom) and characters (left and right) and reports alignment within the margins.

Page Titles and Footers. Controls the number of lines that separate page titles and footers from the body of the report.

Break Columns. Controls the display of break columns. If multiple break variables are specified, they can be in separate columns or in the first column. Placing all break variables in the first column produces a narrower report.

Column Titles. Controls the display of column titles, including title underlining, space between titles and the body of the report, and vertical alignment of column titles.

Data Column Rows and Break Labels. Controls the placement of data column information (data values and/or summary statistics) in relation to the break labels at the start of each break category. The first row of data column information can start either on the same line as the break category label or on a specified number of lines after the break category label. (Not available for column summary reports.)

Report Titles

Report Titles controls the content and placement of report titles and footers. You can specify up to 10 lines of page titles and up to 10 lines of page footers, with left-justified, centered, and right-justified components on each line.

Figure 38-8
Report Titles dialog box

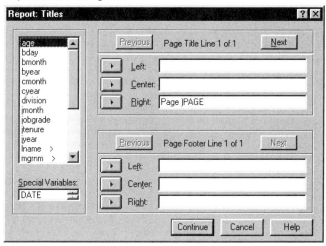

If you insert variables into titles or footers, the current value label or value of the variable is displayed in the title or footer. In titles, the value label corresponding to the value of the variable at the beginning of the page is displayed. In footers, the value label corresponding to the value of the variable at the end of the page is displayed. If there is no value label, the actual value is displayed.

Special Variables. The special variables *DATE* and *PAGE* allow you to insert the current date or the page number into any line of a report header or footer. If your data file contains variables named *DATE* or *PAGE*, you cannot use these variables in report titles or footers.

Report Summaries in Columns

Report Summaries in Columns produces summary reports in which different summary statistics appear in separate columns.

Example. A company with a chain of retail stores keeps records of employee information, including salary, job tenure, and the division in which each employee works. You could generate a report that provides summary salary statistics (for example, mean, minimum, and maximum) for each division.

Data Columns. Lists the report variables for which you want summary statistics and controls the display format and summary statistics displayed for each variable.

Break Columns. Lists optional break variables that divide the report into groups and controls the display formats of break columns. For multiple break variables, there will be a separate group for each category of each break variable within categories of the preceding break variable in the list. Break variables should be discrete categorical variables that divide cases into a limited number of meaningful categories.

Report. Controls overall report characteristics, including display of missing values, page numbering, and titles.

Preview. Displays only the first page of the report. This option is useful for previewing the format of your report without processing the whole report.

Data are already sorted. For reports with break variables, the data file must be sorted by break variable values before generating the report. If your data file is already sorted by values of the break variables, you can save processing time by selecting this option. This option is particularly useful after running a preview report.

Sample Output

Figure 38-9
Summary report with summary statistics in columns

Division	Mean Annual Mean Age	Minimum Annual Salary	Maximum Annual Salary	Salary
Carpeting	30.75	$11,754	$9,200	$19,500
Appliances	31.11	$12,508	$7,500	$28,300
Furniture	36.87	$13,255	$8,975	$17,050
Hardware	36.20	$17,580	$7,450	$22,500

To Obtain a Summary Report: Summaries in Columns

▶ From the menus choose:

Analyze
  Reports
    Report Summaries in Columns...

▶ Select one or more variables for Data Columns. One column in the report is generated for each variable selected.

▶ To change the summary measure for a variable, select the variable in the Data Columns list and click Summary.

▶ To obtain more than one summary measure for a variable, select the variable in the source list and move it into the Data Columns list multiple times, one for each summary measure you want.

▶ To display a column containing the sum, mean, ratio, or other function of existing columns, click Insert Total. This places a variable called *total* into the Data Columns list.

▶ For reports sorted and displayed by subgroups, select one or more variables for Break Columns.

Figure 38-10
Report Summaries in Columns dialog box

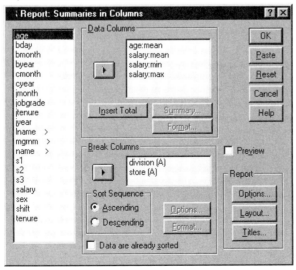

Data Columns Summary Function

Summary Lines controls the summary statistic displayed for the selected data column variable.

Figure 38-11
Report Summary Lines dialog box

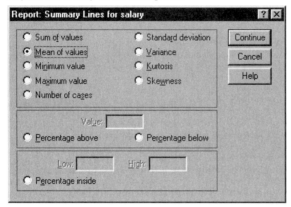

Available summary statistics are sum, mean, minimum, maximum, number of cases, percentage of cases above or below a specified value, percentage of cases within a specified range of values, standard deviation, variance, kurtosis, and skewness.

Data Columns Summary for Total Column

Summary Column controls the total summary statistics that summarize two or more data columns.

Available total summary statistics are sum of columns, mean of columns, minimum, maximum, difference between values in two columns, quotient of values in one column divided by values in another column, and product of columns values multiplied together.

Figure 38-12
Report Summary Column dialog box

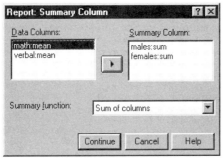

Sum of columns. The *total* column is the sum of the columns in the Summary Column list.

Mean of columns. The *total* column is the average of the columns in the Summary Column list.

Minimum of columns. The *total* column is the minimum of the columns in the Summary Column list.

Maximum of columns. The *total* column is the maximum of the columns in the Summary Column list.

1st column – 2nd column. The *total* column is the difference of the columns in the Summary Column list. The Summary Column list must contain exactly two columns.

1st column / 2nd column. The *total* column is the quotient of the columns in the Summary Column list. The Summary Column list must contain exactly two columns.

% 1st column / 2nd column. The *total* column is the first column's percentage of the second column in the Summary Column list. The Summary Column list must contain exactly two columns.

Product of columns. The *total* column is the product of the columns in the Summary Column list.

Report Column Format

Data and break column formatting options for Report Summaries in Columns are the same as those described for Report Summaries in Rows.

Report Summaries in Columns Break Options

Break Options controls subtotal display, spacing, and pagination for break categories.

Figure 38-13
Report Break Options dialog box

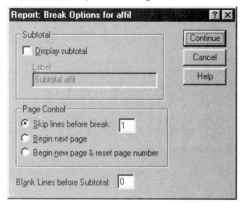

Subtotal. Controls the display subtotals for break categories.

Page Control. Controls spacing and pagination for categories of the selected break variable. You can specify a number of blank lines between break categories or start each break category on a new page.

Blank Lines before Subtotal. Controls the number of blank lines between break category data and subtotals.

Report Summaries in Columns Options

Options controls the display of grand totals, the display of missing values, and pagination in column summary reports.

Figure 38-14
Report Options dialog box

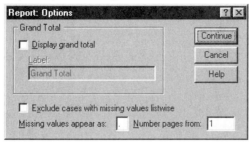

Grand Total. Displays and labels a grand total for each column; displayed at the bottom of the column.

Missing values. You can exclude missing values from the report or select a single character to indicate missing values in the report.

Report Layout for Summaries in Columns

Report layout options for Report Summaries in Columns are the same as those described for Report Summaries in Rows.

REPORT Command Additional Features

The SPSS command language also allows you to:

■ Display different summary functions in the columns of a single summary line.

■ Insert summary lines into data columns for variables other than the data column variable or for various combinations (composite functions) of summary functions.

■ Use Median, Mode, Frequency, and Percent as summary functions.

■ Control more precisely the display format of summary statistics.

■ Insert blank lines at various points in reports.

■ Insert blank lines after every *n*th case in listing reports.

Because of the complexity of the REPORT syntax, you may find it useful, when building a new report with syntax, to approximate the report generated from the dialog boxes, copy and paste the corresponding syntax, and refine that syntax to yield the exact report that you want.

See the *SPSS Command Syntax Reference* for complete syntax information.

Reliability Analysis

Reliability analysis allows you to study the properties of measurement scales and the items that compose the scales. The Reliability Analysis procedure calculates a number of commonly used measures of scale reliability and also provides information about the relationships between individual items in the scale. Intraclass correlation coefficients can be used to compute inter-rater reliability estimates.

Example. Does my questionnaire measure customer satisfaction in a useful way? Using reliability analysis, you can determine the extent to which the items in your questionnaire are related to each other, you can get an overall index of the repeatability or internal consistency of the scale as a whole, and you can identify problem items that should be excluded from the scale.

Statistics. Descriptives for each variable and for the scale, summary statistics across items, inter-item correlations and covariances, reliability estimates, ANOVA table, intraclass correlation coefficients, Hotelling's T^2, and Tukey's test of additivity.

Models. The following models of reliability are available:

- **Alpha (Cronbach).** This model is a model of internal consistency, based on the average inter-item correlation.

- **Split-half.** This model splits the scale into two parts and examines the correlation between the parts.

- **Guttman.** This model computes Guttman's lower bounds for true reliability.

- **Parallel.** This model assumes that all items have equal variances and equal error variances across replications.

- **Strict parallel.** This model makes the assumptions of the Parallel model and also assumes equal means across items.

Data. Data can be dichotomous, ordinal, or interval, but the data should be coded numerically.

Assumptions. Observations should be independent, and errors should be uncorrelated between items. Each pair of items should have a bivariate normal distribution. Scales should be additive, so that each item is linearly related to the total score.

Related procedures. If you want to explore the dimensionality of your scale items (to see whether more than one construct is needed to account for the pattern of item scores), use factor analysis or multidimensional scaling. To identify homogeneous groups of variables, use hierarchical cluster analysis to cluster variables.

To Obtain a Reliability Analysis

▶ From the menus choose:

Analyze
  Scale
    Reliability Analysis...

Figure 39-1
Reliability Analysis dialog box

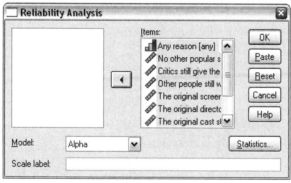

▶ Select two or more variables as potential components of an additive scale.

▶ Choose a model from the Model drop-down list.

Reliability Analysis Statistics

Figure 39-2
Reliability Analysis Statistics dialog box

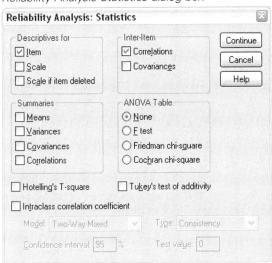

You can select various statistics that describe your scale and items. Statistics that are reported by default include the number of cases, the number of items, and reliability estimates as follows:

- **Alpha models: Coefficient alpha.** For dichotomous data, this is equivalent to the Kuder-Richardson 20 (KR20) coefficient.

- **Split-half models: Correlation between forms, Guttman split-half reliability, Spearman-Brown reliability** (equal and unequal length), and coefficient alpha for each half.

- **Guttman models: Reliability coefficients lambda 1 through lambda 6.**

- **Parallel and Strict parallel models: Test for goodness of fit of model; estimates of error variance, common variance, and true variance; estimated common inter-item correlation; estimated reliability; and unbiased estimate of reliability.**

Descriptives for. Produces descriptive statistics for scales or items across cases.

- **Item.** Produces descriptive statistics for items across cases.

■ **Scale.** Produces descriptive statistics for scales.

■ **Scale if item deleted.** Displays summary statistics comparing each item to the scale that is composed of the other items. Statistics include scale mean and variance if the item were to be deleted from the scale, correlation between the item and the scale that is composed of other items, and Cronbach's alpha if the item were to be deleted from the scale.

Summaries. Provides descriptive statistics of item distributions across all items in the scale.

■ **Means.** Summary statistics for item means. The smallest, largest, and average item means, the range and variance of item means, and the ratio of the largest to the smallest item means are displayed.

■ **Variances.** Summary statistics for item variances. The smallest, largest, and average item variances, the range and variance of item variances, and the ratio of the largest to the smallest item variances are displayed.

■ **Covariances.** Summary statistics for inter-item covariances. The smallest, largest, and average inter-item covariances, the range and variance of inter-item covariances, and the ratio of the largest to the smallest inter-item covariances are displayed.

■ **Correlations.** Summary statistics for inter-item correlations. The smallest, largest, and average inter-item correlations, the range and variance of inter-item correlations, and the ratio of the largest to the smallest inter-item correlations are displayed.

Inter-Item. Produces matrices of correlations or covariances between items.

ANOVA Table. Produces tests of equal means.

■ **F test.** Displays a repeated measures analysis-of-variance table.

■ **Friedman chi-square.** Displays Friedman's chi-square and Kendall's coefficient of concordance. This option is appropriate for data that are in the form of ranks. The chi-square test replaces the usual F test in the ANOVA table.

■ **Cochran chi-square.** Displays Cochran's Q. This option is appropriate for data that are dichotomous. The Q statistic replaces the usual F statistic in the ANOVA table.

Hotelling's T-square. Produces a multivariate test of the null hypothesis that all items on the scale have the same mean.

Tukey's test of additivity. Produces a test of the assumption that there is no multiplicative interaction among the items.

Intraclass correlation coefficient. Produces measures of consistency or agreement of values within cases.

- **Model.** Select the model for calculating the intraclass correlation coefficient. Available models are Two-Way Mixed, Two-Way Random, and One-Way Random. Select Two-Way Mixed when people effects are random and the item effects are fixed, select Two-Way Random when people effects and the item effects are random. Select One-Way Random when people effects are random.

- **Type.** Select the type of index. Available types are Consistency and Absolute Agreement.

- **Confidence interval.** Specify the level for the confidence interval. The default is 95%.

- **Test value.** Specify the hypothesized value of the coefficient for the hypothesis test. This value is the value to which the observed value is compared. The default value is 0.

RELIABILITY Command Additional Features

The SPSS command language also allows you to:

- Read and analyze a correlation matrix.

- Write a correlation matrix for later analysis.

- Specify splits other than equal halves for the split-half method.

See the *SPSS Command Syntax Reference* for complete syntax information.

Multidimensional Scaling

Multidimensional scaling attempts to find the structure in a set of distance measures between objects or cases. This task is accomplished by assigning observations to specific locations in a conceptual space (usually two- or three-dimensional) such that the distances between points in the space match the given dissimilarities as closely as possible. In many cases, the dimensions of this conceptual space can be interpreted and used to further understand your data.

If you have objectively measured variables, you can use multidimensional scaling as a data reduction technique (the Multidimensional Scaling procedure will compute distances from multivariate data for you, if necessary). Multidimensional scaling can also be applied to subjective ratings of dissimilarity between objects or concepts. Additionally, the Multidimensional Scaling procedure can handle dissimilarity data from multiple sources, as you might have with multiple raters or questionnaire respondents.

Example. How do people perceive relationships between different cars? If you have data from respondents indicating similarity ratings between different makes and models of cars, multidimensional scaling can be used to identify dimensions that describe consumers' perceptions. You might find, for example, that the price and size of a vehicle define a two-dimensional space, which accounts for the similarities that are reported by your respondents.

Statistics. For each model: data matrix, optimally scaled data matrix, S-stress (Young's), stress (Kruskal's), RSQ, stimulus coordinates, average stress and RSQ for each stimulus (RMDS models). For individual difference (INDSCAL) models: subject weights and weirdness index for each subject. For each matrix in replicated multidimensional scaling models: stress and RSQ for each stimulus. Plots: stimulus coordinates (two- or three-dimensional), scatterplot of disparities versus distances.

Data. If your data are dissimilarity data, all dissimilarities should be quantitative and should be measured in the same metric. If your data are multivariate data, variables can be quantitative, binary, or count data. Scaling of variables is an important issue—differences in scaling may affect your solution. If your variables have large differences in scaling (for example, one variable is measured in dollars and the other variable is measured in years), consider standardizing them (this process can be done automatically by the Multidimensional Scaling procedure).

Assumptions. The Multidimensional Scaling procedure is relatively free of distributional assumptions. Be sure to select the appropriate measurement level (ordinal, interval, or ratio) in the Multidimensional Scaling Options dialog box so that the results are computed correctly.

Related procedures. If your goal is data reduction, an alternative method to consider is factor analysis, particularly if your variables are quantitative. If you want to identify groups of similar cases, consider supplementing your multidimensional scaling analysis with a hierarchical or *k*-means cluster analysis.

To Obtain a Multidimensional Scaling Analysis

▶ From the menus choose:

Analyze
  Scale
    Multidimensional Scaling...

Figure 40-1
Multidimensional Scaling dialog box

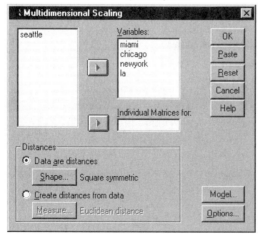

▶ In Distances, select either Data are distances or Create distances from data.

▶ If your data are distances, select at least four numeric variables for analysis. (You can also click Shape to indicate the shape of the distance matrix.)

▶ If you want SPSS to create the distances before analyzing them, select at least one numeric variable. (You can also click Measure to specify the type of distance measure that you want.) Each grouping variable can be either numeric or string, and you can create separate matrices for each category of a grouping variable by moving that variable into the Individual Matrices For list.

Multidimensional Scaling Shape of Data

Figure 40-2
Multidimensional Scaling Shape dialog box

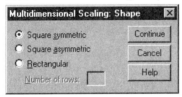

If your active dataset represents distances among a set of objects or represents distances between two sets of objects, specify the shape of your data matrix in order to get the correct results. *Note*: You cannot select Square symmetric if the Model dialog box specifies row conditionality.

Multidimensional Scaling Create Measure

Figure 40-3
Multidimensional Scaling Create Measure from Data dialog box

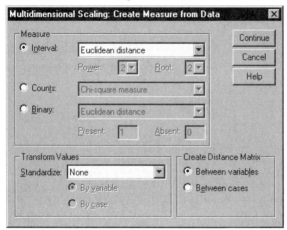

Multidimensional scaling uses dissimilarity data to create a scaling solution. If your data are multivariate data (values of measured variables), you must create dissimilarity data in order to compute a multidimensional scaling solution. You can specify the details of creating dissimilarity measures from your data.

Measure. Allows you to specify the dissimilarity measure for your analysis. Select one alternative from the Measure group corresponding to your type of data, and then choose one of the measures from the drop-down list corresponding to that type of measure. Available alternatives are:

- **Interval**. Euclidean distance, Squared Euclidean distance, Chebychev, Block, Minkowski, or Customized.

- **Counts**. Chi-square measure or Phi-square measure.

- **Binary**. Euclidean distance, Squared Euclidean distance, Size difference, Pattern difference, Variance, or Lance and Williams.

Create Distance Matrix. Allows you to choose the unit of analysis. Alternatives are Between variables or Between cases.

Transform Values. In certain cases, such as when variables are measured on very different scales, you may want to standardize values before computing proximities (not applicable to binary data). Choose a standardization method from the Standardize drop-down list. If no standardization is required, choose None).

Multidimensional Scaling Model

Figure 40-4
Multidimensional Scaling Model dialog box

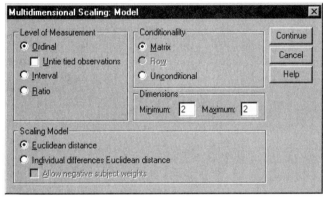

Correct estimation of a multidimensional scaling model depends on aspects of the data and the model itself.

Level of Measurement. Allows you to specify the level of your data. Alternatives are Ordinal, Interval, or Ratio. If your variables are ordinal, selecting Untie tied observations requests that the variables be treated as continuous variables, so that ties (equal values for different cases) are resolved optimally.

Conditionality. Allows you to specify which comparisons are meaningful. Alternatives are Matrix, Row, or Unconditional.

Dimensions. Allows you to specify the dimensionality of the scaling solution(s). One solution is calculated for each number in the range. Specify integers between 1 and 6; a minimum of 1 is allowed only if you select Euclidean distance as the scaling model. For a single solution, specify the same number for minimum and maximum.

Scaling Model. Allows you to specify the assumptions by which the scaling is performed. Available alternatives are Euclidean distance or Individual differences Euclidean distance (also known as INDSCAL). For the Individual differences

Euclidean distance model, you can select Allow negative subject weights, if appropriate for your data.

Multidimensional Scaling Options

Figure 40-5
Multidimensional Scaling Options dialog box

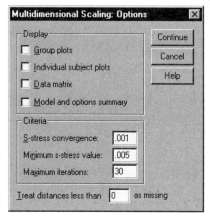

You can specify options for your multidimensional scaling analysis.

Display. Allows you to select various types of output. Available options are Group plots, Individual subject plots, Data matrix, and Model and options summary.

Criteria. Allows you to determine when iteration should stop. To change the defaults, enter values for S-stress convergence, Minimum s-stress value, and Maximum iterations.

Treat distances less than n as missing. Distances that are less than this value are excluded from the analysis.

ALSCAL Command Additional Features

The SPSS command language also allows you to:

- Use three additional model types, known as ASCAL, AINDS, and GEMSCAL in the literature about multidimensional scaling.
- Carry out polynomial transformations on interval and ratio data.
- Analyze similarities (rather than distances) with ordinal data.

- Analyze nominal data.

- Save various coordinate and weight matrices into files and read them back in for analysis.

- Constrain multidimensional unfolding.

See the *SPSS Command Syntax Reference* for complete syntax information.

Ratio Statistics

The Ratio Statistics procedure provides a comprehensive list of summary statistics for describing the ratio between two scale variables.

You can sort the output by values of a grouping variable in ascending or descending order. The ratio statistics report can be suppressed in the output, and the results can be saved to an external file.

Example. Is there good uniformity in the ratio between the appraisal price and sale price of homes in each of five counties? From the output, you might learn that the distribution of ratios varies considerably from county to county.

Statistics. Median, mean, weighted mean, confidence intervals, coefficient of dispersion (COD), median-centered coefficient of variation, mean-centered coefficient of variation, price-related differential (PRD), standard deviation, average absolute deviation (AAD), range, minimum and maximum values, and the concentration index computed for a user-specified range or percentage within the median ratio.

Data. Use numeric codes or short strings to code grouping variables (nominal or ordinal level measurements).

Assumptions. The variables that define the numerator and denominator of the ratio should be scale variables that take positive values.

To Obtain Ratio Statistics

▶ From the menus choose:
Analyze
  Descriptive Statistics
    Ratio...

Figure 41-1
Ratio Statistics dialog box

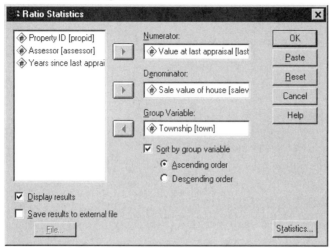

▶ Select a numerator variable.

▶ Select a denominator variable.

▶ Optionally:

■ Select a grouping variable and specify the ordering of the groups in the results.

■ Choose whether to display the results in the Viewer.

■ Choose whether to save the results to an external file for later use, and specify the name of the file to which the results are saved.

Ratio Statistics

Figure 41-2
Ratio Statistics dialog box

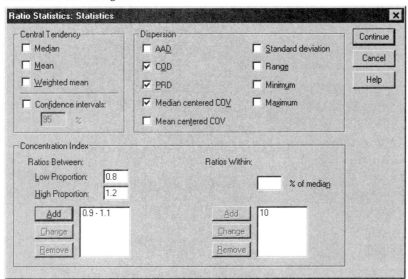

Central Tendency. Measures of central tendency are statistics that describe the distribution of ratios.

■ **Median.** The value such that the number of ratios that are less than this value and the number of ratios that are greater than this value are the same.

■ **Mean.** The result of summing the ratios and dividing the result by the total number of ratios.

■ **Weighted mean.** The result of dividing the mean of the numerator by the mean of the denominator. Weighted mean is also the mean of the ratios weighted by the denominator.

■ **Confidence intervals.** Displays confidence intervals for the mean, the median, and the weighted mean (if requested). Specify a value that is greater than or equal to 0 and less than 100 as the confidence level.

Dispersion. These statistics measure the amount of variation, or spread, in the observed values.

- **AAD.** The average absolute deviation is the result of summing the absolute deviations of the ratios about the median and dividing the result by the total number of ratios.

- **COD.** The coefficient of dispersion is the result of expressing the average absolute deviation as a percentage of the median.

- **PRD.** The price-related differential, also known as the index of regressivity, is the result of dividing the mean by the weighted mean.

- **Median centered COV.** The median-centered coefficient of variation is the result of expressing the root mean squares of deviation from the median as a percentage of the median.

- **Mean centered COV.** The mean-centered coefficient of variation is the result of expressing the standard deviation as a percentage of the mean.

- **Standard deviation.** The standard deviation is the result of summing the squared deviations of the ratios about the mean, dividing the result by the total number of ratios minus one, and taking the positive square root.

- **Range.** The range is the result of subtracting the minimum ratio from the maximum ratio.

- **Minimum.** The minimum is the smallest ratio.

- **Maximum.** The maximum is the largest ratio.

Concentration Index. The coefficient of concentration measures the percentage of ratios that fall within an interval. It can be computed in two different ways:

- **Ratios Between.** Here the interval is defined explicitly by specifying the low and high values of the interval. Enter values for the low proportion and high proportion, and click Add to obtain an interval.

- **Ratios Within.** Here the interval is defined implicitly by specifying the percentage of the median. Enter a value between 0 and 100, and click Add. The lower end of the interval is equal to $(1 - 0.01 \times \text{value}) \times \text{median}$, and the upper end is equal to $(1 + 0.01 \times \text{value}) \times \text{median}$.

ROC Curves

This procedure is a useful way to evaluate the performance of classification schemes in which there is one variable with two categories by which subjects are classified.

Example. It is in a bank's interest to correctly classify customers into those customers who will and will not default on their loans, so special methods are developed for making these decisions. ROC curves can be used to evaluate how well these methods perform.

Statistics. Area under the ROC curve with confidence interval and coordinate points of the ROC curve. Plots: ROC curve.

Methods. The estimate of the area under the ROC curve can be computed either nonparametrically or parametrically using a binegative exponential model.

Data. Test variables are quantitative. Test variables are often composed of probabilities from discriminant analysis or logistic regression or composed of scores on an arbitrary scale indicating a rater's "strength of conviction" that a subject falls into one category or another category. The state variable can be of any type and indicates the true category to which a subject belongs. The value of the state variable indicates which category should be considered *positive*.

Assumptions. It is assumed that increasing numbers on the rater scale represent the increasing belief that the subject belongs to one category, while decreasing numbers on the scale represent the increasing belief that the subject belongs to the other category. The user must choose which direction is *positive*. It is also assumed that the *true* category to which each subject belongs is known.

Chapter 42

Figure 42-1
ROC Curve output

Case Processing Summary

ACTUAL	Valid N (listwise)
Positive[1]	74
Negative	76

Larger values of the test result variable(s) indicate stronger evidence for a positive actual state.

[1]. The positive actual state is 1.00.

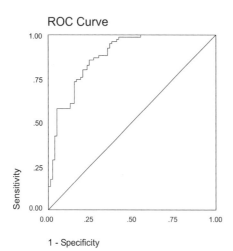

ROC Curve

Area Under the Curve

Test Result Variable(s): PROBS

Area	Std. Error[1]	Asymptotic Sig.[2]	Asymptotic 95% Confidence Interval	
			Lower Bound	Upper Bound
.877	.028	.000	.823	.931

[1]. Under the nonparametric assumption

[2]. Null hypothesis: true area = 0.5

To Obtain an ROC Curve

▶ From the menus choose:

Analyze
 ROC Curve...

Figure 42-2
ROC Curve dialog box

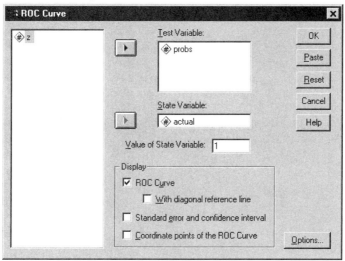

▶ Select one or more test probability variables.

▶ Select one state variable.

▶ Identify the *positive* value for the state variable.

Chapter 42

ROC Curve Options

Figure 42-3
ROC Curve Options dialog box

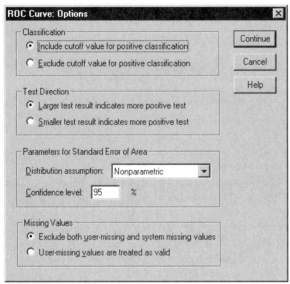

You can specify the following options for your ROC analysis:

Classification. Allows you to specify whether the cutoff value should be included or excluded when making a *positive* classification. This setting currently has no effect on the output.

Test Direction. Allows you to specify the direction of the scale in relation to the *positive* category.

Parameters for Standard Error of Area. Allows you to specify the method of estimating the standard error of the area under the curve. Available methods are nonparametric and binegative exponential. Also allows you to set the level for the confidence interval. The available range is 50.1% to 99.9%.

Missing Values. Allows you to specify how missing values are handled.

Overview of the Chart Facility

High-resolution charts and plots are created by the procedures on the Graphs menu and by many of the procedures on the Analyze menu. This chapter provides an overview of the chart facility.

Creating and Modifying a Chart

Before you can create a chart, you need to have your data in the Data Editor. You can enter the data directly into the Data Editor, open a previously saved data file, or read a spreadsheet, tab-delimited data file, or database file. The Tutorial selection on the Help menu has online examples of creating and modifying a chart, and the online Help system provides information about creating and modifying all chart types.

Creating the Chart

▶ After you get your data into the Data Editor, select Chart Builder from the Graphs menu. This opens the Chart Builder dialog box.

Figure 43-1
Chart Builder dialog box

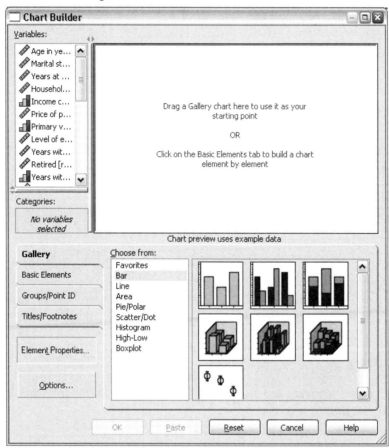

The Gallery tab provides a number of common charts, organized by chart types.

▶ In the Chart Builder, drag the icon representing the chart onto the "canvas," which is the large area that appears above the gallery.

▶ Drag variables from the Variables list to the axis drop zones. (For more information about the Chart Builder, click Help.)

When the chart definition is complete, it will look similar to the following chart.

Figure 43-2
Chart Builder dialog box with completed drop zones

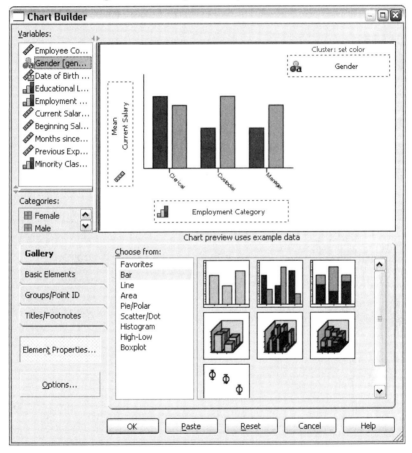

▶ If you need to change statistics or modify attributes of the axes or legends, click Element Properties.

Figure 43-3
Element Properties window

▶ In the Edit Properties of list, select the item you want to change. (For information about the specific properties, click Help.)

▶ After making any changes, click Apply.

▶ In the Chart Builder dialog box, click OK to create the chart. The chart is displayed in the Viewer.

Figure 43-4
Bar chart displayed in Viewer window

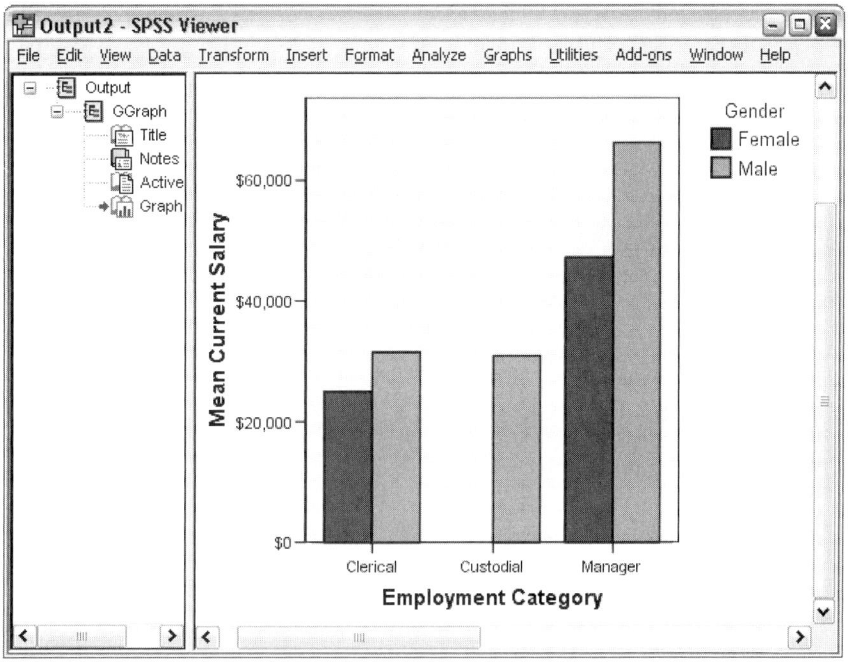

Modifying the Chart

To modify a chart, double-click anywhere on the chart that is displayed in the Viewer. This displays the chart in the Chart Editor.

Figure 43-5
Original chart in the Chart Editor

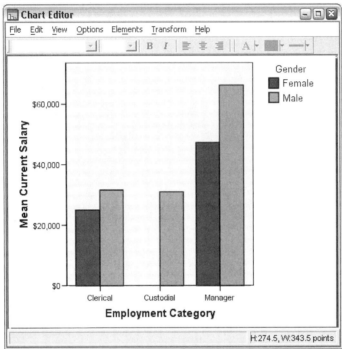

You can modify any part of the chart or change to another type of chart that illustrates the same data. You add items, show items, or hide items by using the menus in the Chart Editor.

To Modify a Chart Item

▶ Select the item that you want to modify.

▶ From the menus choose:
Edit
  Properties...

This opens the Properties window. The tabs that appear in the Properties window are specific to your selection. The online Help describes how to display the tabs that you need.

Figure 43-6
Properties window

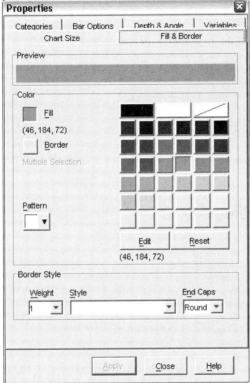

Some typical modifications include:

- Editing text in the chart.

- Changing the color and fill pattern of the bars.

- Adding text to the chart, such as a title or an annotation.

- Changing the location of the bar origin line.

Following is a modified chart.

Figure 43-7
Modified chart

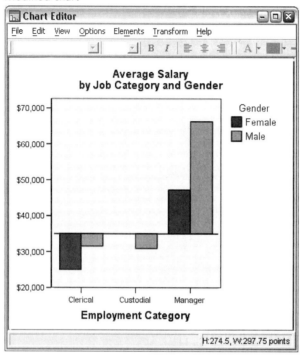

Chart modifications are saved when you close the chart window, and the modified chart is displayed in the Viewer.

Chart Definition Options

When you are defining a chart, you can add titles and change options for the chart creation. In the Chart Builder, click the Optional Elements tab to specify titles, subtitles, and footnotes. You can click Options to control various chart options, such as the treatment of missing values or the use of templates. The next few sections describe how to define these characteristics at the time that you define the chart.

Titles, Subtitles, and Footnotes

In any chart, you can define two title lines, one subtitle line, and two footnote lines as part of your original chart definition. To specify titles or footnotes while defining a chart, click the Titles/Footnotes tab in the Chart Builder dialog box. Then select one or more of the titles and footnotes. You can modify the text that is associated with the item by clicking Element Properties and selecting the item from the Edit Properties Of list.

Each line can be up to 72 characters. The number of characters that will actually fit in the chart depends on the font and size. Most titles are left-justified by default and are cropped on the right if they are too long. Pie chart titles are center-justified by default and are cropped at both ends if they are too long.

Titles, subtitles, and footnotes are rendered as text boxes in the Chart Editor. Within the Chart Editor, you can add, delete, or revise text boxes as well as change their font, size, and justification.

Options

The Options dialog box provides options for the chart that you are creating. This dialog box is available by clicking Options.

Figure 43-8
Options dialog box

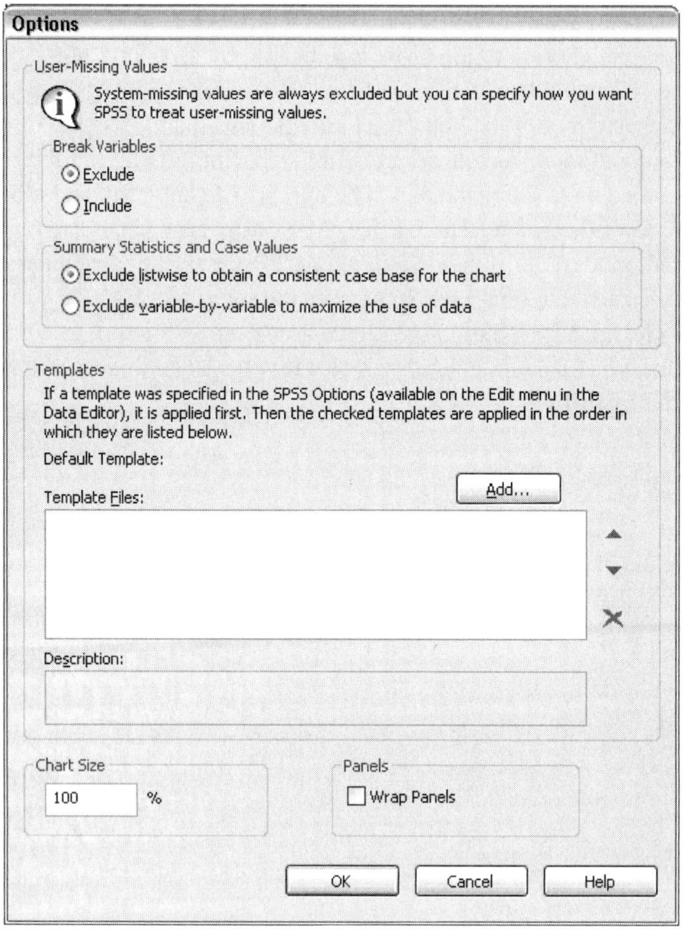

Missing Values

Break Variables

If there are missing values in the data for variables that are used to define categories or subgroups, select Include so that the category or categories of user-missing values (values that are identified as missing by the user) are included in the chart. These categories also act as break variables in calculating the statistic. "Missing" categories are displayed on the category axis or in the legend, adding, for example, an extra bar or a slice to a pie chart. If there are no missing values, the "missing" categories are not displayed.

If you select this option and want to suppress display after the chart is drawn, select the chart, and then choose Properties from the Edit menu. Click the Categories tab and move the categories that you want suppressed to the Excluded list.

Summary Statistics and Case Values

You can choose one of the following alternatives to exclude cases that have missing values:

- **Exclude listwise to obtain a consistent case base for the chart**. If any of the variables in the chart has a missing value for a given case, the whole case is excluded from the chart.

- **Exclude variable-by-variable to maximize the use of data**. If a selected variable has any missing values, the cases that have those missing values are excluded when the variable is analyzed.

To see the difference between listwise and variable-by-variable exclusion of missing values, consider the following figures, which show a bar chart for each of the two options.

Figure 43-9
Listwise exclusion of missing values

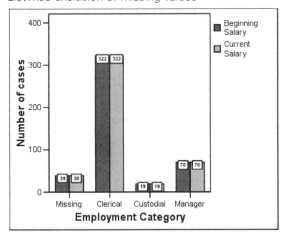

Figure 43-10
Variable-by-variable exclusion of missing values

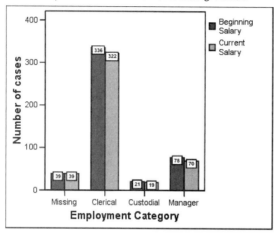

The charts were created from a version of the *Employee data.sav* file that was edited to have some system-missing (blank) values in the variables for current salary and job category. In some other cases, the value 0 was entered and defined as missing. For both charts, the option Display groups defined by missing values is selected, which adds the category *Missing* to the other displayed job categories. In each chart, the values of the summary function, *Number of cases*, are displayed in the bar labels.

In both charts, 26 cases have a system-missing value for the job category, and 13 cases have the user-missing value (0). In the listwise chart, the number of cases is the same for both variables in each bar cluster because whenever a value was missing, the case was excluded for all variables. In the variable-by-variable chart, the number of nonmissing cases for each variable in a category is plotted without regard to missing values in other variables.

Chart Templates

You can apply many of the attributes and text elements from one chart to another chart. This capability allows you to modify one chart, save that chart as a template, and then use the template to create a number of other similar charts.

To use a template when creating a chart, click Add in the Templates group. This opens a standard file selection dialog box. If you add multiple templates, the templates are applied in the order in which they appear in Template Files list. You can change the order if needed.

To apply a template to a chart that is already in the Chart Editor, from the menus choose:
File
  Apply Chart Template...

This opens a standard file selection dialog box. Select a file to use as a template. If you are creating a new chart, the filename that you select is displayed in the Template group when you return to the chart definition dialog box.

A template borrows the format from one chart and applies it to the new chart that you are generating. In general, any formatting information from the old chart that can apply to the new chart will automatically apply. For example, if the old chart is a clustered bar chart with bar colors modified to yellow and green, and the new chart is a multiple line chart, the lines will be yellow and green. If the old chart is a simple bar chart with drop shadows, and the new chart is a simple line chart, the lines will not have drop shadows because drop shadows don't apply to line charts. If there are titles in the template chart but not in the new chart, you will get the titles from the template chart. If there are titles defined in the new chart, they will override the titles in the template chart.

To Create a Chart Template

▶ Create a chart.

▶ Edit the chart to contain the attributes that you want to have in a template.

▶ From the Chart Editor menus choose:
File
  Save Chart Template...

▶ In the Save Chart Template dialog box, specify which characteristics of the chart you want to save in the template. The online Help describes the settings in detail.

▶ Click Continue.

▶ Enter a filename and location for the new template. The template's extension is *.sgt*.

Chart Size and Panels

You can use the Groups/Point ID tab to add paneling to the chart. A paneled chart is like a table of charts, with the rows and columns determined by categorical variables. Depending on the number of categories in the paneled variables, each panel may be very small. To improve this behavior, you can choose to increase the chart size or to force the panels to wrap. You may also want to change the chart size for other reasons, like decreasing the size for publication.

Chart Size. Specify a percentage greater than 100 to enlarge the chart or less than 100 to shrink it. The percentage is relative to the default chart size.

Panels. When there are many panel columns, select Wrap Panels to allow panels to wrap across rows rather than being forced to fit in a specific row. Unless this option is selected, the panels are shrunk to force them to fit in a row.

Utilities

This chapter describes the functions found on the Utilities menu and the ability to reorder target variable lists using the Windows system menus.

Variable Information

The Variables dialog box displays variable definition information for the currently selected variable, including:

- Data format
- Variable label
- User-missing values
- Value labels

Figure 44-1
Variables dialog box

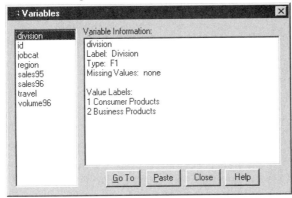

Go To. Goes to the selected variable in the Data Editor window.

Paste. Pastes the selected variables into the designated syntax window at the cursor location.

To modify variable definitions, use the Variable view in the Data Editor.

To Obtain Variable Information

▶ From the menus choose:
Utilities
  Variables...

▶ Select the variable for which you want to display variable definition information.

Data File Comments

You can include descriptive comments with a data file. For SPSS-format data files, these comments are saved with the data file.

To Add, Modify, Delete, or Display Data File Comments

▶ From the menus choose:
Utilities
  Data File Comments...

▶ To display the comments in the Viewer, select Display comments in output.

Comments can be any length but are limited to 80 bytes (typically 80 characters in single-byte languages) per line; lines will automatically wrap at 80 characters. Comments are displayed in the same font as text output to accurately reflect how they will appear when displayed in the Viewer.

A date stamp (the current date in parentheses) is automatically appended to the end of the list of comments whenever you add or modify comments. This may lead to some ambiguity concerning the dates associated with comments if you modify an existing comment or insert a new comment between existing comments.

Variable Sets

You can restrict the variables that are displayed in the Data Editor and in dialog variable lists by defining and using variable sets. This is particularly useful for data files with a large number of variables. Small variable sets make it easier to find and select the variables for your analysis.

Define Variable Sets

Define Variable Sets creates subsets of variables to display in the Data Editor and in dialog box variable lists. Defined variable sets are saved with SPSS-format data files.

Figure 44-2
Define Variable Sets dialog box

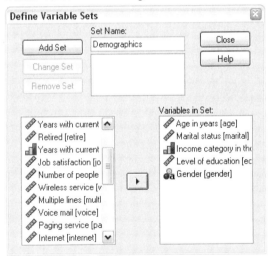

Set Name. Set names can be up to 64 bytes long. Any characters, including blanks, can be used.

Variables in Set. Any combination of numeric and string variables can be included in a set. The order of variables in the set has no effect on the display order of the variables in the Data Editor or in dialog box variable lists. A variable can belong to multiple sets.

To Define Variable Sets

▶ From the menus choose:
Utilities
  Define Variable Sets...

▶ Select the variables that you want to include in the set.

▶ Enter a name for the set (up to 64 bytes).

▶ Click Add Set.

Use Variable Sets

Use Variable Sets restricts the variables displayed in the Data Editor and in dialog box variable lists to the variables in the selected (checked) sets.

Figure 44-3
Use Variable Sets dialog box

- The set of variables displayed in the Data Editor and in dialog box variable lists is the union of all selected sets.
- A variable can be included in multiple selected sets.

■ The order of variables in the selected sets and the order of selected sets have no effect on the display order of variables in the Data Editor or in dialog box variable lists.

■ Although the defined variable sets are saved with SPSS-format data files, the list of currently selected sets is reset to the default, built-in sets each time you open the data file.

The list of available variable sets includes any variable sets defined for the active dataset, plus two built-in sets:

■ **ALLVARIABLES.** This set contains all variables in the data file, including new variables created during a session.

■ **NEWVARIABLES.** This set contains only new variables created during the session.

Note: Even if you save the data file after creating new variables, those new variables are still included in the *NEWVARIABLES* set until you close and reopen the data file.

At least one variable set must be selected. If *ALLVARIABLES* is selected, any other selected sets will not have any visible effect, since this set contains all variables.

To Select Variable Sets to Display

▶ From the menus choose:
Utilities
  Use Variable Sets...

▶ Select the defined variable sets that contain the variables that you want to appear in the Data Editor and in dialog box variable lists.

To Display All Variables

▶ From the menus choose:
Utilities
  Show All Variables

Reordering Target Variable Lists

Variables appear on dialog box target lists in the order in which they are selected from the source list. If you want to change the order of variables on a target list—but you don't want to deselect all of the variables and reselect them in the new order—you can move variables up and down on the target list using the system menu in the upper left corner of the dialog box (accessed by clicking the left side of the dialog box title bar).

Figure 44-4
Windows system menu with target list reordering

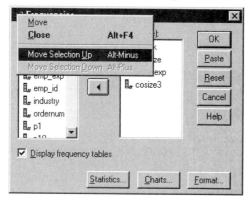

Move Selection Up. Moves the selected variable(s) up one position on the target list.

Move Selection Down. Moves the selected variable(s) down one position on the target list.

You can move multiple variables simultaneously if they are contiguous (grouped together). You cannot move noncontiguous groups of variables.

Options

Options control a wide variety of settings, including:

- Session journal, which keeps a record of all commands run in every session
- Display order for variables in dialog box source lists
- Items displayed and hidden in new output results
- TableLook for new pivot tables and ChartLook for new interactive charts
- Custom currency formats
- Autoscript files and autoscript functions to customize output

To Change Options Settings

▶ From the menus choose:
Edit
  Options...

▶ Click the tabs for the settings that you want to change.

▶ Change the settings.

▶ Click OK or Apply.

General Options

Figure 45-1
Options dialog box, General tab

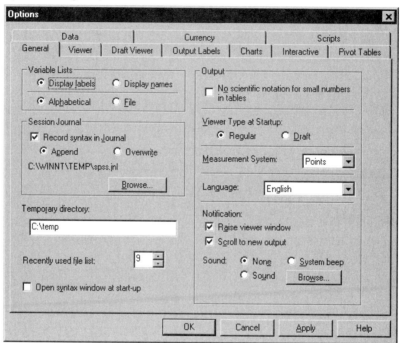

Variable Lists. Controls display of variables in dialog box lists. You can display variable names or variable labels. Names or labels can be displayed in alphabetical order or in file order, which is the order in which they actually occur in the data file (and are displayed in the Data Editor window). Display order affects only source variable lists. Target variable lists always reflect the order in which variables were selected.

Session Journal. The journal file of all commands run in a session. This includes commands entered and run in syntax windows and commands generated by dialog box choices. You can edit the journal file and use the commands again in other sessions. You can turn journaling off and on, append or overwrite the journal file, and select the journal filename and location. You can copy command syntax from the journal file and save it in a syntax file for use with the automated Production Facility.

Temporary directory. Controls the location of temporary files created during a session. In distributed mode (available with the server version), this does not affect the location of temporary data files. In distributed mode, the location of temporary data files is controlled by the environment variable *SPSSTMPDIR*, which can be set only on the computer running the server version of the software. If you need to change the location of the temporary directory, contact your system administrator.

Recently used file list. Controls the number of recently used files that appear on the File menu.

Open syntax window at start-up. Syntax windows are text file windows used to enter, edit, and run commands. If you frequently work with command syntax, select this option to automatically open a syntax window at the beginning of each session. This is useful primarily for experienced users who prefer to work with command syntax instead of dialog boxes. (Not available with the Student Version.)

No scientific notation for small numbers in tables. Suppresses the display of scientific notation for small decimal values in output. Very small decimal values will be displayed as 0 (or 0.000).

Viewer Type at Start-up. Controls the type of Viewer used and the output format. The Viewer produces interactive pivot tables and interactive charts. The Draft Viewer converts pivot tables to text output and charts to metafiles.

Measurement System. The measurement system used (points, inches, or centimeters) for specifying attributes such as pivot table cell margins, cell widths, and space between tables for printing.

Language. Controls the language used in the output. Does not apply to simple text output, interactive graphics, or maps (available with the Maps add-on module). The list of available languages depends on the currently installed language files.

Note: Custom scripts that rely on language-specific text strings in the output may not run correctly when you change the output language. For more information, see "Script Options" on p. 655.

Notification. Controls the manner in which the program notifies you that it has finished running a procedure and that the results are available in the Viewer.

Viewer Options

Viewer output display options affect only new output produced after you change the settings. Output already displayed in the Viewer is not affected by changes in these settings.

Figure 45-2
Options dialog box, Viewer tab

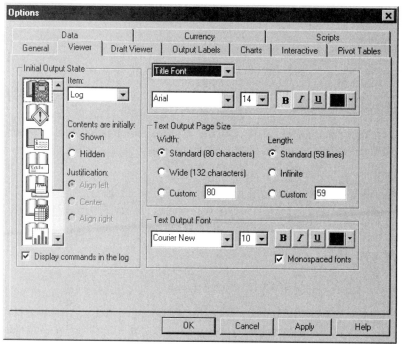

Initial Output State. Controls which items are automatically displayed or hidden each time you run a procedure and how items are initially aligned. You can control the display of the following items: log, warnings, notes, titles, pivot tables, charts, and text output (output not displayed in pivot tables). You can also turn the display of commands in the log on or off. You can copy command syntax from the log and save it in a syntax file for use with the automated Production Facility.

Note: All output items are displayed left-aligned in the Viewer. Only the alignment of printed output is affected by the justification settings. Centered and right-aligned items are identified by a small symbol above and to the left of the item.

Title Font. Controls the font style, size, and color for new output titles.

Page Title Font. Controls the font style, size, and color for new page titles and page titles generated by TITLE and SUBTITLE command syntax or created by New Page Title on the Insert menu.

Text Output Page Size. For text output, controls the page width (expressed in number of characters) and page length (expressed in number of lines). For some procedures, some statistics are displayed only in wide format.

Text Output Font. Font used for text output. Text output is designed for use with a monospaced (fixed-pitch) font. If you select a nonmonospaced font, tabular output will not align properly. The text output font is also used in the Text Wizard to display the file contents, which will not be aligned if the font is not monospaced.

Draft Viewer Options

Draft Viewer output display options affect only new output produced after you change the settings. Output already displayed in the Draft Viewer is not affected by changes in these settings.

Figure 45-3
Options dialog box, Draft Viewer tab

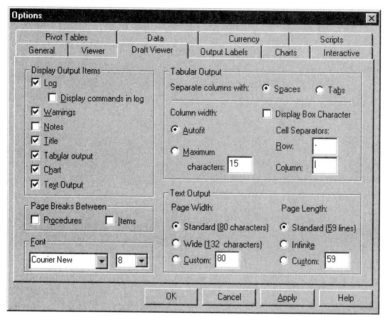

Display Output Items. Controls which items are automatically displayed each time that you run a procedure. You can control the display of the following items: log, warnings, notes, titles, tabular output (pivot tables converted to text output), charts, and text output (space-separated output). You can also turn the display of commands in the log on or off. You can copy command syntax from the log and save it in a syntax file for use with the automated Production Facility.

Page Breaks Between. Inserts page breaks between output from different procedures and/or between individual output items.

Font. The font used for new output. Only fixed-pitch (monospaced) fonts are available because space-separated text output will not align properly with a proportional font.

Tabular Output. Controls settings for pivot table output converted to tabular text output. Column width and column separator specifications are available only if you select Spaces for the column separator. For space-separated tabular output, by default all line wrapping is removed and each column is set to the width of the longest label or value in the column. To limit the width of columns and wrap long labels, specify a number of characters for the column width.

Note: Tab-separated tabular output will not align properly in the Draft Viewer. This format is useful for copying and pasting results to word-processing applications where you can use any font that you want (not only fixed-pitch fonts) and set the tabs to align output properly.

Text Output. For text output other than converted pivot table output, controls the page width (expressed in number of characters) and page length (expressed in number of lines). For some procedures, some statistics are displayed only in wide format.

Output Label Options

Output label options control the display of variable and data value information in the outline and pivot tables. You can display variable names, defined variable labels and actual data values, defined value labels, or a combination.

Descriptive variable and value labels (Variable view in the Data Editor, *Label* and *Values* columns) often make it easier to interpret your results. However, long labels can be awkward in some tables.

Figure 45-4
Options dialog box, Output Labels tab

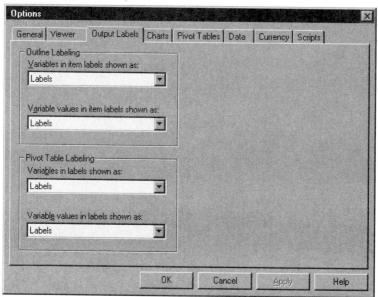

Output label options affect only new output produced after you change the settings. Output already displayed in the Viewer is not affected by changes in these settings. These settings affect only pivot table output. Text output is not affected by these settings.

Chart Options

Figure 45-5
Options dialog box, Charts tab

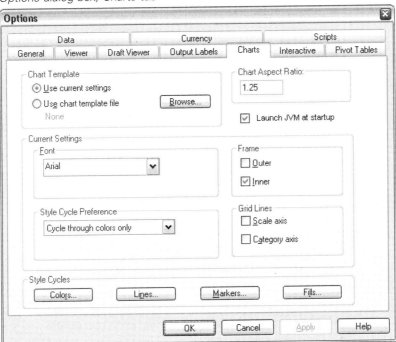

Chart Template. New charts can use either the settings selected here or the settings from a chart template file. Click Browse to select a chart template file. To create a chart template file, create a chart with the attributes that you want and save it as a template (choose Save Chart Template from the File menu).

Chart Aspect Ratio. The width-to-height ratio of the outer frame of new charts. You can specify a width-to-height ratio from 0.1 to 10.0. Values less than 1 make charts that are taller than they are wide. Values greater than 1 make charts that are wider than they are tall. A value of 1 produces a square chart. Once a chart is created, its aspect ratio cannot be changed.

Launch JVM at startup. Charting features require the Java Virtual Machine (JVM). By default, the JVM is launched when you start an SPSS session. If you deselect (uncheck) this option, SPSS may start faster, but there will be slight delays while the JVM is launched the first time you use the Chart Builder or create a chart in a session.

Font. Font used for all text in new charts.

Style Cycle Preference. The initial assignment of colors and patterns for new charts. Cycle through colors only uses only colors to differentiate chart elements and does not use patterns. Cycle through patterns only uses only line styles, marker symbols, or fill patterns to differentiate chart elements and does not use color.

Frame. Controls the display of inner and outer frames on new charts.

Grid Lines. Controls the display of scale and category axis grid lines on new charts.

Style Cycles. Customizes the colors, line styles, marker symbols, and fill patterns for new charts. You can change the order of the colors and patterns that are used when a new chart is created.

Note: These settings have no effect on interactive charts (the Graphs menu's Interactive submenu).

Data Element Colors

Specify the order in which colors should be used for the data elements (such as bars and markers) in your new chart. Colors are used whenever you select a choice that includes *color* in the Style Cycle Preference group in the main Chart Options dialog box.

For example, if you create a clustered bar chart with two groups and you select Cycle through colors, then patterns in the main Chart Options dialog box, the first two colors in the Grouped Charts list are used as the bar colors on the new chart.

To Change the Order in Which Colors Are Used

▶ Select Simple Charts and then select a color that is used for charts without categories.

▶ Select Grouped Charts to change the color cycle for charts with categories. To change a category's color, select a category and then select a color for that category from the palette.

Optionally, you can:

■ Insert a new category above the selected category.

■ Move a selected category.

■ Remove a selected category.

- Reset the sequence to the default sequence.

- Edit a color by selecting its well and then clicking Edit.

Data Element Lines

Specify the order in which styles should be used for the line data elements in your new chart. Line styles are used whenever your chart includes line data elements and you select a choice that includes *patterns* in the Style Cycle Preference group in the main Chart Options dialog box.

For example, if you create a line chart with two groups and you select Cycle through patterns only in the main Chart Options dialog box, the first two styles in the Grouped Charts list are used as the line patterns on the new chart.

To Change the Order in Which Line Styles Are Used

▶ Select Simple Charts and then select a line style that is used for line charts without categories.

▶ Select Grouped Charts to change the pattern cycle for line charts with categories. To change a category's line style, select a category and then select a line style for that category from the palette.

Optionally, you can:

- Insert a new category above the selected category.

- Move a selected category.

- Remove a selected category.

- Reset the sequence to the default sequence.

Data Element Markers

Specify the order in which symbols should be used for the marker data elements in your new chart. Marker styles are used whenever your chart includes marker data elements and you select a choice that includes *patterns* in the Style Cycle Preference group in the main Chart Options dialog box.

For example, if you create a scatterplot chart with two groups and you select Cycle through patterns only in the main Chart Options dialog box, the first two symbols in the Grouped Charts list are used as the markers on the new chart.

To Change the Order in Which Marker Styles Are Used

▶ Select Simple Charts and then select a marker symbol that is used for charts without categories.

▶ Select Grouped Charts to change the pattern cycle for charts with categories. To change a category's marker symbol, select a category and then select a symbol for that category from the palette.

Optionally, you can:

■ Insert a new category above the selected category.

■ Move a selected category.

■ Remove a selected category.

■ Reset the sequence to the default sequence.

Data Element Fills

Specify the order in which fill styles should be used for the bar and area data elements in your new chart. Fill styles are used whenever your chart includes bar or area data elements, and you select a choice that includes *patterns* in the Style Cycle Preference group in the main Chart Options dialog box.

For example, if you create a clustered bar chart with two groups and you select Cycle through patterns only in the main Chart Options dialog box, the first two styles in the Grouped Charts list are used as the bar fill patterns on the new chart.

To Change the Order in Which Fill Styles Are Used

▶ Select Simple Charts and then select a fill pattern that is used for charts without categories.

▶ Select Grouped Charts to change the pattern cycle for charts with categories. To change a category's fill pattern, select a category and then select a fill pattern for that category from the palette.

Optionally, you can:

- Insert a new category above the selected category.

- Move a selected category.

- Remove a selected category.

- Reset the sequence to the default sequence.

Interactive Chart Options

Figure 45-6
Options dialog box, Interactive tab

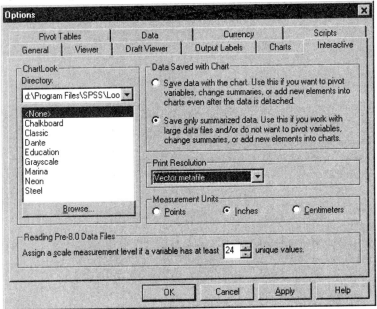

For interactive charts (Graphs menu, Interactive submenu), the following options
are available:

ChartLook. Select a ChartLook from the list of files and click OK or Apply. By default,
the list displays the ChartLooks saved in the *Looks* directory of the directory in which
the program is installed. You can use one of the ChartLooks provided with the
program, or you can create your own in the Interactive Graphics Editor (in an activated
chart, choose ChartLooks from the Format menu).

- **Directory.** Allows you to select a ChartLook directory. Use Browse to add directories to the list.

- **Browse.** Allows you to select a ChartLook from another directory.

Data Saved with Chart. Controls information saved with interactive charts once the charts are no longer attached to the data file that created them (for example, if you open a Viewer file saved in a previous session). Saving data with the chart enables you to perform most of the interactive functions available for charts attached to the data file that created them (except adding variables that weren't included in the original chart). However, this can substantially increase the size of Viewer files, particularly for large data files.

Print Resolution. Controls the print resolution of interactive charts. In most cases, Vector metafile will print faster and provide the best results. For bitmaps, lower resolution charts print faster; higher resolution charts look better.

Measurement Units. Measurement system used (points, inches, or centimeters) for specifying attributes, such as the size of the data region in a chart.

Reading Pre-8.0 Data Files. For data files created in older versions of SPSS, data read from external file formats, and new variables created in a session, you can specify the minimum number of data values for a numeric variable used to classify the variable as scale or nominal. Variables with fewer than the specified number of unique values are classified as nominal.

Note: With the exception of measurement level, these settings affect only interactive charts (the Graphs menu's Interactive submenu).

Pivot Table Options

Pivot Table options sets the default TableLook used for new pivot table output. TableLooks can control a variety of pivot table attributes, including the display and width of grid lines; font style, size, and color; and background colors.

Figure 45-7
Options dialog box, Pivot Tables tab

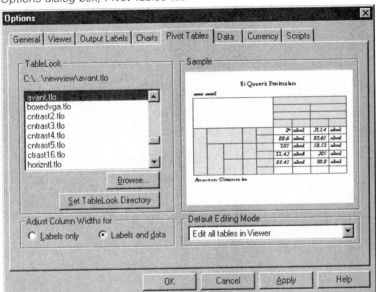

TableLook. Select a TableLook from the list of files and click OK or Apply. By default, the list displays the TableLooks saved in the *Looks* directory of the directory in which the program is installed. You can use one of the TableLooks provided with the program, or you can create your own in the Pivot Table Editor (choose TableLooks from the Format menu).

■ **Browse.** Allows you to select a TableLook from another directory.

■ **Set TableLook Directory.** Allows you to change the default TableLook directory.

Adjust Column Widths for. Controls the automatic adjustment of column widths in pivot tables.

■ **Labels only.** Adjusts column width to the width of the column label. This produces more compact tables, but data values wider than the label will not be displayed (asterisks indicate values too wide to be displayed).

■ **Labels and data.** Adjusts column width to whichever is larger, the column label or the largest data value. This produces wider tables, but it ensures that all values will be displayed.

Default Editing Mode. Controls activation of pivot tables in the Viewer window or in a separate window. By default, double-clicking a pivot table activates the table in the Viewer window. You can choose to activate pivot tables in a separate window or select a size setting that will open smaller pivot tables in the Viewer window and larger pivot tables in a separate window.

Data Options

Figure 45-8
Options dialog box, Data tab

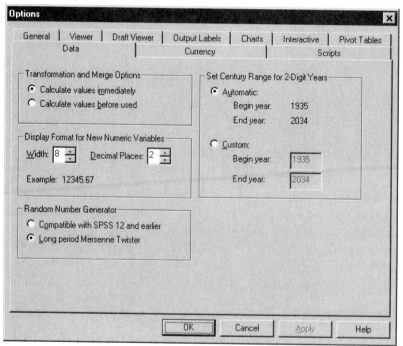

Transformation and Merge Options. Each time the program executes a command, it reads the data file. Some data transformations (such as Compute and Recode) and file transformations (such as Add Variables and Add Cases) do not require a separate pass of the data, and execution of these commands can be delayed until the program reads the data to execute another command, such as a statistical or charting procedure.

■ For large data files, where reading the data can take some time, you may want to select Calculate values before used to delay execution and save processing time. When this option is selected, the results of transformations you make using dialog boxes such as Compute Variable will not appear immediately in the Data Editor; new variables created by transformations will be displayed without any data values; and data values in the Data Editor cannot be changed while there are pending transformations. Any command that reads the data, such as a statistical or charting procedure, will execute the pending transformations and update the data displayed in the Data Editor. Alternatively, you can use Run Pending Transforms on the Transform menu.

■ With the default setting of Calculate values immediately, when you paste command syntax from dialogs, an EXECUTE command is pasted after each transformation command. For more information, see "Multiple Execute Commands" in Chapter 13 on p. 352.

Display Format for New Numeric Variables. Controls the default display width and number of decimal places for new numeric variables. There is no default display format for new string variables. If a value is too large for the specified display format, first decimal places are rounded and then values are converted to scientific notation. Display formats do not affect internal data values. For example, the value 123456.78 may be rounded to 123457 for display, but the original unrounded value is used in any calculations.

Set Century Range for 2-Digit Years. Defines the range of years for date-format variables entered and/or displayed with a two-digit year (for example, 10/28/86, 29-OCT-87). The automatic range setting is based on the current year, beginning 69 years prior to and ending 30 years after the current year (adding the current year makes a total range of 100 years). For a custom range, the ending year is automatically determined based on the value that you enter for the beginning year.

Random Number Generator. Two different random number generators are available:

■ **SPSS 12 Compatible.** The random number generator used in SPSS 12 and previous releases. If you need to reproduce randomized results generated in previous releases based on a specified seed value, use this random number generator.

■ **Mersenne Twister.** A newer random number generator that is more reliable for simulation purposes. If reproducing randomized results from SPSS 12 or earlier is not an issue, use this random number generator.

Currency Options

You can create up to five custom currency display formats that can include special prefix and suffix characters and special treatment for negative values.

The five custom currency format names are CCA, CCB, CCC, CCD, and CCE. You cannot change the format names or add new ones. To modify a custom currency format, select the format name from the source list and make the changes that you want.

Figure 45-9
Options dialog box, Currency tab

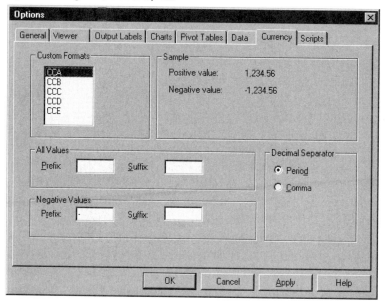

Prefixes, suffixes, and decimal indicators defined for custom currency formats are for display purposes only. You cannot enter values in the Data Editor using custom currency characters.

To Create Custom Currency Formats

▶ Click the Currency tab.

▶ Select one of the currency formats from the list (CCA, CCB, CCC, CCD, and CCE).

▶ Enter the prefix, suffix, and decimal indicator values.

▶ Click OK or Apply.

Script Options

Use the Scripts tab to specify your global procedures file and autoscript file, and select the autoscript subroutines that you want to use. You can use scripts to automate many functions, including customizing pivot tables.

Global Procedures. A global procedures file is a library of script subroutines and functions that can be called by script files, including autoscript files.

Note: The global procedures file that comes with the program is selected by default. Many of the available scripts use functions and subroutines in this global procedures file and will not work if you specify a different global procedures file.

Autoscripts. An autoscript file is a collection of script subroutines that run automatically each time you run procedures that create certain types of output objects.

Figure 45-10
Options dialog box, Scripts tab

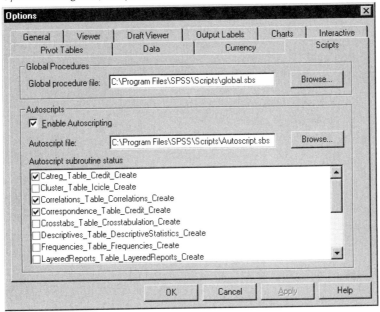

All of the subroutines in the current autoscript file are displayed, allowing you to enable and disable individual subroutines.

To Specify Global Procedure File and Autoscript File

▶ Click the Scripts tab.

▶ Select Enable Autoscripting.

▶ Select the autoscript subroutines that you want to enable.

You can also specify a different autoscript file or global procedure file.

46

Customizing Menus and Toolbars

Menu Editor

You can use the Menu Editor to customize your menus. With the Menu Editor you can:

■ Add menu items that run customized scripts.

■ Add menu items that run command syntax files.

■ Add menu items that launch other applications and automatically send data to other applications.

You can send data to other applications in the following formats: SPSS, Excel 4.0, Lotus 1-2-3 release 3, SYLK, tab-delimited, and dBASE IV.

To Add Items to Menus

▶ From the menus choose:
Utilities
  Menu Editor...

▶ In the Menu Editor dialog box, double-click the menu to which you want to add a new item.

▶ Select the menu item above which you want the new item to appear.

▶ Click Insert Item to insert a new menu item.

▶ Select the file type for the new item (script file, command syntax file, or external application).

▶ Click Browse to select a file to attach to the menu item.

Figure 46-1
Menu Editor dialog box

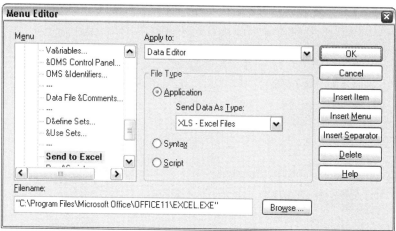

You can also add entirely new menus and separators between menu items.

Optionally, you can automatically send the contents of the Data Editor to another application when you select that application on the menus.

Customizing Toolbars

You can customize toolbars and create new toolbars. Toolbars can contain any of the available tools, including tools for all menu actions. They can also contain custom tools that launch other applications, run command syntax files, or run script files.

Show Toolbars

Use Show Toolbars to show or hide toolbars, customize toolbars, and create new toolbars. Toolbars can contain any of the available tools, including tools for all menu actions. They can also contain custom tools that launch other applications, run command syntax files, or run script files.

Figure 46-2
Show Toolbars dialog box

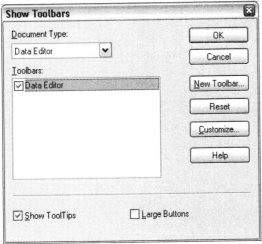

To Customize Toolbars

▶ From the menus choose:
View
  Toolbars...

▶ Select the toolbar you want to customize and click Customize, or click New Toolbar
to create a new toolbar.

▶ For new toolbars, enter a name for the toolbar, select the windows in which you want
the toolbar to appear, and click Customize.

▶ Select an item in the Categories list to display available tools in that category.

▶ Drag and drop the tools you want onto the toolbar displayed in the dialog box.

▶ To remove a tool from the toolbar, drag it anywhere off the toolbar displayed in the
dialog box.

To create a custom tool to open a file, to run a command syntax file, or to run a script:

▶ Click New Tool in the Customize Toolbar dialog box.

▶ Enter a descriptive label for the tool.

▶ Select the action you want for the tool (open a file, run a command syntax file, or run a script).

▶ Click Browse to select a file or application to associate with the tool.

New tools are displayed in the User-Defined category, which also contains user-defined menu items.

Toolbar Properties

Use Toolbar Properties to select the window types in which you want the selected toolbar to appear. This dialog box is also used for creating names for new toolbars.

Figure 46-3
Toolbar Properties dialog box

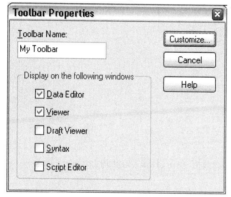

To Set Toolbar Properties

▶ From the menus choose:
View
Toolbars...

▶ For existing toolbars, click Customize, and then click Properties in the Customize Toolbar dialog box.

▶ For new toolbars, click New Tool.

▶ Select the window types in which you want the toolbar to appear. For new toolbars, also enter a toolbar name.

Customize Toolbar

Use the Customize Toolbar dialog box to customize existing toolbars and create
new toolbars. Toolbars can contain any of the available tools, including tools for all
menu actions. They can also contain custom tools that launch other applications, run
command syntax files, or run script files.

Figure 46-4
Customize Toolbar dialog box

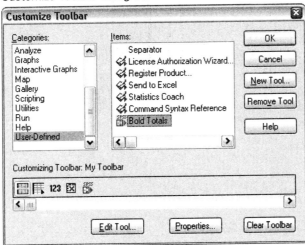

Create New Tool

Use the Create New Tool dialog box to create custom tools to launch other applications,
run command syntax files, and run script files.

Figure 46-5
Create New Tool dialog box

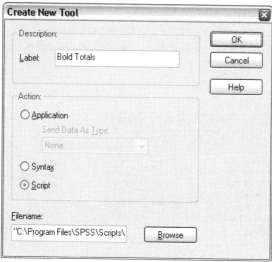

Toolbar Bitmap Editor

Use the Bitmap Editor to create custom icons for toolbar buttons. This is particularly useful for custom tools you create to run scripts, syntax, and other applications.

Figure 46-6
Bitmap Editor

To Edit Toolbar Bitmaps

▶ From the menus choose:
View
  Toolbars...

▶ Select the toolbar you want to customize and click Customize.

▶ Click the tool with the bitmap icon you want to edit on the example toolbar.

▶ Click Edit Tool.

▶ Use the toolbox and the color palette to modify the bitmap or create a new bitmap icon.

Production Facility

The Production Facility provides the ability to run the program in an automated fashion. The program runs unattended and terminates after executing the last command, so you can perform other tasks while it runs. Production mode is useful if you often run the same set of time-consuming analyses, such as weekly reports.

The Production Facility uses command syntax files to tell the program what to do. A command syntax file is a simple text file containing command syntax. You can use any text editor to create the file. You can also generate command syntax by pasting dialog box selections into a syntax window or by editing the journal file.

After you create syntax files and include them in a production job, you can view and edit them from the Production Facility.

Figure 47-1
Production Facility

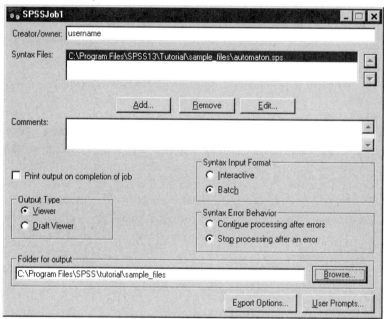

Syntax Input Format. Controls the form of the syntax rules used for the job:

- **Interactive.** Each command must end with a period. Periods can appear anywhere within the command, and commands can continue on multiple lines, but a period as the last non-blank character on a line is interpreted as the end of the command. Continuation lines and new commands can start anywhere on a new line. These are the "interactive" rules in effect when you select and run commands in a syntax window.

- **Batch.** Each command must start at the beginning of a new line (no blank spaces before the start of the command), and continuation lines must be indented at least one space. If you want to indent new commands, you can use a plus sign, dash, or period as the first character at the start of the line and then indent the actual command. The period at the end of the command is optional. This setting is compatible with the syntax rules for command files included with the INCLUDE command.

Syntax Error Behavior. Controls the treatment of error conditions in the job:

- **Continue.** Errors in the job do not automatically stop command processing. The commands in the production job files are treated as part of the normal command stream, and command processing continues in the normal fashion.

- **Stop.** Command processing stops when the first error in a production job file is encountered. This is compatible with the behavior of command files included with the INCLUDE command.

Production job results. Each production run creates an output file with the same name as the production job and the extension *.spo*. For example, a production job file named *prodjob.spp* creates an output file named *prodjob.spo*. The output file is a Viewer document.

Output Type. Viewer output produces pivot tables and high-resolution, interactive charts. Draft Viewer output produces text output and metafile pictures of charts. Text output can be edited in the Draft Viewer, but charts cannot be edited in the Draft Viewer.

Production Jobs with OUTPUT Commands

Production jobs honor SPSS OUTPUT commands, such as OUTPUT SAVE, OUTPUT ACTIVATE, and OUTPUT NEW. OUTPUT SAVE commands executed during the course of a production job will write the contents of the specified output documents to the specified locations. This is in addition to the output file created by the Production Facility, as in the output file *prodjob.spo* for a production job file named *prodjob.spp*. When using OUTPUT NEW to create a new output document, it is recommended that you explicitly save it with the OUTPUT SAVE command.

The Production Facility output file consists of the contents of the active output document as of the end of the job. For jobs containing OUTPUT commands, the output file may not contain all output created in the session. For example, suppose the production job consists of a number of SPSS procedures followed by an OUTPUT NEW command, followed by more SPSS procedures but no more OUTPUT commands. The OUTPUT NEW command defines a new active output document. At the end of the production job, it will contain output from only the procedures executed after the OUTPUT NEW command.

Using the Production Facility

▶ Create a command syntax file.

▶ Start the Production Facility, available on the Start menu.

▶ Specify the syntax files that you want to use in the production job. Click Add to select the syntax files.

▶ Save the production job file.

▶ Run the production job file. Click the Run button on the toolbar, or from the menus choose:

Run
  Production Job

Export Options

Export Options saves pivot tables and text output in HTML, text, Word/RTF, and Excel format, and it saves charts in a variety of common formats used by other applications.

Figure 47-2
Export Options dialog box

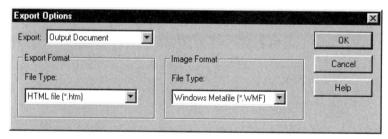

Export

This drop-down list specifies what you want to export.

Output Document. Exports any combination of pivot tables, text output, and charts.

■ For HTML and text formats, charts are exported in the currently selected chart export format. For HTML document format, charts are embedded by reference, and you should export charts in a suitable format for inclusion in HTML documents. For text document format, a line is inserted in the text file for each chart, indicating the filename of the exported chart.

- For Word/RTF format, charts are exported in Windows metafile format and embedded in the Word document.

- Charts are not included in Excel documents.

Output Document (No Charts). Exports pivot tables and text output. Any charts in the Viewer are ignored.

Charts Only. Exports charts only. For HTML and text documents, export formats include: Enhanced metafile (EMF), Windows metafile (WMF), Windows bitmap (BMP), encapsulated PostScript (EPS), JPEG, TIFF, PNG, and Macintosh PICT. For Word/RTF documents, charts are always exported in Windows metafile format.

Export Format. For output documents, the available options are HTML, text, Excel, Word/RTF, PowerPoint, and PDF; for HTML and text format, charts are exported in the currently selected chart format in the Options dialog box for the selected format. For Charts Only, choose a chart export format from the drop-down list. For output documents, pivot tables and text are exported in the following manner:

- **HTML file (*.htm).** Pivot tables are exported as HTML tables. Text output is exported as preformatted HTML.

- **Text file (*.txt).** Pivot tables can be exported in tab-separated or space-separated format. All text output is exported in space-separated format.

- **Excel file (*.xls).** Pivot table rows, columns, and cells are exported as Excel rows, columns, and cells, with all formatting attributes intact—for example, cell borders, font styles, and background colors. Text output is exported with all font attributes intact. Each line in the text output is a row in the Excel file, with the entire contents of the line contained in a single cell.

- **Word/RTF file (*.doc).** Pivot tables are exported as Word tables with all formatting attributes intact—for example, cell borders, font styles, background colors, and so on. Text output is exported as formatted RTF. Text output in SPSS is always displayed in a fixed-pitch (monospaced) font and is exported with the same font attributes. A fixed-pitch font is required for proper alignment of space-separated text output. (*Note*: Microsoft Word may not display extremely wide tables properly.)

- **PowerPoint file (*.ppt).** Pivot tables are exported as Word tables and are embedded on separate slides in the PowerPoint file, with one slide for each pivot table. All formatting attributes of the pivot table are retained—for example, cell borders, font styles, and background colors. Text output is exported as formatted RTF. Text output in SPSS is always displayed in a fixed-pitch (monospaced) font and is exported with the same font attributes. A fixed-pitch font is required for proper alignment of space-separated text output. (*Note*: Export to PowerPoint is not available in the Student Version.)

- **Portable Document Format (*.pdf).** All output is exported as it appears in Print Preview, with all formatting attributes intact.

Image Format

Image Format controls the export format for charts. Charts can be exported in the following formats: Enhanced metafile (EMF), Windows metafile (WMF), Windows bitmap (BMP), encapsulated PostScript (EPS), JPEG, TIFF, PNG, or Macintosh PICT.

Exported chart names are based on the production job filename, a sequential number, and the extension of the selected format. For example, if the production job *prodjob.spp* exports charts in Windows metafile format, the chart names would be *prodjob1.wmf, prodjob2.wmf, prodjob3.wmf*, and so on.

Text and Image Options

Text export options (for example, tab-separated or space-separated) and chart export options (for example, color settings, size, and resolution) are set in SPSS and cannot be changed in the Production Facility. Use Export on the File menu in SPSS to change text and chart export options.

Draft Viewer Export

The only Export option available for Draft Viewer output is to export the output in simple text format. Charts for Draft Viewer output cannot be exported.

User Prompts

Macro symbols defined in a production job file and used in a command syntax file simplify tasks such as running the same analysis for different data files or running the same set of commands for different sets of variables. For example, you could define the macro symbol *@datfile* to prompt you for a data filename each time you run a production job that uses the string *@datfile* in place of a filename in the command syntax file.

Figure 47-3
User Prompts dialog box

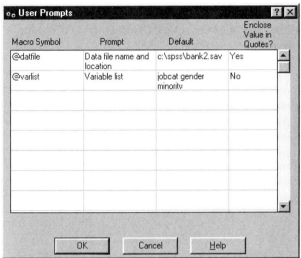

Macro Symbol. The macro name used in the command syntax file to invoke the macro that prompts the user to enter information. The macro symbol name must begin with an @. *Note*: These "macro" symbols are unrelated to macros created by the SPSS macro facility using DEFINE-!ENDDEFINE.

Prompt. The descriptive label that is displayed when the production job prompts you to enter information. For example, you could use the phrase "What data file do you want to use?" to identify a field that requires a data filename.

Default. The value that the production job supplies by default if you don't enter a different value. This value is displayed when the production job prompts you for information. You can replace or modify the value at runtime.

Enclose Value in Quotes? Enter Y or Yes if you want the value enclosed in quotes. Otherwise, leave the field blank or enter N or No. For example, you should enter Yes for a filename specification because filename specifications should be enclosed in quotes.

Figure 47-4
Macro prompts in a command syntax file

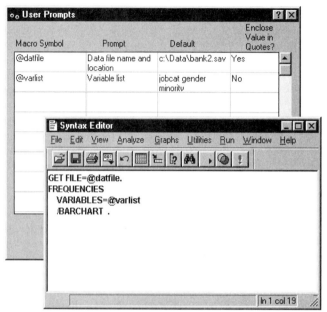

Production Macro Prompting

The Production Facility prompts you for values whenever you run a production job
that contains defined macro symbols. You can replace or modify the default values that
are displayed. Those values are then substituted for the macro symbols in all command
syntax files associated with the production job.

Figure 47-5
Production macro prompting dialog box

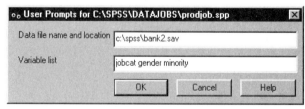

Note: These "macro" symbols are unrelated to macros created by the SPSS macro facility using DEFINE-!ENDDEFINE, and the production job will prompt you for these values even if none of the command syntax files in the job include references to the defined "macros."

Production Options

Production Options enable you to:

- Specify a default text editor for syntax files accessed with the Edit button in the main dialog box.

- Run the production job as an invisible background process or display the results it generates as the job runs.

- Specify a remote server, domain name, user ID, and password for distributed analysis (applicable only if you have network access to the server version of SPSS). If you don't specify these settings, the default settings in the SPSS Server Login dialog box are used. You can select only remote servers that you have previously defined in the Add Server dialog box in SPSS (File menu, Switch Server, Add).

Figure 47-6
Options dialog box

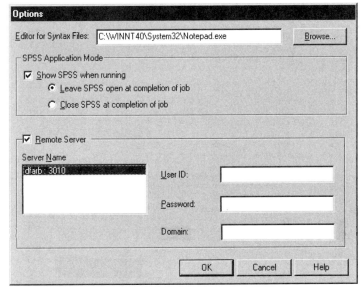

Changing Production Options

From the Production Facility menus choose:
Edit
  Options...

Format Control for Production Jobs

A number of settings in SPSS can help to ensure the best format for pivot tables created in production jobs.

TableLooks. By editing and saving TableLooks (Format menu in an activated pivot table), you can control many pivot table attributes. You can specify font sizes and styles, colors, and borders. To ensure that wide tables do not split across pages, select Rescale wide table to fit page on the Table Properties General tab.

Output labels. Output label options (Edit menu, Options, Output Labels tab) control the display of variable and data value information in pivot tables. You can display variable names and/or defined variable labels, actual data values and/or defined value labels. Descriptive variable and value labels often make it easier to interpret your results; however, long labels can be awkward in some tables.

Column width. Pivot table options (Edit menu, Options, Pivot Tables tab) control the default TableLook and the automatic adjustment of column widths in pivot tables.

- **Labels only.** Adjusts the column width to the width of the column label. This produces more compact tables, but data values wider than the label will not be displayed (asterisks indicate values too wide to be displayed).

- **Labels and data.** Adjusts the column width to whichever is larger, the column label or the largest data value. This produces wider tables, but it ensures that all values will be displayed.

Production jobs use the current TableLook and Options settings in effect. You can set the TableLook and Options settings before running your production job, or you can use SET commands in your syntax files to control them. Using SET commands in syntax files enables you to use multiple TableLooks and Options settings in the same job.

Creating a Custom Default TableLook

▶ Activate a pivot table (double-click anywhere in the table).

▶ From the menus choose:
Format
    TableLook...

▶ Select a TableLook from the list and click Edit Look.

▶ Adjust the table properties for the attributes that you want.

▶ Click Save Look or Save As to save the TableLook and click OK.

▶ From the menus choose:
Edit
    Options...

▶ Click the Pivot Tables tab.

▶ Select the TableLook from the list and click OK.

Setting Options for Production Jobs

▶ From the menus choose:
Edit
    Options...

▶ Select the options that you want.

▶ Click OK.

You can set the default TableLook, output label settings, and automatic column width adjustment with Options. Options settings are saved with the program. When you run a production job, the Options settings in effect the last time that you ran the program are applied to the production job.

Controlling Pivot Table Format with Command Syntax

SET TLOOK. Controls the default TableLook for new pivot tables, as in:

■   SET TLOOK = 'c:\prodjobs\mytable.tlo'.

SET TVARS. Controls the display of variable names and labels in new pivot tables.

■   SET TVARS = LABELS displays variable labels.

- SET TVARS = NAMES displays variable names.

- SET TVARS = BOTH displays both variable names and labels.

SET ONUMBER. Controls the display of data values or value labels in new pivot tables.

- SET ONUMBER = LABELS displays value labels.

- SET ONUMBER = VALUES displays data values.

- SET ONUMBER = BOTH displays data values and value labels.

SET TFIT. Controls automatic column width adjustment for new pivot tables.

- SET TFIT = LABELS adjusts column width to the width of the column label.

- SET TFIT = BOTH adjusts column width to the width of the column label or the largest data value, whichever is wider.

Running Production Jobs from a Command Line

Command line switches enable you to schedule production jobs to run at certain times with scheduling utilities like the one available in Microsoft Plus!. You can run production jobs from a command line with the following switches:

-r. Runs the production job. If the production job contains any user prompts, you must supply the requested information before the production job will run.

-s. Runs the production job and suppresses any user prompts or alerts. The default user prompt values are used automatically.

Distributed analysis. If you have network access to the server version of SPSS, you can also use the following switches to run the Production Facility in distributed analysis mode:

-x. Name or IP address of the remote server.

-n. Port number.

-d. Domain name.

-u. User ID for remote server access.

-p. Password for remote server access.

If you specify any of the command lines switches for distributed analysis, you must specify all of the distributed analysis command line switches (-x, -n, -d, -u, and -p).

You should provide the full path for both the Production Facility (*spssprod.exe*) and the production job, and both should be enclosed in quotes, as in:

```
"c:\program files\spss\spssprod.exe"  "c:\spss\datajobs\prodjob.spp" -s -r
```

For command line switches that require additional specifications, the switch must be followed by an equals sign followed immediately by the specification. If the specification contains spaces (such as a two-word server name), enclose the value in quotes or apostrophes, as in:

```
-x="HAL 9000" -u="secret word"
```

Default server. If you have network access to the server version of SPSS, the default server and related information (if not specified in command line switches) is the default server specified in the SPSS Server Login dialog box. If no default is specified there, the job runs in local mode.

If you want to run a production job in local mode but your local computer is not your default server, specify null quoted strings for all of the distributed analysis command line switches, as in:

```
"c:\program files\spss\spssprod.exe" "c:\spss\datajobs\prodjob.spp" -x="" -n="" -d="" -u="" -p=""
```

Running Multiple Production Jobs

If you use a batch (*.bat*) file or similar facility to run multiple production jobs, use the Windows Start command with the /wait switch to control execution of each job, preventing subsequent jobs from starting before the previous job ends, as in:

```
cd \program files\spss
start /wait spssprod.exe prodjob1.spp -s
start /wait spssprod.exe prodjob2.spp -s
```

Publish to Web

Publish to Web exports output for publishing to SmartViewer Web Server. Tables and reports published in SmartViewer can be viewed and manipulated over the Web, in real time, using a standard browser.

■ Pivot tables are published as dynamic tables that can be manipulated over the Web to obtain different views of the data.

■ Charts are published as JPEG or PNG graphic files.

■ Text output is published as preformatted HTML. (By default, most Web browsers use a fixed-pitch font for preformatted text.)

Publish. Allows you to specify the output that you want to publish:

■ **Output Document.** Publishes the entire output document, including hidden or collapsed items.

■ **Output Document (No Notes).** Publishes everything but the Notes tables that are automatically produced for each procedure.

■ **Tables Only.** Excludes charts. All pivot tables and all text tables are published.

■ **Tables Only (No Notes).** Excludes charts and Notes tables.

■ **Charts Only.** Publishes only the charts in the document.

■ **Nothing.** Turns off publishing to the Web. Since all settings are saved with the production job (*.spp* file), results will be published every time that you run the production job unless you select Nothing. This turns off publishing while still generating other types of output (Viewer files, HTML files) specified in the production job.

Publish Tables as. Controls how pivot tables are published:

■ **Interactive.** Tables are dynamic objects that can be manipulated over the Web to obtain different views of the data.

■ **Static.** Tables are static and cannot be manipulated after publishing.

Configure. Opens the SmartViewer Web Server "Configure Automated Publishing" page in a browser window. This is required when you create a new production job to publish to the Web.

A user ID and password are also required to access the SmartViewer Web Server. When you create a new production job to publish to the Web, you will be prompted for your user ID and password. This information is stored in the production job in encrypted format.

Note: Publish to Web is available only for sites with SmartViewer Web Server installed and requires a plug-in to activate the publishing feature. Contact your system administrator or Webmaster for instructions on downloading the plug-in. If SmartViewer is unavailable at your site, use Export Output to save output in HTML format.

SmartViewer Web Server Login

Publishing to SmartViewer Web Server requires a valid SmartViewer Web Server user name (user ID) and password.

Contact your system administrator or Webmaster for more information.

SPSS Scripting Facility

The scripting facility allows you to automate tasks, including:

- Opening and saving data files.
- Displaying and manipulating dialog boxes.
- Running data transformations and statistical procedures using command syntax.
- Exporting charts as graphic files in a variety of formats.
- Customizing output in the Viewer.

A number of scripts are included with the software, including autoscripts that run automatically every time a specific type of output is produced. You can use these scripts as they are or you can customize them to your needs. If you want to create your own scripts, you can begin by choosing from a number of starter scripts.

To Run a Script

▶ From the menus choose:
Utilities
  Run Script...

Figure 48-1
Run Script dialog box

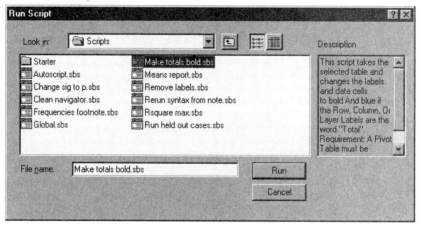

▶ Select the *Scripts* folder.

▶ Select the script you want.

For more information, see "Customizing Menus and Toolbars" in Chapter 46 on p. 657.

Scripts Included with SPSS

The following scripts are included with the program:

Analyze held out cases. Repeats a factor or discriminant analysis using cases not selected in a previous analysis. A Notes table produced by a previous run of a factor or discriminant analysis must be selected before running the script.

Change significance to p. Change *Sig.* to *p=* in the column labels of any pivot table. The table must be selected before running the script.

Clean navigator. Delete all Notes tables from an output document. The document must be open in the designated Viewer window before running the script.

Frequencies footnote. Insert statistics displayed in a Frequencies Statistics table as footnotes in the corresponding frequency table for each variable. The Frequencies Statistics table must be selected before running the script.

Make totals bold. Apply the bold format and blue color to any row, column, or layer of data labeled *Total* in a pivot table. The table must be selected before running the script.

Means report. Extract information from a Means table and write results to several output ASCII files. The Means table must be selected before running the script.

Remove labels. Delete all row and column labels from the selected pivot table. The table must be selected before running the script.

Rerun syntax from note. Resubmit the command found in the selected Notes table using the active data file. If no data file is open, the script attempts to read the data file used originally. The Notes table must be selected before running the script.

Rsquare max. In a Regression Model Summary table, apply the bold format and blue color to the row corresponding to the model that maximizes adjusted R^2. The Model Summary table must be selected before running the script.

For more information, see "Options" in Chapter 45 on p. 637.

Note: This list may not be complete.

Autoscripts

Autoscripts run automatically when triggered by the creation of a specific piece of output by a given procedure. For example, there is an autoscript that automatically removes the upper diagonal and highlights correlation coefficients below a certain significance whenever a Correlations table is produced by the Bivariate Correlations procedure.

The Scripts tab of the Options dialog box (Edit menu) displays the autoscripts that are available on your system and allows you to enable or disable individual scripts.

Figure 48-2
Scripts tab of Options dialog box

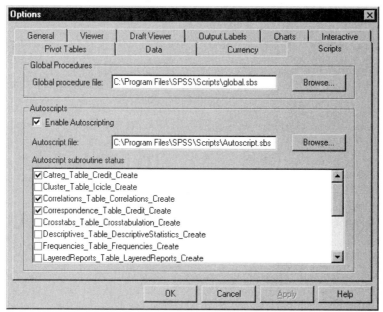

Autoscripts are specific to a given procedure and output type. An autoscript that formats the ANOVA tables produced by One-Way ANOVA is not triggered by ANOVA tables produced by other statistical procedures (although you could use global procedures to create separate autoscripts for these other ANOVA tables that share much of the same code). However, you can have a separate autoscript for each type of output produced by the same procedure. For example, Frequencies produces both a frequency table and a table of statistics, and you can have a different autoscript for each.

For more information, see "Options" in Chapter 45 on p. 637.

Creating and Editing Scripts

You can customize many of the scripts included with the software for your specific needs. For example, there is a script that removes all Notes tables from the designated output document. You can easily modify this script to remove output items of any type and label you want.

Figure 48-3
Modifying a script in the script window

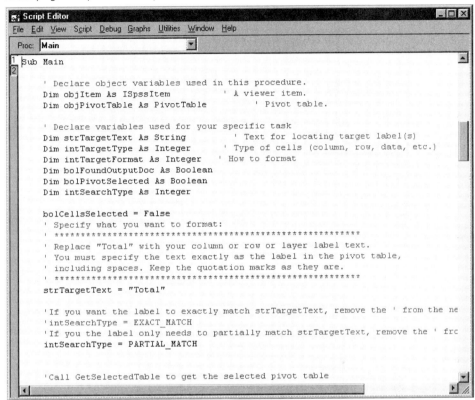

If you prefer to create your own scripts, you can begin by choosing from a number of starter scripts.

To Edit a Script

▶ From the menus choose:

File
  Open
    Script...

Figure 48-4
Opening a script file

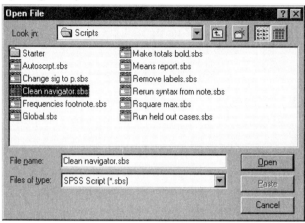

▶ Select the *Scripts* folder.

▶ Under Files of Type, select SPSS Script (*.sbs).

▶ Select the script you want.

If you open more than one script, each opens in its own window.

Script Window

The script window is a fully featured programming environment that uses the Sax BASIC language and includes a dialog box editor, object browser, debugging features, and context-sensitive Help.

Figure 48-5
Script window

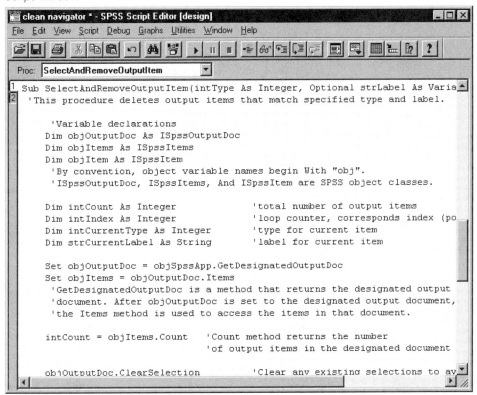

- As you move the cursor, the name of the current procedure is displayed at the top of the window.

- Terms colored blue are reserved words in BASIC (for example `Sub`, `End Sub`, and `Dim`). You can access context-sensitive Help on these terms by clicking them and pressing F1.

- Terms colored magenta are objects, properties, or methods. You can also click these terms and press F1 for Help, but only where they appear in valid statements and are colored magenta. (Clicking the name of an object in a comment will not work because it brings up Help on the Sax BASIC language rather than on SPSS objects.)

- Comments are displayed in green.

- Press F2 at any time to display the object browser, which displays objects, properties, and methods.

Script Editor Properties (Script Window)

Code elements in the script window are color-coded to make them easier to distinguish. By default, comments are green, Sax BASIC terms are blue, and names of valid objects, properties, and methods are magenta. You can specify different colors for these elements and change the size and font for all text.

To Set Script Editor Properties

▶ From the menus choose:

Script
  Editor Properties...

Figure 48-6
Editor Properties dialog box

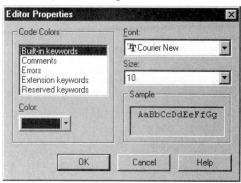

▶ To change the color of a code element type, select the element and choose a color from the drop-down palette.

Starter Scripts

When you create a new script, you can begin by choosing from a number of starter scripts.

Figure 48-7
Use Starter Script dialog box

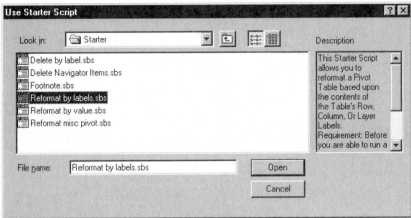

Each starter script supplies code for one or more common procedures and is commented with hints on how to customize the script to your particular needs.

Delete by label. Delete rows or columns in a pivot table based on the contents of the RowLabels or ColumnLabels. In order for this script to work, the Hide empty rows and columns option must be selected in the Table Properties dialog box.

Delete navigator items. Delete items from the Viewer based on a number of different criteria.

Footnote. Reformat a pivot table footnote, change the text in a footnote, or add a footnote.

Reformat by labels. Reformat a pivot table based upon the row, column, or layer labels.

Reformat by value. Reformat a pivot table based upon the value of data cells or a combination of data cells and labels.

Reformat misc pivot. Reformat or change the text in a pivot table title, corner text, or caption.

In addition, you can use any of the other available scripts as starter scripts, although they may not be as easy to customize. Just open the script and save it with a different filename.

Creating a Script

▶ From the menus choose:
New
  Script...

▶ Select a starter script if you want to begin with one.

▶ If you do not want to use a starter script, click Cancel.

Creating Autoscripts

You can create an autoscript by starting with the output object that you want to serve as the trigger. For example, to create an autoscript that runs whenever a frequency table is produced, create a frequency table in the usual manner and single-click the table in the Viewer to select it. You can then right-click or use the Utilities menu to create a new autoscript triggered whenever that type of table is produced.

Figure 48-8
Creating a new autoscript

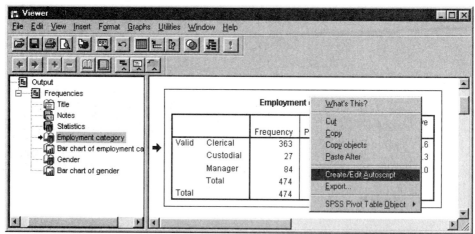

By default, each autoscript you create is added to the current autoscript file (*autoscript.sbs*) as a new procedure. The name of the procedure references the event that serves as the trigger. For example, if you create an autoscript triggered whenever the Explore procedure creates a Descriptives table, the name of the autoscript subroutine would be Explore_Table_Descriptives_Create.

Figure 48-9
New autoscript procedure displayed in script window

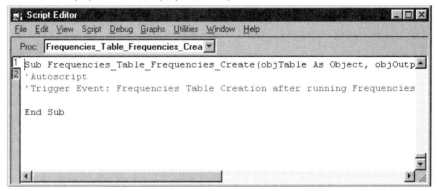

This makes autoscripts easier to develop because you do not need to write code to get the object you want to operate on, but it requires that autoscripts are specific to a given piece of output and statistical procedure.

To Create an Autoscript

▶ Select the object you want to serve as a trigger in the Viewer.

▶ From the menus choose:
Utilities
  Create/Edit Autoscript...

If no autoscript exists for the selected object, a new autoscript is created. If an autoscript already exists, the existing script is displayed.

▶ Type the code.

▶ From the Edit menu, choose Options to enable or disable the autoscript.

Events that Trigger Autoscripts

The name of the autoscript procedure references the event that serves as the trigger. The following events can trigger autoscripts:

Creation of pivot table. The name of the procedure references both the table type and the procedure that created it—for example, `Correlations_Table_Correlations_Create`.

Figure 48-10
Autoscript procedure for Correlations table

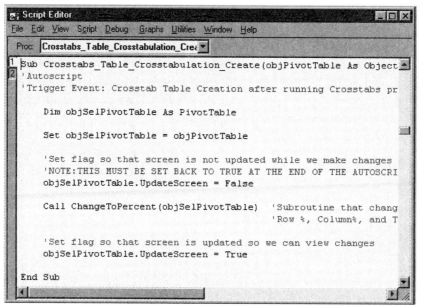

Creation of title. Referenced to the statistical procedure that created it: `Correlations_Title_Create`.

Creation of notes. Referenced to the procedure that created it: `Correlations_Notes_Create`.

Creation of warnings. Referenced by the procedure that created it.

You can also use a script to trigger an autoscript indirectly. For example, you could write a script that invokes the Correlations procedure, which in turn triggers the autoscript registered to the resulting Correlations table.

Autoscript File

All autoscripts are saved in a single file (unlike other scripts, each of which is saved in a separate file). Any new autoscripts you create are also added to this file. The name of the current autoscript file is displayed in the Scripts tab of the Options dialog box (Edit menu).

Figure 48-11
Autoscript subroutines displayed in Options dialog box

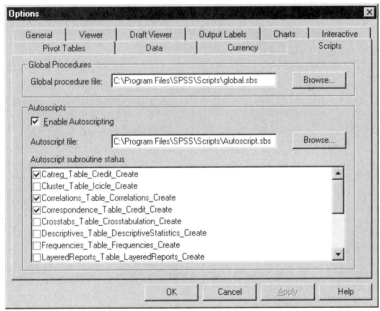

The Options dialog box also displays all of the autoscripts in the currently selected autoscript file, allowing you to enable and disable individual scripts.

The default autoscript file is *autoscript.sbs*. You can specify a different autoscript file, but only one can be active at any one time.

How Scripts Work

Scripts work by manipulating objects using properties and methods. For example, pivot tables are a class of objects. With objects of this class, you can use the SelectTable method to select all of the elements in the table, and you can use the

TextColor property to change the color of selected text. Each object class has specific properties and methods associated with it. The collection of all SPSS object classes (or types) is called the SPSS type library.

Figure 48-12
Tree view of object hierarchy

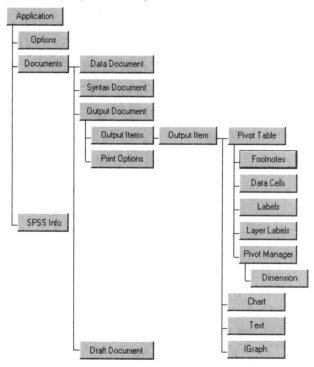

Using objects is a two-step process. First, you create a reference to the object (called *getting* the object). Then, you use properties and methods to do something. You get objects by navigating the hierarchy of objects, at each step using properties or methods of objects higher in the hierarchy to get at the objects beneath. For example, to get a pivot table object, you have to first get the output document that contains the pivot table and then get the items in that output document.

Each object that you get is stored in a variable. (Remember that all you are really storing in the variable is a reference to the object.) One of the first steps in creating a script is often to declare variables for the objects that you need.

Tip: It is difficult to understand how scripts work if you do not understand how the program works. Before writing a script, use the mouse to perform the task several times as you normally would. At each step, consider what objects you are manipulating and what properties of each object you are changing.

Variable Declarations (Scripting)

Although not always required, it is a good idea to declare all variables before using them. This is most often done using `Dim` declaration statements:

```
Dim objOutputDoc As ISpssOutputDoc
Dim objPivotTable As PivotTable
Dim intType As Integer
Dim strLabel As String
```

Each declaration specifies the variable name and type. For example, the first declaration above creates an object variable named `objOutputDoc` and assigns this variable to the `ISpssOutputDoc` object class. The variable does not yet have a value because it has not been set to a particular output document. All the statement does is declare that the variable exists. (This process has been referred to as "renaming the objects you want to use.")

Variable naming conventions. By convention, the name of each variable indicates its type. Object variable names begin with `obj`, integer variables begin with `int`, and string variables begin with `str`. These are only conventions—you can name your variables anything you want—but following them makes it much easier to understand your code.

SPSS object classes. `ISpssOutputDoc` and `PivotTable` are names of SPSS object classes. Each class represents a type of object that the program can create, such as an output document or pivot table. Each object class has specific properties and methods associated with it. The collection of all SPSS object classes (or types) is referred to as the SPSS type library.

Table of Object Classes and Naming Conventions

The following variable names are used in the sample scripts included with the program and are recommended for all scripts. Notice that with the exception of pivot tables, object classes have names beginning with `ISpss`.

Object	Type or Class	Variable Name
SPSS application	IspssApp	objSpssApp—variable is global and does not require declaration
SPSS options	ISpssOptions	objSpssOptions
SPSS file information	ISpssInfo	objSpssInfo
Documents	ISpssDocuments	objDocuments
Data document	ISpssDataDoc	objDataDoc
Syntax document	ISpssSyntaxDoc	objSyntaxDoc
Viewer document	ISpssOutputDoc	objOutputDoc
Print options	ISpssPrintOptions	objPrintOptions
Output items collection	ISpssItems	objOutputItems
Output item	ISpssItem	objOutputItem
Chart	ISpssChart	objSPSSChart
Text	ISpssRtf	objSPSSText
Pivot table	PivotTable	objPivotTable
Footnotes	ISpssFootnotes	objFootnotes
Data cells	ISpssDataCells	objDataCells
Layer labels	ISpssLayerLabels	objLayerLabels
Column labels	ISpssLabels	objColumnLabels
Row labels	ISpssLabels	objRowLabels
Pivot manager	ISpssPivotMgr	objPivotMgr
Dimension	ISpssDimension	objDimension

Getting SPSS Automation Objects (Scripting)

To *get* an object means to create a reference to the object so that you can use properties and methods to do something. Each object reference that you get is stored in a variable. To get an object, first declare an object variable of the appropriate class, then set the variable to the specific object. For example, to get the designated output document:

```
Dim objOutputDoc As ISpssOutputDoc
Set objOutputDoc = objSpssApp.GetDesignatedOutputDoc
```

you use properties and methods of objects higher in the object hierarchy to get at the objects beneath. The second statement above gets the designated output document using `GetDesignatedOutputDoc`, a method associated with the application object, which is the highest-level object. Similarly, to get a pivot table object, you first get the output document that contains the pivot table, and then get the collection of items in that output document, and so on.

Example: Getting an Output Object

This script gets the third output item in the designated output document and activates it. If that item is not an OLE object, the script produces an error.

```
Sub Main

Dim objOutputDoc As ISpssOutputDoc'declare object variables
Dim objOutputItems As ISpssItems
Dim objOutputItem As ISpssItem

Set objOutputDoc = objSpssApp.GetDesignatedOutputDoc'get reference to designated output doc
Set objOutputItems = objOutputDoc.Items() 'get collection of items in doc
Set objOutputItem = objOutputItems.GetItem(2) 'get third output item
'(item numbers start at 0 so "2" gets third)

objOutputItem.Activate 'activate output item

End sub
```

Example: Getting the First Pivot Table

This script gets the first pivot table in the designated output document and activates it.

```
Sub Main

Dim objOutputDoc As ISpssOutputDoc 'declare object variables
Dim objOutputItems As ISpssItems
Dim objOutputItem As ISpssItem
Dim objPivotTable As PivotTable

Set objOutputDoc = objSpssApp.GetDesignatedOutputDoc'get reference to designated output doc
Set objOutputItems = objOutputDoc.Items()'get collection of items in doc

Dim intItemCount As Integer'number of output items
Dim intItemType As Integer'type of item (defined by SpssType property)

intItemCount = objOutputItems.Count()'get number of output items
   For index = 0 To intItemCount'loop through output items
   Set objOutputItem = objOutputItems.GetItem(index)'get current item
   intItemType = objOutputItem.SPSSType()'get type of current item
   If intItemType = SPSSPivot Then
      Set objPivotTable = objOutputItem.Activate()'if item is a pivot table, activate it
      Exit For
   End If
Next index

End sub
```

Examples are also available in the online Help. You can try them yourself by pasting the code from Help into the script window.

Properties and Methods (Scripting)

Like real world objects, OLE automation objects have features and uses. In programming terminology, the features are referred to as properties, and the uses are referred to as methods. Each object class has specific methods and properties that determine what you can do with that object.

Object	Property	Method
Pencil (real world)	Hardness Color	Write Erase
Pivot table (SPSS)	TextFont DataCellWidths CaptionText	SelectTable ClearSelection HideFootnotes

Example: Using Properties (Scripting)

Properties set or return attributes of objects, such as color or cell width. When a property appears to the left side of an equals sign, you are writing to it. For example, to set the caption for an activated pivot table (`objPivotTable`) to `"Anita's results"`:

objPivotTable.CaptionText = "Anita's results"

When a property appears on the right side, you are reading from it. For example, to get the caption of the activated pivot table and save it in a variable:

strFontName = objPivotTable.CaptionText

Example: Using Methods (Scripting)

Methods perform actions on objects, such as selecting all the elements in a table:

objPivotTable.SelectTable

or removing a selection:

objPivotTable.ClearSelection

Some methods return another object. Such methods are extremely important for navigating the object hierarchy. For example, the `GetDesignatedOutputDoc` method returns the designated output document, allowing you to access the items in that output document:

```
Set objOutputDoc = objSpssApp.GetDesignatedOutputDoc
Set objItems = objOutputDoc.Items
```

Object Browser

The object browser displays all object classes and the methods and properties associated with each. You can also access Help on individual properties and methods and paste selected properties and methods into your script.

Figure 48-13
Object browser

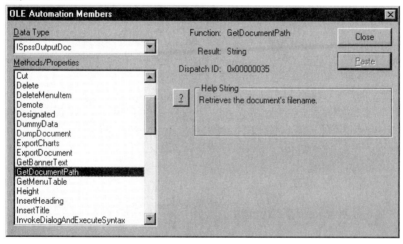

Using the Object Browser

▶ From the script window menus choose:
Debug
  Object Browser...

▶ Select an object class from the Data Type list to display the methods and properties for that class.

▶ Select properties and methods for context-sensitive Help or to paste them into your script.

New Procedure (Scripting)

A procedure is a named sequence of statements that are executed as a unit. Organizing code in procedures makes it easier to manage and reuse pieces of code. Scripts must have at least one procedure (the `Main` subroutine) and often they have several. The `Main` procedure may contain few statements, aside from calls to subroutines that do most of the work.

Figure 48-14
New Procedure dialog box

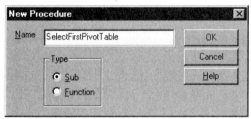

Procedures can be subroutines or functions. A procedure begins with a statement that specifies the type of procedure and the name (for example, `Sub Main` or `Function DialogMonitor( )`) and concludes with the appropriate End statement (`End Sub` or `End Function`).

As you scroll through the script window, the name of the current procedure is displayed at the top of the script window. Within a script, you can call any procedure as many times as you want. You can also call any procedure in the global script file, which makes it possible to share procedures between scripts.

To Add a New Procedure in a Script

▶ From the menus choose:
Script
  New Procedure...

▶ Type a name for the procedure.

▶ Select Subroutine or Function.

Alternatively, you can create a new procedure by typing the statements that define the procedure directly in the script.

Global Procedures (Scripting)

If you have a procedure or function that you want to use in a number of different scripts, you can add it to the global script file. Procedures in the global script file can be called by all other scripts.

Figure 48-15
Global script file

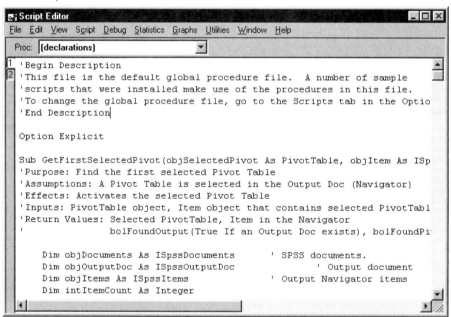

The default global script file is *global.sbs*. You can freely add procedures to this file. You can also specify a different global file on the Scripts tab in the Options dialog box (Edit menu), but only one file can be active as the global file at any given time. That means that if you create a new global file and specify it as the global file, the procedures and functions in *global.sbs* are no longer available.

You can view the global script file in any script window (click the #2 tab on the left side of the window just below the toolbar), but you can edit it in only one window at a time.

Global procedures must be called by other script procedures. You cannot run a global script directly from the Utilities menu or a script window.

Adding a Description to a Script

You can add a description to be displayed in the Run Script and Use Starter Script dialog boxes. Just add a comment on the first line of the script that starts with `Begin Description`, followed by the desired comment (one or more lines), followed by `End Description`. For example:

```
'Begin Description
'This script changes "Sig." to "p=" in the column labels of any pivot table.
'Requirement: The Pivot Table that you want to change must be selected.
'End Description
```

The description must be formatted as a comment (each line beginning with an apostrophe).

Scripting Custom Dialog Boxes

There are two steps to implementing a custom dialog box: first create the dialog box using the UserDialog Editor, and then create a dialog monitor function (`DialogFunc`) that monitors the dialog box and defines its behavior.

The dialog box itself is defined by a `Begin Dialog...End Dialog` block. You do not need to type this code directly—the UserDialog Editor provides an easy, graphical way to define the dialog box.

Figure 48-16
Creating a dialog box in the UserDialog Editor

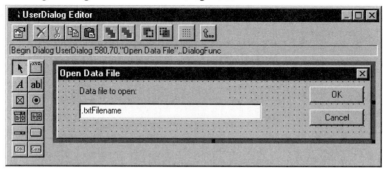

The Editor initially displays a blank dialog box form. You can add controls, such as radio buttons and check boxes, by selecting the appropriate tool and dragging with the mouse. (Hold the mouse over each tool for a description.) You can also drag the sides and corners to resize the dialog box. After adding a control, right-click the control to set properties for that control.

Dialog monitor function. To create the dialog monitor function, right-click the dialog box form (make sure no control is selected on the form) and enter a name for the function in the DialogFunc field. The statements that define the function are added to your script, although you will have to edit the function manually to define the behavior for each action.

When finished, click the Save and Exit button (far right on the toolbar) to add the code for the dialog box to your script.

To Create a Custom Dialog Box

▶ In the script window, click the cursor in the script where you want to insert the code for the dialog box.

▶ From the menus choose:
  Script
    Dialog Editor...

▶ Select tools from the palette and drag in the new dialog box form to add controls, such as buttons and check boxes.

▶ Resize the dialog box by dragging the handles on the sides and corners.

▶ Right-click the form (with no control selected) and enter a name for the dialog monitor function in the DialogFunc field.

▶ Click the Save and Exit button (far right on the toolbar) when you are finished.

You have to edit your dialog monitor function manually to define the behavior of the dialog box.

Dialog Monitor Functions (Scripting)

A dialog monitor function defines the behavior of a dialog box for each of a number of specified cases. The function takes the following (generic) form:

```
Function DialogFunc(strDlgItem as String, intAction as Integer, intSuppValue as Integer)
    Select Case intAction
        Case 1 ' dialog box initialization
        ...    'statements to execute when dialog box is initialized
        Case 2 ' value changing or button pressed
        ...    'statements...
        Case 3 ' TextBox or ComboBox text changed ...
        Case 4 ' focus changed ...
        Case 5 ' idle ...
    End Select
End Function
```

Parameters. The function must be able to pass three parameters: one string (strDlgItem) and two integers (intAction and intSuppValue). The parameters are values passed between the function and the dialog box, depending on what action is taken.

For example, when a user clicks a control in the dialog box, the name of the control is passed to the function as strDlgItem (the field name is specified in the dialog box definition). The second parameter (intAction) is a numeric value that indicates what action took place in the dialog box. The third parameter is used for additional information in some cases. You must include all three parameters in the function definition even if you do not use all of them.

Select Case intAction. The value of `intAction` indicates what action took place in the dialog box. For example, when the dialog box initializes, `intAction` = 1. If the user presses a button, `intAction` changes to 2, and so on. There are five possible actions, and you can specify statements that execute for each action as indicated below. You do not need to specify all five possible cases—only the ones that apply. For example, if you do not want any statements to execute on initialization, omit Case 1.

- **Case intAction = 1.** Specify statements to execute when the dialog box is initialized. For example, you could disable one or more controls or add a beep. The string `strDlgItem` is a null string; `intSuppValue` is 0.

- **Case 2.** Executes when a button is pushed or when a value changes in a `CheckBox`, `DropListBox`, `ListBox`, or `OptionGroup` control. If a button is pushed, `strDlgItem` is the button, `intSuppValue` is meaningless, and you must set `DialogFunc = True` to prevent the dialog box from closing. If a value changes, `strDlgItem` is the item whose value has changed, and `intSuppValue` is the new value.

- **Case 3.** Executes when a value changes in a `TextBox` or `ComboBox` control. The string `strDlgItem` is the control whose text changed and is losing focus; `intSuppValue` is the number of characters.

- **Case 4.** Executes when the focus changes in the dialog box. The string `strDlgItem` is gaining focus, and `intSuppValue` is the item that is losing focus (the first item is 0, second is 1, and so on).

- **Case 5.** Idle processing. The string `strDlgItem` is a null string; `intSuppValue` is 0. Set `DialogFunc = True` to continue receiving idle actions.

For more information, see the examples and the `DialogFunc` prototype in the Sax BASIC Language Reference Help file.

Example: Scripting a Simple Dialog Box

This script creates a simple dialog box that opens a data file. See related sections for explanations of the `BuildDialog` subroutine and dialog monitor function.

Figure 48-17
Open Data File dialog box created by script

```
Sub Main
    Call BuildDialog
End Sub

 'define dialog box
Sub BuildDialog
    Begin Dialog UserDialog 580,70,"Open Data File",.DialogFunc
        Text 40,7,280,21,"Data file to open:",.txtDialogTitle
        TextBox 40,28,340,21,.txtFilename
        OKButton 470,7,100,21,.cmdOK
        CancelButton 470,35,100,21,.cmdCancel
    End Dialog
    Dim dlg As UserDialog
    Dialog dlg
End Sub

 'define function that determines behavior of dialog box
Function DialogFunc(strDlgItem As String, intAction As Integer, intSuppValue As Integer) As Boolean
    Select Case intAction
        Case 1' beep when dialog is initialized
          Beep
        Case 2' value changing or button pressed
          Select Case strDlgItem
          Case "cmdOK"'if user clicks OK, open data file with specified filename
          strFilename = DlgText("txtFilename")
          Call OpenDataFile(strFilename)
          DialogFunc = False
          Case "cmdCancel"'If user clicks Cancel, close dialog
          DialogFunc = False
          End Select
    End Select
End Function

Sub OpenDataFile(strFilename As Variant)'Open data file with specified filename
```

```
Dim objDataDoc As ISpssDataDoc
Set objDataDoc = objSpssApp.OpenDataDoc(strFilename)
End Sub
```

Examples are also available in the online Help. You can try them yourself by pasting the code from Help into the script window.

Debugging Scripts

The Debug menu allows you to step through your code, executing one line or subroutine at a time and viewing the result. You can also insert a break point in the script to pause the execution at the line that contains the break point.

To debug an autoscript, open the autoscript file in a script window, insert break points in the procedure that you want to debug, and then run the statistical procedure that triggers the autoscript.

Step Into. Execute the current line. If the current line is a subroutine or function call, stop on the first line of that subroutine or function.

Step Over. Execute to the next line. If the current line is a subroutine or function call, execute the subroutine or function completely.

Step Out. Step out of the current subroutine or function call.

Step to Cursor. Execute to the current line.

Toggle Break. Insert or remove a break point. The script pauses at the break point, and the debugging pane is displayed.

Quick Watch. Display the value of the current expression.

Add Watch. Add the current expression to the watch window.

Object Browser. Display the object browser.

Set Next Statement. Set the next statement to be executed. Only statements in the current subroutine/function can be selected.

Show Next Statement. Display the next statement to be executed.

To Step through a Script

▶ From the Debug menu, choose any of the Step options to execute code, one line or subroutine at a time.

The Immediate, Watch, Stack, and Loaded tabs are displayed in the script window, along with the debugging toolbar.

▶ Use the toolbar (or hot keys) to continue stepping through the script.

▶ Alternatively, select Toggle Break to insert a break point at the current line.

The script pauses at the break point.

Debugging Pane (Scripting)

When you step through code, the Immediate, Watch, Stack, and Loaded tabs are displayed.

Figure 48-18
Debugging pane displayed in script window

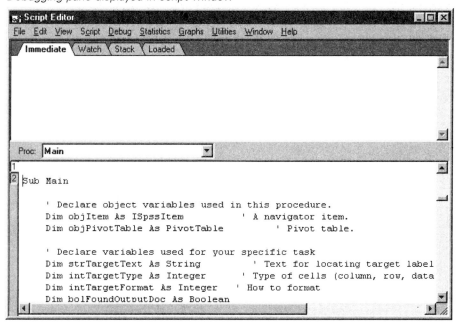

Immediate tab. Click the name of any variable and click the eyeglass button to display the current value of the variable. You can also evaluate an expression, assign a variable, or call a subroutine.

■ Type ?expr and press Enter to show the value of *expr*.

■ Type var = expr and press Enter to change the value of *var*.

■ Type subname args and press Enter to call a subroutine or built-in instruction.

■ Type Trace and press Enter to toggle trace mode. Trace mode prints each statement in the immediate window when a script is running.

Watch tab. To display a variable, function, or expression, click it and choose Add Watch from the Debug menu. Displayed values are updated each time execution pauses. You can edit the expression to the left of ->. Press Enter to update all of the values immediately. Press Ctrl-Y to delete the line.

Stack tab. Displays the lines that called the current statement. The first line is the current statement, the second line is the one that called the first, and so on. Click any line to highlight that line in the edit window.

Loaded tab. List the currently active scripts. Click a line to view that script.

Script Files and Syntax Files

Syntax files (*.sps*) are not the same as script files (*.sbs*). Syntax files have commands written in the command language that allows you to run statistical procedures and data transformations. While scripts allow you to manipulate output and automate other tasks that you normally perform using the graphical interface of menus and dialog boxes, the command language provides an alternative method for communicating directly with the program's back end, the part of the system that handles statistical computations and data transformations.

You can combine scripts and syntax files for even greater flexibility, by running a script from within command syntax, or by embedding command syntax within a script.

Running Command Syntax from a Script

You can run command syntax from within an automation script using the ExecuteCommands method. Command syntax allows you to run data transformations and statistical procedures and to produce charts. Much of this functionality cannot be automated directly from command scripts.

The easiest way to build a command syntax file is to make selections in dialog boxes and paste the syntax for the selections into the script window.

Figure 48-19
Pasting command syntax into a script

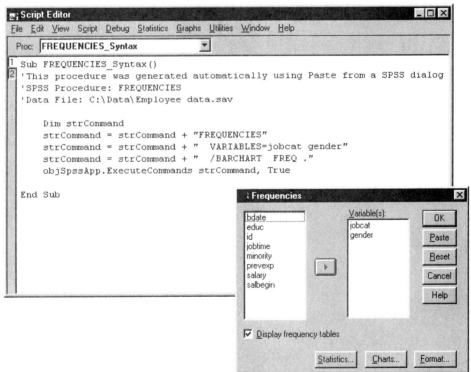

When you open dialog boxes using the script window menus, the Paste button pastes all of the code needed to run commands from within a script.

Note: You must use the script window menus to open the dialog box; otherwise, commands will be pasted to a syntax window rather than the scripting window.

Pasting SPSS Command Syntax into a Script

▶ From the script window menus, choose commands from the Statistics, Graphs, and Utilities menus to open dialog boxes.

▶ Make selections in the dialog box.

▶ Click Paste.

Note: You must use the script window menus to open the dialog box; otherwise, commands will be pasted to a syntax window rather than the scripting window.

Running a Script from Command Syntax

You can use the SCRIPT command to run a script from within command syntax. Specify the name of the script you want to run, with the filename enclosed in quotes, as follows:

```
SCRIPT 'C:\PROGRAM FILES\SPSS\CLEAN NAVIGATOR.SBS'.
```

See the *SPSS Command Syntax Reference* for complete syntax information.

Output Management System

The Output Management System (OMS) provides the ability to automatically write selected categories of output to different output files in different formats. Formats include:

- **SPSS data file format (.sav).** Output that would be displayed in pivot tables in the Viewer can be written out in the form of an SPSS data file, making it possible to use output as input for subsequent commands.

- **XML.** Tables, text output, and even many charts can be written out in XML format.

- **HTML.** Tables and text output can be written out in HTML format. Standard (not interactive) charts and tree model diagrams (Classification Tree option) can be included as image files. The image files are saved in a separate subdirectory (folder).

- **Text.** Tables and text output can be written out as tab-delimited or space-separated text.

To Use the Output Management System Control Panel

▶ From the menus choose:
Utilities
  OMS Control Panel...

Figure 49-1
Output Management System Control Panel

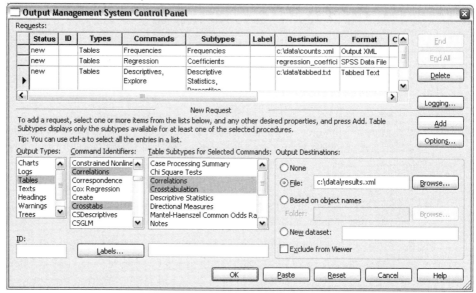

You can use the control panel to start and stop the routing of output to various destinations.

- Each OMS request remains active until explicitly ended or until the end of the session.

- A destination file that is specified on an OMS request is unavailable to other SPSS procedures and other applications until the OMS request is ended.

- While an OMS request is active, the specified destination files are stored in memory (RAM), so active OMS requests that write a large amount of output to external files may consume a large amount of memory.

- Multiple OMS requests are independent of each other. The same output can be routed to different locations in different formats, based on the specifications in different OMS requests.

- The order of the output objects in any particular destination is the order in which they were created, which is determined by the order and operation of the procedures that generate the output.

- OMS cannot route charts or warnings objects created by interactive graphics procedures (Graphs menu, Interactive submenu) or maps created by the mapping procedures (Graphs menu, Map submenu).

To Add New OMS Requests

▶ Select the output types (tables, charts, and so on) that you want to include. (For more information, see "Output Object Types" on p. 717.)

▶ Select the commands to include. If you want to include all output, select all items in the list. (For more information, see "Command Identifiers and Table Subtypes" on p. 719.)

▶ For commands that produce pivot table output, select the specific table types to include.

The list displays only the tables that are available in the selected commands; any table type that is available in one or more of the selected commands is displayed in the list. If no commands are selected, all table types are displayed. (For more information, see "Command Identifiers and Table Subtypes" on p. 719.)

▶ To select tables based on text labels instead of subtypes, click Labels. (For more information, see "Labels" on p. 720.)

▶ Click Options to specify the output format (for example, SPSS data file, XML, or HTML). (By default, Output XML format is used. For more information, see "OMS Options" on p. 721.)

▶ Specify an output destination:

- **File.** All selected output is routed to a single file.

- **Based on object names.** Output is routed to multiple destination files based on object names. A separate file is created for each output object, with a filename based on either table subtype names or table labels. Enter the destination folder name.

- **New dataset.** For SPSS data file format output, you can route the output to a dataset. The dataset is available for subsequent use in the same session but is not saved unless you explicitly save it as a file prior to the end of the session. This option is available only for SPSS data file format output. Dataset names must conform to SPSS variable-naming rules. For more information, see "Variable Names" in Chapter 5 on p. 106.

▶ Optionally:

■ Exclude the selected output from the Viewer. If you select Exclude from Viewer, the output types in the OMS request will not be displayed in the Viewer window. If multiple active OMS requests include the same output types, the display of those output types in the Viewer is determined by the most recent OMS request that contains those output types. For more information, see "Excluding Output Display from the Viewer" on p. 727.

■ Assign an ID string to the request. All requests are automatically assigned an ID value, and you can override the system default ID string with a descriptive ID, which can be useful if you have multiple active requests that you want to identify easily. ID values that you assign cannot start with a dollar sign ($).

The following tips are for selecting multiple items in a list:

■ Press Ctrl-A to select all items in a list.

■ Use Shift-click to select multiple contiguous items.

■ Use Ctrl-click to select multiple noncontiguous items.

To End and Delete OMS Requests

Active and new OMS requests are displayed in the Requests list, with the most recent request at the top. You can change the widths of the information columns by clicking and dragging the borders, and you can scroll the list horizontally to see more information about a particular request.

An asterisk (*) after the word Active in the *Status* column indicates an OMS request that was created with command syntax that includes features that are not available in the Control Panel.

To end a specific, active OMS request:

▶ In the Requests list, click any cell in the row for the request.

▶ Click End.

To end all active OMS requests:

▶ Click End All.

To delete a new request (a request that has been added but is not yet active):

▶ In the Requests list, click any cell in the row for the request.

▶ Click Delete.

Note: Active OMS requests are not ended until you click OK.

Output Object Types

There are seven different types of output objects:

Charts. Charts (except "interactive" charts and maps). Chart objects are included only with XML and HTML destination formats. For HTML format, the chart image files are saved in a separate subdirectory (folder).

Logs. Log text objects. Log objects contain certain types of error and warning messages. Depending on your Options settings (Edit menu, Options, Viewer tab), log objects may also contain the command syntax that is executed during the session. Log objects are labeled *Log* in the outline pane of the Viewer.

Tables. Output objects that are pivot tables in the Viewer (includes Notes tables). Tables are the only output objects that can be routed to SPSS data file (*.sav*) format.

Text. Text objects that aren't logs or headings (includes objects labeled *Text Output* in the outline pane of the Viewer).

Trees. Tree model diagrams that are produced by the Classification Tree option. Tree objects are included only with XML and HTML destination formats.

Headings. Text objects that are labeled *Title* in the outline pane of the Viewer. For Output XML format, heading text objects are not included.

Warnings. Warnings objects. Warnings objects contain certain types of error and warning messages.

Figure 49-2
Output object types

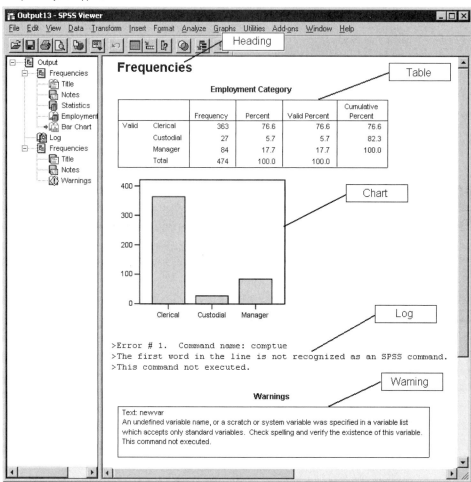

Command Identifiers and Table Subtypes

Command Identifiers

Command identifiers are available for all statistical and charting procedures and any other commands that produce blocks of output with their own identifiable heading in the outline pane of the Viewer. These identifiers are usually (but not always) the same or similar to the procedure names on the menus and dialog box titles, which are usually (but not always) similar to the underlying SPSS command names. For example, the command identifier for the Frequencies procedure is "Frequencies," and the underlying command name is also the same.

There are, however, some cases where the procedure name and the command identifier and/or the command name are not all that similar. For example, all of the procedures on the Nonparametric Tests submenu (from the Analyze menu) use the same underlying command, and the command identifier is the same as the underlying command name: Npar Tests.

Table Subtypes

Table subtypes are the different types of pivot tables that can be produced. Some subtypes are produced by only one command; other subtypes can be produced by multiple commands (although the tables may not look similar). Although table subtype names are generally descriptive, there can be many names to choose from (particularly if you have selected a large number of commands); also, two subtypes may have very similar names.

To Find Command Identifiers and Table Subtypes

When in doubt, you can find command identifiers and table subtype names in the Viewer window:

▶ Run the procedure to generate some output in the Viewer.

▶ Right-click the item in the outline pane of the Viewer.

▶ Choose Copy OMS Command Identifier or Copy OMS Table Subtype.

▶ Paste the copied command identifier or table subtype name into any text editor (such as an SPSS syntax window).

Labels

As an alternative to table subtype names, you can select tables based on the text that is displayed in the outline pane of the Viewer. You can also select other object types based on their labels. Labels are useful for differentiating between multiple tables of the same type in which the outline text reflects some attribute of the particular output object, such as the variable names or labels. There are, however, a number of factors that can affect the label text:

■ If split-file processing is on, split-file group identification may be appended to the label.

■ Labels that include information about variables or values are affected by your current output label options settings (Edit menu, Options, Output Labels tab).

■ Labels are affected by the current output language setting (Edit menu, Options, General tab).

To Specify Labels to Use to Identify Output Objects

▶ In the Output Management System Control Panel, select one or more output types and then select one or more commands.

▶ Click Labels.

Figure 49-3
OMS Labels dialog box

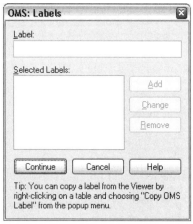

▶ Enter the label *exactly* as it appears in the outline pane of the Viewer window. (You can also right-click the item in the outline, choose Copy OMS Label, and paste the copied label into the Label text field.)

▶ Click Add.

▶ Repeat the process for each label that you want to include.

▶ Click Continue.

Wildcards

You can use an asterisk (*) as the last character of the label string as a wildcard character. All labels that begin with the specified string (except for the asterisk) will be selected. This process works only when the asterisk is the last character, because asterisks can appear as valid characters inside a label.

OMS Options

You can use the OMS Options dialog box to:

- Specify the output format.
- Include or exclude chart and tree model diagram output and specify the graphic format.
- Specify what table dimension elements should go into the row dimension.
- For SPSS data file format, include a variable that identifies the sequential table number that is the source for each case.

To Specify OMS Options

▶ Click Options in the Output Management System Control Panel.

Figure 49-4
OMS Options dialog box

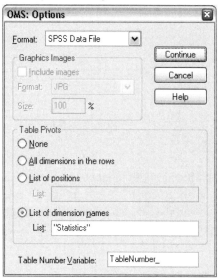

Format

Output XML. XML that conforms to the *spss-output* schema. Standard charts are included as XML that conforms to the *vizml* schema. You can also export all charts and maps as separate files in the selected graphics format.

HTML. Output objects that would be pivot tables in the Viewer are converted to simple HTML tables. No TableLook attributes (font characteristics, border styles, colors, and so on) are supported. Text output objects are tagged <PRE> in the HTML. If you choose to include charts, they are exported as separate files in the selected graphics format and are embedded by reference (<IMG SRC='filename.ext'>) in the HTML document. The image files are saved in a separate subdirectory (folder).

SPSS Data File. This format is a binary file format. All output object types other than tables are excluded. Each column of a table becomes a variable in the data file. To use a data file that is created with OMS in the same session, you must end the active OMS request before you can open the data file. For more information, see "Routing Output to SPSS Data Files" on p. 727.

SVWS XML. XML used by SmartViewer Web Server. This is actually a JAR/ZIP file containing XML, CSV, and other files. SmartViewer Web Server is a separate, server-based product.

Text. Space-separated text. Output is written as text, with tabular output aligned with spaces for fixed-pitch fonts. All charts and maps are excluded.

Tabbed Text. Tab-delimited text. For output that is displayed as pivot tables in the Viewer, tabs delimit table column elements. Text block lines are written as is; no attempt is made to divide them with tabs at useful places. All charts and maps are excluded.

Graphics Images

For HTML format, you can include charts (excluding interactive charts) and tree model diagrams as image files. A separate image file is created for each chart and/or tree, and standard <IMG SRC='filename'> tags are included in the HTML for each image file.

Image files are saved in a separate subdirectory (folder). The subdirectory name is the name of the HTML destination file, without any extension and with *files* appended to the end. For example, if the HTML destination file is *julydata.htm*, the images subdirectory will be named *julydata_files*.

- **Format.** The available image formats are PNG, JPG, EMF, and BMP.

- **Size.** You can scale the image size from 10% to 200%.

Table Pivots

For pivot table output, you can specify the dimension element(s) that should appear in the columns. All other dimension elements appear in the rows. For SPSS data file format, table columns become variables, and rows become cases.

- If you specify multiple dimension elements for the columns, they are nested in the columns in the order in which they are listed. For SPSS data file format, variable names are constructed by nested column elements. For more information, see "Variable Names in OMS-Generated Data Files" on p. 735.

- If a table doesn't contain any of the listed dimension elements, all dimension elements for that table will appear in the rows.

- Table pivots that are specified here have no effect on tables that are displayed in the Viewer.

Each dimension of a table—row, column, layer—may contain zero or more elements. For example, a simple two-dimensional crosstabulation contains a single row dimension element and a single column dimension element, each of which contains one of the variables that are used in the table. You can use either positional arguments or dimension element "names" to specify the dimension elements that you want to put into the column dimension.

All dimensions in rows. Creates a single row for each table. For SPSS format data files, this means each table is a single case, and all the table elements are variables.

List of positions. The general form of a positional argument is a letter indicating the default position of the element—C for column, R for row, or L for layer—followed by a positive integer indicating the default position within that dimension. For example, R1 would indicate the outermost row dimension element.

- To specify multiple elements from multiple dimensions, separate each dimension with a space—for example, R1 C2.

- The dimension letter followed by ALL indicates all elements in that dimension in their default order. For example, CALL is the same as the default behavior (using all column elements in their default order to create columns).

- CALL RALL LALL (or RALL CALL LALL, and so on) will put all dimension elements into the columns. For SPSS data file format, this creates one row/case per table in the data file.

Figure 49-5
Row and column positional arguments

Model				Previous Experience (months)	Months since Hire	Beginning Salary	Date of Birth
1	Correlations	Previous Experience (months)		1.000	.067	-.087	.805
		Months since Hire		.067	1.000	.012	.085
		Beginning Salary		-.087	.012	1.000	-.075
		Date of Birth		.805	.085	-.075	1.000
	Covariances	Previous Experience (months)		31.307	12.940	-.022	7.096E-06
		Months since Hire		12.940	1205.248	.019	4.635E-06
		Beginning Salary		-.022	.019	.002	-5.236E-09
		Date of Birth		7.096E-06	4.635E-06	-5.236E-09	2.485E-12

List of dimension names. As an alternative to positional arguments, you can use dimension element "names," which are the text labels that appear in the table. For example, a simple two-dimensional crosstabulation contains a single row dimension element and a single column dimension element, each with labels based on the variables in those dimensions, plus a single layer dimension element labeled *Statistics* (if English is the output language).

■ Dimension element names may vary, based on the output language and/or settings that affect the display of variable names and/or labels in tables.

■ Each dimension element name must be enclosed in single or double quotation marks. To specify multiple dimension element names, include a space between each quoted name.

The labels that are associated with the dimension elements may not always be obvious.

To See All Dimension Elements and Their Labels for a Pivot Table

▶ Activate (double-click) the table in the Viewer.

▶ From the menus choose:
View
  Show All

and/or

▶ If the pivoting trays aren't displayed, from the menus choose:
Pivot
  Pivoting Trays

▶ Hover the pointer over each icon in the pivoting trays to display the label.

Figure 49-6

Dimension element names displayed in table and pivoting trays

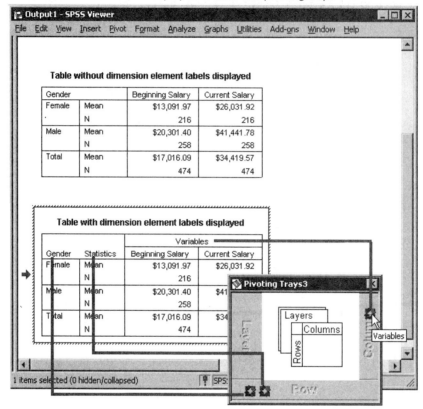

Logging

You can record OMS activity in a log in XML or text format.

- ■ The log tracks all new OMS requests for the session but does not include OMS requests that were already active before you requested a log.

- ■ The current log file ends if you specify a new log file or if you deselect (uncheck) Log OMS activity.

To specify OMS logging:

▶ Click Logging in the Output Management System Control Panel.

Excluding Output Display from the Viewer

The Exclude from Viewer check box affects all output that is selected in the OMS request by suppressing the display of that output in the Viewer window. This process is often useful for production jobs that generate a lot of output and when you don't need the results in the form of a Viewer document (*.spo* file). You can also use this functionality to suppress the display of particular output objects that you simply never want to see, without routing any other output to some external file and format.

To suppress the display of certain output objects without routing other output to an external file:

▶ Create an OMS request that identifies the unwanted output.

▶ Select Exclude from Viewer.

▶ For the output destination, select File—but leave the File field blank.

▶ Click Add.

The selected output will be excluded from the Viewer while all other output will be displayed in the Viewer in the normal fashion.

Routing Output to SPSS Data Files

An SPSS data file consists of variables in the columns and cases in the rows, which is essentially the format in which pivot tables are converted to data files:

- Columns in the table are variables in the data file. Valid variable names are constructed from the column labels.

- Row labels in the table become variables with generic variable names (*Var1*, *Var2*, *Var3*, and so on) in the data file. The values of these variables are the row labels in the table.

- Three table-identifier variables are automatically included in the data file: *Command_*, *Subtype_*, and *Label_*. All three are string variables. The first two variables correspond to the command and subtype identifiers. For more information, see "Command Identifiers and Table Subtypes" on p. 719. *Label_* contains the table title text.

- Rows in the table become cases in the data file.

Example: Single Two-Dimensional Table

In the simplest case—a single two-dimensional table—the table columns become variables, and the rows become cases in the data file.

Figure 49-7
Single two-dimensional table

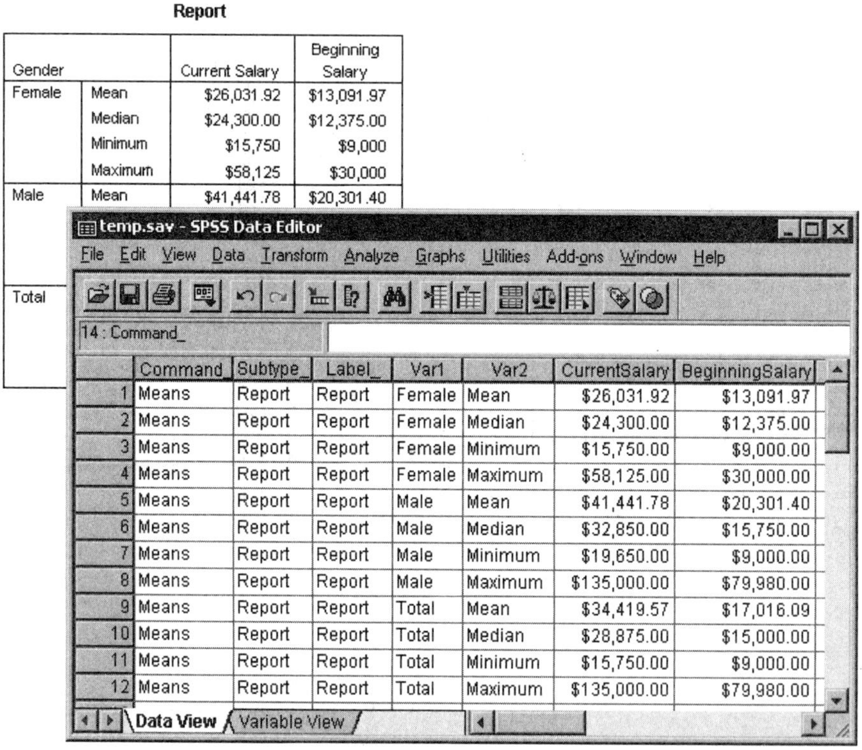

Report

Gender		Current Salary	Beginning Salary
Female	Mean	$26,031.92	$13,091.97
	Median	$24,300.00	$12,375.00
	Minimum	$15,750	$9,000
	Maximum	$58,125	$30,000
Male	Mean	$41,441.78	$20,301.40
Total			

temp.sav - SPSS Data Editor

File Edit View Data Transform Analyze Graphs Utilities Add-ons Window Help

14 : Command_

	Command	Subtype_	Label_	Var1	Var2	CurrentSalary	BeginningSalary
1	Means	Report	Report	Female	Mean	$26,031.92	$13,091.97
2	Means	Report	Report	Female	Median	$24,300.00	$12,375.00
3	Means	Report	Report	Female	Minimum	$15,750.00	$9,000.00
4	Means	Report	Report	Female	Maximum	$58,125.00	$30,000.00
5	Means	Report	Report	Male	Mean	$41,441.78	$20,301.40
6	Means	Report	Report	Male	Median	$32,850.00	$15,750.00
7	Means	Report	Report	Male	Minimum	$19,650.00	$9,000.00
8	Means	Report	Report	Male	Maximum	$135,000.00	$79,980.00
9	Means	Report	Report	Total	Mean	$34,419.57	$17,016.09
10	Means	Report	Report	Total	Median	$28,875.00	$15,000.00
11	Means	Report	Report	Total	Minimum	$15,750.00	$9,000.00
12	Means	Report	Report	Total	Maximum	$135,000.00	$79,980.00

Data View / Variable View /

- The first three variables identify the source table by command, subtype, and label.

- The two elements that defined the rows in the table—values of the variable *Gender* and statistical measures—are assigned the generic variable names *Var1* and *Var2*. These variables are both string variables.

- The column labels from the table are used to create valid variable names. In this case, those variable names are based on the variable labels of the three scale variables that are summarized in the table. If the variables didn't have defined variable labels, or you chose to display variable names instead of variable labels as the column labels in the table, the variable names in the new data file would be the same as in the source data file.

Example: Tables with Layers

In addition to rows and columns, a table can also contain a third dimension: the layer dimension.

Figure 49-8
Table with layers

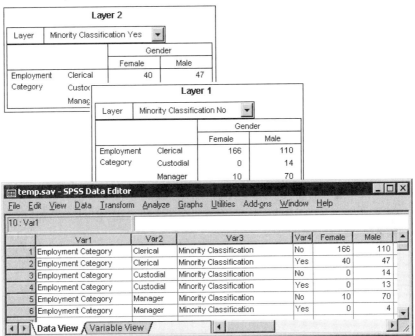

- In the table, the variable labeled *Minority Classification* defines the layers. In the data file, two additional variables are created: one variable that identifies the layer element and one variable that identifies the categories of the layer element.

- As with the variables that are created from the row elements, the variables that are created from the layer elements are string variables with generic variable names (the prefix *Var* followed by a sequential number).

Data Files Created from Multiple Tables

When multiple tables are routed to the same data file, each table is added to the data file in a fashion that is similar to merging data files by adding cases from one data file to another data file (Data menu, Merge Files, Add Cases).

- Each subsequent table will always add cases to the data file.

- If column labels in the tables differ, each table may also add variables to the data file, with missing values for cases from other tables that don't have an identically labeled column.

Example: Multiple Tables with the Same Column Labels

Multiple tables that contain the same column labels will typically produce the most immediately useful data files (data files that don't require additional manipulation). For example, two or more frequency tables from the Frequencies procedure will all have identical column labels.

Figure 49-9
Two tables with identical column labels

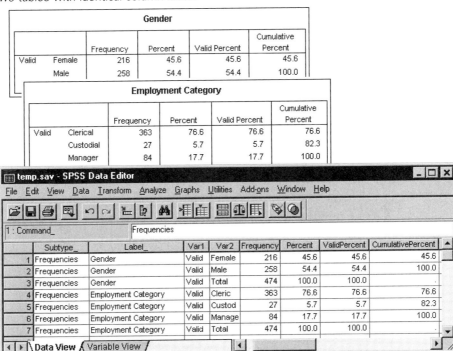

- The second table contributes additional cases (rows) to the data file but contributes no new variables because the column labels are exactly the same, so there are no large patches of missing data.

- Although the values for *Command_* and *Subtype_* are the same, the *Label_* value identifies the source table for each group of cases because the two frequency tables have different titles.

Example: Multiple Tables with Different Column Labels

A new variable is created in the data file for each unique column label in the tables that are routed to the data file. This process results in blocks of missing values if the tables contain different column labels.

Figure 49-10
Two tables with different column labels

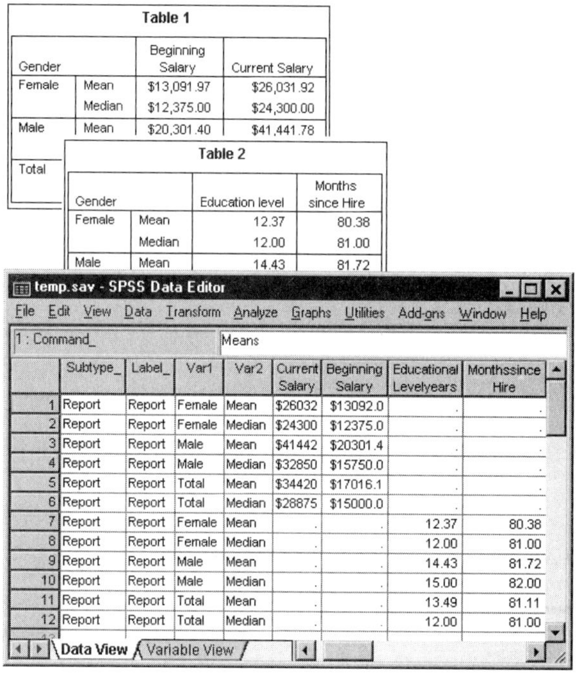

Table 1

Gender		Beginning Salary	Current Salary
Female	Mean	$13,091.97	$26,031.92
	Median	$12,375.00	$24,300.00
Male	Mean	$20,301.40	$41,441.78
Total			

Table 2

Gender		Education level	Months since Hire
Female	Mean	12.37	80.38
	Median	12.00	81.00
Male	Mean	14.43	81.72

temp.sav - SPSS Data Editor

File Edit View Data Transform Analyze Graphs Utilities Add-ons Window Help

1 : Command_   Means

	Subtype_	Label_	Var1	Var2	Current Salary	Beginning Salary	Educational Levelyears	Monthssince Hire
1	Report	Report	Female	Mean	$26032	$13092.0	.	.
2	Report	Report	Female	Median	$24300	$12375.0	.	.
3	Report	Report	Male	Mean	$41442	$20301.4	.	.
4	Report	Report	Male	Median	$32850	$15750.0	.	.
5	Report	Report	Total	Mean	$34420	$17016.1	.	.
6	Report	Report	Total	Median	$28875	$15000.0	.	.
7	Report	Report	Female	Mean	.	.	12.37	80.38
8	Report	Report	Female	Median	.	.	12.00	81.00
9	Report	Report	Male	Mean	.	.	14.43	81.72
10	Report	Report	Male	Median	.	.	15.00	82.00
11	Report	Report	Total	Mean	.	.	13.49	81.11
12	Report	Report	Total	Median	.	.	12.00	81.00

Data View Variable View

- The first table has columns labeled *Beginning Salary* and *Current Salary*, which are not present in the second table, resulting in missing values for those variables for cases from the second table.

- Conversely, the second table has columns labeled *Education level* and *Months since Hire*, which are not present in the first table, resulting in missing values for those variables for cases from the first table.

- Mismatched variables like the variables in this example can occur even with tables of the same subtype. In this example, both tables are the same subtype.

Example: Data Files Not Created from Multiple Tables

If any tables do not have the same number of row elements as the other tables, no data file will be created. The number of rows doesn't have to be the same; the number of row *elements* that become variables in the data file must be the same. For example, a two-variable crosstabulation and a three-variable crosstabulation contain different numbers of row elements, because the "layer" variable is actually nested within the row variable in the default three-variable crosstabulation display.

Figure 49-11
Tables with different numbers of row elements

Employment Category * Gender Crosstabulation

		Gender		Total
		Female	Male	
Employment Category	Clerical	206	157	363
	Custodial	0	27	27
	Manager	10	74	84
Total		216	258	474

Employment Category * Gender * Minority Classification Crosstabulation

Minority Classification			Gender		Total
			Female	Male	
No	Employment Category	Clerical	166	110	276
		Custodial	0	14	14
		Manager	10	70	80
	Total		176	194	370
Yes	Employment Category	Clerical	40	47	87
		Custodial	0	13	13
		Manager	0	4	4
	Total		40	64	104

Controlling Column Elements to Control Variables in the Data File

In the Options dialog box of the Output Management System Control Panel, you can specify which dimension elements should be in the columns and therefore will be used to create variables in the generated data file. This process is equivalent to pivoting the table in the Viewer.

For example, the Frequencies procedure produces a descriptive statistics table with statistics in the rows, while the Descriptives procedure produces a descriptive statistics table with statistics in the columns. To include both table types in the same data file in a meaningful fashion, you need to change the column dimension for one of the table types.

Because both table types use the element name "Statistics" for the statistics dimension, we can put the statistics from the Frequencies statistics table in the columns simply by specifying "Statistics" (in quotation marks) in the list of dimension names in the OMS Options dialog box.

Figure 49-12
OMS Options dialog box

Figure 49-13
Combining different table types in a data file by pivoting dimension elements

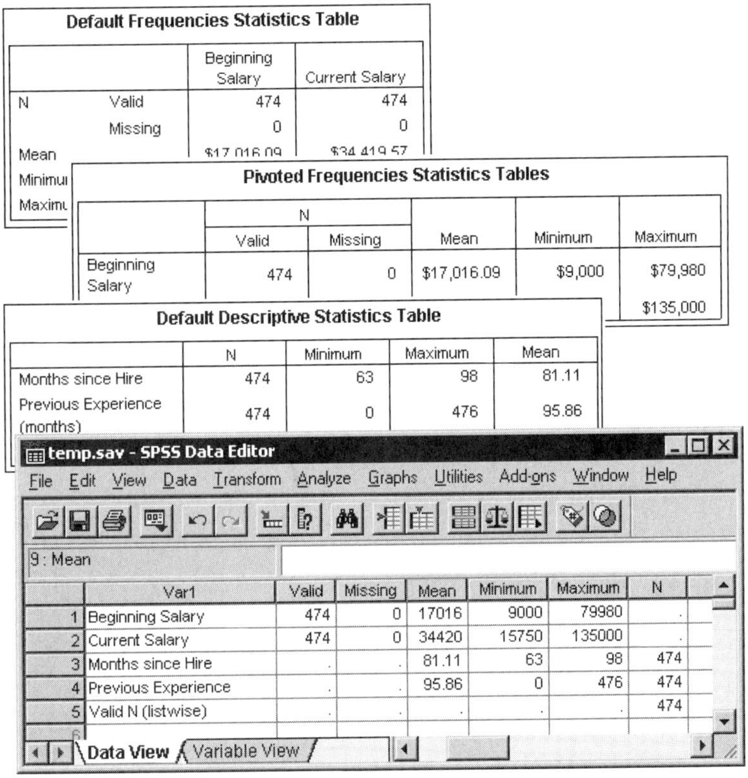

Some of the variables will have missing values, because the table structures still aren't exactly the same with statistics in the columns.

Variable Names in OMS-Generated Data Files

OMS constructs valid, unique variable names from column labels:

■ Row and layer elements are assigned generic variable names—the prefix *Var* followed by a sequential number.

■ Characters that aren't allowed in variable names (space, parenthesis, etc.) are removed. For example, "This (Column) Label" would become a variable named *ThisColumnLabel*.

- If the label begins with a character that is allowed in variable names but not allowed as the first character (for example, a number), "@" is inserted as a prefix. For example, "2nd" would become a variable named @2nd.

- Underscores or periods at the end of labels are removed from the resulting variable names. (The underscores at the end of the automatically generated variables *Command_*, *Subtype_*, and *Label_* are not removed.)

- If more than one element is in the column dimension, variable names are constructed by combining category labels with underscores between category labels. Group labels are not included. For example, if *VarB* is nested under *VarA* in the columns, you would get variables like *CatA1_CatB1*, not *VarA_CatA1_VarB_CatB1*.

Figure 49-14

Variable names constructed from table elements

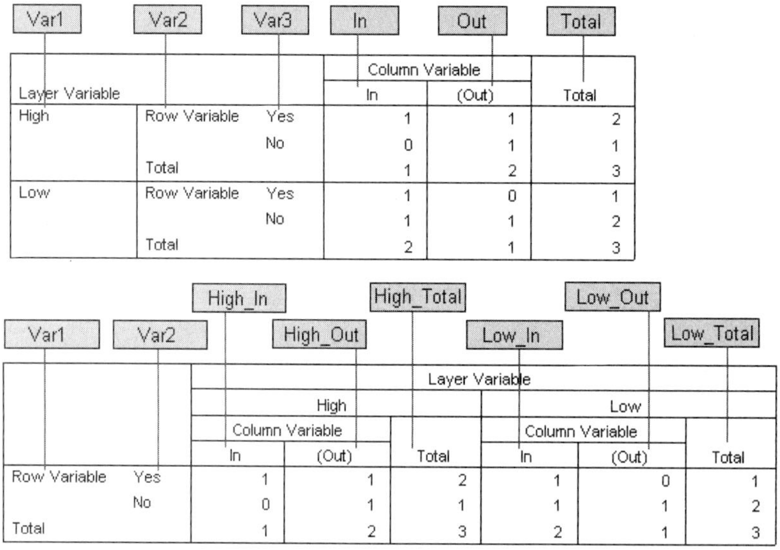

OXML Table Structure

Output XML (OXML) is XML that conforms to the *spss-output* schema. For a detailed description of the schema, see the Output Schema section of the Help system.

- OMS command and subtype identifiers are used as values of the command and subType attributes in OXML. An example is as follows:

```
<command text="Frequencies" command="Frequencies"...>
 <pivotTable text="Gender" label="Gender" subType="Frequencies"...>
```

- OMS command and subType attribute values are not affected by output language or display settings for variable names/labels or values/value labels.

- XML is case-sensitive. A subType attribute value of "frequencies" is *not* the same as a subType attribute value of "Frequencies."

- All information that is displayed in a table is contained in attribute values in OXML. At the individual cell level, OXML consists of "empty" elements that contain attributes but no "content" other than the content that is contained in attribute values.

- Table structure in OXML is represented row by row; elements that represent columns are nested within the rows, and individual cells are nested within the column elements:

```
<pivotTable...>
 <dimension axis='row'...>
  <dimension axis='column'...>
   <category...>
    <cell text='...' number='...' decimals='...'/>
   </category>
   <category...>
    <cell text='...' number='...' decimals='...'/>
   </category>
  </dimension>
 </dimension>
   ...
 </pivotTable>
```

The preceding example is a simplified representation of the structure that shows the descendant/ancestor relationships of these elements. However, the example does not necessarily show the parent/child relationships, because there are typically intervening nested element levels.

The following figure shows a simple frequency table and the complete output XML representation of that table.

Figure 49-15
Simple frequency table

Gender

		Frequency	Percent	Valid Percent	Cumulative Percent
Valid	Female	216	45.6	45.6	45.6
	Male	258	54.4	54.4	100.0
	Total	474	100.0	100.0	

Figure 49-16
Output XML for the simple frequency table

```
<?xml version="1.0" encoding="UTF-8" ?>
<outputTreeoutputTree xmlns="http://xml.spss.com/spss/oms"
 xmlns:xsi="http://www.w3.org/2001/XMLSchema-instance"
 xsi:schemaLocation="http://xml.spss.com/spss/oms
 http://xml.spss.com/spss/oms/spss-output-1.0.xsd">
<command text="Frequencies" command="Frequencies"
 displayTableValues="label" displayOutlineValues="label"
 displayTableVariables="label" displayOutlineVariables="label">
<pivotTable text="Gender" label="Gender" subType="Frequencies"
 varName="gender" variable="true">
 <dimension axis="row" text="Gender" label="Gender"
 varName="gender" variable="true">
 <group text="Valid">
  <group hide="true" text="Dummy">
  <category text="Female" label="Female" string="f"
 varName="gender">
  <dimension axis="column" text="Statistics">
  <category text="Frequency">
  <cell text="216" number="216"/>
  </category>
  <category text="Percent">
  <cell text="45.6" number="45.569620253165" decimals="1"/>
  </category>
  <category text="Valid Percent">
  <cell text="45.6" number="45.569620253165" decimals="1"/>
  </category>
  <category text="Cumulative Percent">
  <cell text="45.6" number="45.569620253165" decimals="1"/>
```

```
  </category>
 </dimension>
 </category>
 <category text="Male" label="Male" string="m" varName="gender">
 <dimension axis="column" text="Statistics">
 <category text="Frequency">
  <cell text="258" number="258"/>
 </category>
 <category text="Percent">
  <cell text="54.4" number="54.430379746835" decimals="1"/>
 </category>
 <category text="Valid Percent">
  <cell text="54.4" number="54.430379746835" decimals="1"/>
 </category>
 <category text="Cumulative Percent">
  <cell text="100.0" number="100" decimals="1"/>
 </category>
 </dimension>
 </category>
 </group>
 <category text="Total">
 <dimension axis="column" text="Statistics">
 <category text="Frequency">
  <cell text="474" number="474"/>
 </category>
 <category text="Percent">
  <cell text="100.0" number="100" decimals="1"/>
 </category>
 <category text="Valid Percent">
  <cell text="100.0" number="100" decimals="1"/>
 </category>
 </dimension>
 </category>
 </group>
 </dimension>
 </pivotTable>
 </command>
</outputTree>
```

As you may notice, a simple, small table produces a substantial amount of XML. That's partly because the XML contains some information that is not readily apparent in the original table, some information that might not even be available in the original table, and a certain amount of redundancy.

- The table contents as they are (or would be) displayed in a pivot table in the Viewer are contained in text attributes. An example is as follows:

```
<command text="Frequencies" command="Frequencies"...>
```

- Text attributes can be affected by both output language and settings that affect the display of variable names/labels and values/value labels. In this example, the text attribute value will differ, depending on the output language, whereas the command attribute value remains the same, regardless of output language.

- Wherever variables or values of variables are used in row or column labels, the XML will contain a text attribute and one or more additional attribute values. An example is as follows:

```
<dimension axis="row" text="Gender" label="Gender" varName="gender">
  ...<category text="Female" label="Female" string="f" varName="gender">
```

- For a numeric variable, there would be a number attribute instead of a string attribute. The label attribute is present only if the variable or values have defined labels.

- The <cell> elements that contain cell values for numbers will contain the text attribute and one or more additional attribute values. An example is as follows:

```
<cell text="45.6" number="45.569620253165" decimals="1"/>
```

The number attribute is the actual, unrounded numeric value, and the decimals attribute indicates the number of decimal positions that are displayed in the table.

- Because columns are nested within rows, the category element that identifies each column is repeated for each row. For example, because the statistics are displayed in the columns, the element <category text="Frequency"> appears three times in the XML: once for the male row, once for the female row, and once for the total row.

OMS Identifiers

The OMS Identifiers dialog box is designed to assist you in writing OMS command syntax. You can use this dialog box to paste selected command and subtype identifiers into a command syntax window.

Figure 49-17
OMS Identifiers dialog box

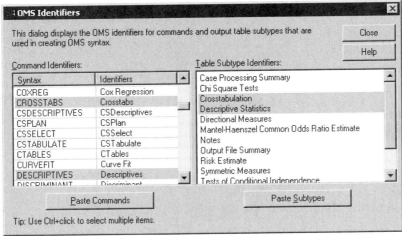

To Use the OMS Identifiers Dialog Box

▶ From the menus choose:

Utilities
    OMS Identifiers...

▶ Select one or more command or subtype identifiers. (Use Ctrl-click to select multiple identifiers in each list.)

▶ Click Paste Commands and/or Paste Subtypes.

■ The list of available subtypes is based on the currently selected command(s). If multiple commands are selected, the list of available subtypes is the union of all subtypes that are available for any of the selected commands. If no commands are selected, all subtypes are listed.

■ The identifiers are pasted into the designated command syntax window at the current cursor location. If there are no open command syntax windows, a new syntax window is automatically opened.

■ Each command and/or subtype identifier is enclosed in quotation marks when pasted, because OMS command syntax requires these quotation marks.

■ Identifier lists for the COMMANDS and SUBTYPES keywords must be enclosed in brackets, as in:

```
/IF COMMANDS=['Crosstabs' 'Descriptives']
    SUBTYPES=['Crosstabulation' 'Descriptive Statistics']
```

Copying OMS Identifiers from the Viewer Outline

You can copy and paste OMS command and subtype identifiers from the Viewer outline pane.

▶ In the outline pane, right-click the outline entry for the item.

▶ Choose Copy OMS Command Identifier or Copy OMS Table Subtype.

This method differs from the OMS Identifiers dialog box method in one respect: The copied identifier is not automatically pasted into a command syntax window. The identifier is simply copied to the clipboard, and you can then paste it anywhere you want. Because command and subtype identifier values are identical to the corresponding command and subtype attribute values in XML-format output (OXML), you might find this copy/paste method useful if you write XSLT transformations.

Copying OMS Labels

Instead of identifiers, you can copy labels for use with the LABELS keyword. Labels can be used to differentiate between multiple graphs or multiple tables of the same type in which the outline text reflects some attribute of the particular output object, such as the variable names or labels. There are, however, a number of factors that can affect the label text:

■ If split-file processing is on, split-file group identification may be appended to the label.

- Labels that include information about variables or values are affected by the settings for the display of variable names/labels and values/value labels in the outline pane (Edit menu, Options, Output Labels tab).

- Labels are affected by the current output language setting (Edit menu, Options, General tab).

To copy OMS labels

▶ In the outline pane, right-click the outline entry for the item.

▶ Choose Copy OMS Label.

As with command and subtype identifiers, the labels must be in quotation marks, and the entire list must be enclosed in square brackets, as in:

```
/IF LABELS=['Employment Category' 'Education Level']
```

Database Access Administrator

The Database Access Administrator is a utility designed to simplify large or confusing data sources for use with the Database Wizard. The Database Access Administrator allows users and administrators to customize their data sources in the following ways:

- Create aliases for database tables and fields.

- Create variable names for fields.

- Hide extraneous tables and fields.

The Database Access Administrator does not actually change your database. Instead, it generates files that hold all of your information. These files act as database *views*.

You can use the Database Access Administrator to specify up to three different views per database: enterprise level, department level, and personal level. Both the Database Access Administrator and the Database Wizard recognize these files by the following names:

- Enterprise level: *dba01.inf*

- Department level: *dba02.inf*

- Personal level: *dba03.inf*

Each file contains level-specific information about any number of data sources. For example, your *dba03.inf* file could contain personal view information for a corporate accounts database, your company's hourlog database, and a database that you use to keep track of your CD collection.

When you open the Database Access Administrator, it will search your system's path for these files and automatically display information for all three views of any data source that you have configured.

Inheritance and priorities. Whenever you use the Database Wizard, it presents the lowest-level view of your data source that it can find on your system's path, where the levels are (from highest to lowest) enterprise, department, and personal. Each level's file holds information about all of your data sources for that level. For example, your marketing department will have one file, *dba02.inf*, that contains the aliasing information for all of the database views that are established for the marketing department. Each person in the marketing department will have a file, *dba03.inf*, that contains customized views of all of the databases that he or she uses.

In the Database Access Administrator, aliases, variable names, and hide orders are inherited from the top down.

Examples

■ If the Regions table is hidden at the enterprise level, it is invisible at both the department and personal levels. This table would not be displayed in the Database Wizard.

■ The field *JOBCAT* in the Employee Sales table is not aliased at the enterprise level, but it is aliased as Job Categories at the department level. It appears as Job Categories at the personal level. Additionally, if the Employee Sales table is aliased as Employee Information at the personal level, the original field (*EmployeeSales.JOBCAT*) would appear in the Database Wizard as *'Employee Information'.'Job Categories'*.

To start the Database Access Administrator, run the file *spssdbca.exe*, which is installed in your SPSS directory. For more information about the Database Access Administrator, see the online Help.

Customizing HTML Documents

You can automatically add customized HTML code to documents exported in HTML format, including:

- HTML document titles
- Document type specification
- Meta tags and script code (for example, JavaScript)
- Text displayed before and after exported output

To Add Customized HTML Code to Exported Output Documents

▶ In a text editor, open the file *htmlfram.txt* (located in the directory in which SPSS is installed).

▶ In the "fields" on the lines between the double open brackets (<<), replace the comments with the text or HTML code that you want to insert into your exported HTML documents.

▶ Save the file as a text file.

Note: If you change the name or location of the text file, you have to modify the system registry to use the file to customize your exported HTML output.

Content and Format of the Text File for Customized HTML

The HTML code that you want to add automatically to your HTML documents must be specified in a simple text file that contains six fields, each field delimited by two open angle brackets on the preceding line (<<):

<<

Text or code that you want to insert at the top of the document before the <HTML> specification (for example, comments that include document type specifications)

<<

Text that is used as the document title (displayed in the title bar)

<<

Meta tags or script code (for example, JavaScript code)

<<

HTML code that modifies the <BODY> tag (for example, code that specifies background color)

<<

Text and/or HTML code that is inserted after the exported output (for example, copyright notice)

<<

Text and/or HTML code that is inserted before the exported output (for example, company name and logo)

To Use a Different File or Location for Custom HTML Code

If you change the name or location of *htmlfram.txt*, you must modify the system registry to use the file in customized HTML output.

▶ From the Windows Start menu choose Run, type regedit, and then click OK.

▶ In the left pane of the Registry Editor, choose:

```
HHKEY_CURRENT_USER
  Software
   SPSS
    SPSS for Windows
     15.0
      Spsswin
```

▶ In the right pane, double-click the string HTMLFormatFile.

▶ For Value data, enter the full path and name of the text file containing the custom HTML specifications (for example, *c:\myfiles\htmlstuf.txt*).

Sample Text File for Customized HTML

```
<<
<!DOCTYPE HTML PUBLIC "-//W3C//DTD HTML 3.2//EN">
<<
NVI, Inc.
<<
<META NAME="keywords" CONTENT="gizmos, gadgets, gimcracks">
<<
bgcolor="#FFFFFF"
<<
<H4 align=center>This page made possible by...
<br><br>
<IMG SRC="spss2.gif" align=center></H4>
<<
<h2 align=center>NVI Sales</h2>
<h3 align=center>Regional Data</h3>
```

Sample HTML Source for Customized HTML

```
<!DOCTYPE HTML PUBLIC "-//W3C//DTD HTML 3.2//EN">
<HTML>
<HEAD>
<TITLE>
NVI Sales, Inc.
</TITLE>
<META NAME="keywords" CONTENT="gizmos, gadgets, gimcracks">
</HEAD>
<BODY bgcolor="#FFFFFF">
<h2 align=center>NVI Sales</h2>
<h3 align=center>Regional Data</h3>

[Exported output]

<H4 align=center>This page made possible by...
<br><br>
<IMG SRC="spss2.gif" align=center></H4>
</BODY>
</HTML>
```

Index

Access (Microsoft), 26
active file, 86, 88
    caching, 88
    creating a temporary active file, 88
    virtual active file, 86
active window, 6
ActiveX objects, 274
adding group labels, 314
adjusted R^2
    in Linear Regression, 468
aggregating data, 229
    aggregate functions, 232
    variable names and labels, 233
algorithms, 13
alignment, 114, 269, 336, 640
    in cells, 336
    in Data Editor, 114
    output, 269, 640
alpha coefficient
    in Reliability Analysis, 595, 597
alpha factoring, 505
analysis of variance
    in Curve Estimation, 483
    in Linear Regression, 468
    in Means, 390
    in One-Way ANOVA, 411
Anderson-Rubin factor scores, 508
Andrews' wave estimator
    in Explore, 366
ANOVA
    in GLM Univariate, 421
    in Means, 390

    in One-Way ANOVA, 411
    model, 425
aspect ratio, 645
attributes
    custom variable attributes, 116
automated production, 665
automation objects, 693, 696–697, 699–700
    methods, 699
    object browser, 700
    properties, 699
    types, 696
    using in scripts, 693, 697, 700
    variable-naming conventions, 696
autoscripts, 655, 690, 692
    autoscript file, 693
    creating, 690
    trigger events, 692
average absolute deviation (AAD)
    in Ratio Statistics, 611

backward elimination
    in Linear Regression, 462
banding, 160
bar charts
    in Frequencies, 357
Bartlett factor scores, 508
Bartlett's test of sphericity
    in Factor Analysis, 504
beta coefficients
    in Linear Regression, 468
binning, 160

Binomial Test, 544
    command additional features, 546
    dichotomies, 544
    missing values, 546
    options, 546
    statistics, 546
Bivariate Correlations
    command additional features, 443
    correlation coefficients, 439
    missing values, 442
    options, 442
    significance level, 439
    statistics, 442
block distance
    in Distances, 453
Blom estimates, 186
BMP files, 277, 286, 288, 668
    exporting charts, 277, 286, 288, 668
Bonferroni
    in GLM, 430
    in One-Way ANOVA, 415
bookmarking pivot table views, 319
bookmarks, 319
borders, 303, 329–330
    displaying hidden borders, 330
    Draft Viewer, 303
boxplots
    comparing factor levels, 367
    comparing variables, 367
    in Explore, 367
Box's M test
    in Discriminant Analysis, 493
break points, 708
    in scripts, 708
break variables
    in Aggregate Data, 229

Brown-Forsythe statistic
    in One-Way ANOVA, 417
build terms, 426, 480
buttons, 662
    editing toolbar bitmap icons, 662

caching, 88
    active file, 88
Cancel button, 9
captions, 340
    adding to a table, 340
case-control study
    Paired-Samples T Test, 405
cases, 124, 126, 241
    finding duplicates, 157
    finding in Data Editor, 126
    inserting new cases, 124
    restructuring into variables, 241
    selecting subsets, 235, 237–238
    sorting, 219
    weighting, 239
casewise diagnostics
    in Linear Regression, 468
categorical data, 140
    converting interval data to discrete categories, 160
cell borders, 303
    Draft Viewer, 303
cells in pivot tables, 322, 327, 331–332, 334, 336–337, 340
    alignment, 336
    fonts, 331
    formats, 327
    hiding, 320
    margins, 337
    modifying text, 340
    outlines, 337

shading, 337

showing, 320

value formats, 334

widths, 332

centered moving average function, 210

centering output, 269, 640

chart options, 645

charts, 267, 273, 275, 277, 342, 645, 668

aspect ratio, 645

case labels, 483

copying, 273

copying into other applications, 273

creating, 617

creating from pivot tables, 342

exporting, 277, 668

footnotes, 625

hiding, 267

in ROC Curve, 613

missing values, 626

modifying, 621

overview, 617

pasting into other applications, 275

size, 630

subtitles, 625

templates, 628, 645

titles, 625

Chebychev distance

in Distances, 453

chi-square, 540

expected range, 542

expected values, 542

Fisher's exact test, 374

for independence, 374

in Crosstabs, 374

likelihood-ratio, 374

linear-by-linear association, 374

missing values, 543

one-sample test, 540

options, 543

Pearson, 374

statistics, 543

Yates' correction for continuity, 374

chi-square distance

in Distances, 453

classification

in ROC Curve, 613

cluster analysis

efficiency, 535

Hierarchical Cluster Analysis, 523

K-Means Cluster Analysis, 531

cluster frequencies

in TwoStep Cluster Analysis, 521

clustering

choosing a procedure, 511

Cochran's Q

in Tests for Several Related Samples, 566

Cochran's statistic

in Crosstabs, 374

coefficient of dispersion (COD)

in Ratio Statistics, 611

coefficient of variation (COV)

in Ratio Statistics, 611

Cohen's kappa

in Crosstabs, 374

collapsing categories, 160

collinearity diagnostics

in Linear Regression, 468

colors in pivot tables, 329, 331, 337

borders, 329

cell background, 337

cell foreground, 337

font, 331

column format, 335

changing in pivot tables, 335

column percentages
    in Crosstabs, 377
column summary reports, 587
column width, 114, 325, 333, 650
    controlling default width, 650
    controlling maximum width, 325
    controlling width for wrapped text, 325
    in Data Editor, 114
    pivot tables, 333
columns, 333, 339
    changing width in pivot tables, 333
    selecting in pivot tables, 339
COMMA format, 109–110
comma-delimited files, 43
command identifiers, 719
command language, 345
command line switches, 676
    Production Facility, 676
command syntax, 345, 351, 638, 640–641, 657,
    661, 665
    accessing *SPSS Command Syntax Reference*, 13
    adding to menus, 657
    journal file, 350, 352, 638
    log, 640–641
    output log, 348
    pasting, 347
    Production Facility formatting, 675
    Production Facility rules, 665
    running, 351
    running with toolbar buttons, 661
    syntax rules, 345
comparing groups
    in OLAP Cubes, 399
comparing variables
    in OLAP Cubes, 399
compound model
    in Curve Estimation, 487

computing variables, 171
    computing new string variables, 174
concentration index
    in Ratio Statistics, 611
conditional transformations, 173
confidence intervals
    in Explore, 366
    in GLM, 427, 435
    in Independent-Samples T Test, 405
    in Linear Regression, 468
    in One-Sample T Test, 410
    in One-Way ANOVA, 417
    in Paired-Samples T Test, 407
    in ROC Curve, 616
    saving in Linear Regression, 465
context-sensitive Help, 315
    finding label definitions in pivot tables, 315
contingency coefficient
    in Crosstabs, 374
contingency tables, 371
continuation text, 331
    for pivot tables, 331
contrasts
    in GLM, 427–428
    in One-Way ANOVA, 414
control variables
    in Crosstabs, 373
convergence
    in Factor Analysis, 505, 507
    in K-Means Cluster Analysis, 536
Cook's distance
    in GLM, 433
    in Linear Regression, 465
copying, 268, 273
    charts, 273
    output, 268
    pivot tables, 273

correlation matrix
 in Discriminant Analysis, 493
 in Factor Analysis, 499, 504
 in Ordinal Regression, 476
correlations
 in Bivariate Correlations, 439
 in Crosstabs, 374
 in Partial Correlations, 445
 zero-order, 448
counting occurrences, 176
covariance matrix
 in Discriminant Analysis, 493, 496
 in GLM, 433
 in Linear Regression, 468
 in Ordinal Regression, 476
covariance ratio
 in Linear Regression, 465
Cox and Snell R^2
 in Ordinal Regression, 476
Cramér's V
 in Crosstabs, 374
Cronbach's alpha
 in Reliability Analysis, 597
Cronbach's alpha
 in Reliability Analysis, 595
Crosstabs, 371
 cell display, 377
 clustered bar charts, 374
 control variables, 373
 formats, 379
 fractional weights, 239
 layers, 373
 statistics, 374
 suppressing tables, 371
crosstabulation
 in Crosstabs, 371
 multiple response, 574

CSV format
 reading data, 43
 saving data, 60
cubic model
 in Curve Estimation, 487
cumulative frequencies
 in Ordinal Regression, 476
cumulative sum function, 210
currency formats, 654
Curve Estimation, 483
 analysis of variance, 483
 forecast, 488
 including constant, 483
 models, 487
 saving predicted values, 488
 saving prediction intervals, 488
 saving residuals, 488
custom attributes, 116
custom currency formats, 109, 654
custom models
 in GLM, 425
custom variable attributes, 116

d
 in Crosstabs, 374
data analysis, 11
 basic steps, 11
data dictionary
 applying from another file, 147
Data Editor, 103, 106, 114, 120–128, 657
 alignment, 114
 changing data type, 125
 column width, 114
 data value restrictions, 122
 Data View, 104
 defining variables, 106
 display options, 127

editing data, 122–123
entering data, 120
entering non-numeric data, 121
entering numeric data, 121
filtered cases, 126
finding cases, 126
inserting new cases, 124
inserting new variables, 124
moving variables, 125
multiple open data files, 131
multiple views/panes, 127
printing, 128
sending data to other applications, 657
Variable View, 105
data entry, 120
data files, 22, 43, 58–60, 67, 88, 96–98, 241
adding comments, 632
dictionary information, 58
Dimensions, 53
file information, 58
flipping, 220
improving performance for large files, 88
mrInterview, 53
multiple open data files, 131
opening, 22
protecting, 86
Quancept, 53
Quanvert, 53
remote servers, 96–98
restructuring, 241
saving, 59–60
saving output as SPSS-format data files, 713
saving subsets of variables, 67
text, 43
transposing, 220
DATA LIST, 86
versus GET DATA command, 86

data transformations, 652
computing variables, 171
conditional transformations, 173
delaying execution, 652
functions, 174
ranking cases, 184
recoding values, 179–182, 188
string variables, 174
time series, 206, 209
data types, 109–110, 125, 654
changing, 125
custom currency, 109, 654
defining, 109
display formats, 110
input formats, 110
Data View, 104
databases, 25–26, 30–34, 37, 40, 42
adding new fields to a table, 78
appending records (cases) to a table, 79
conditional expressions, 34
converting strings to numeric variables, 40
creating a new table, 81
creating relationships, 32
database security, 30
defining variables, 40
logging into a database, 30
Microsoft Access, 26
parameter queries, 34, 37
Prompt for Value, 37
random sampling, 34
reading, 25–26, 31
relationship properties, 33
replacing a table, 81
replacing values in existing fields, 77
saving, 68
saving queries, 42
selecting a data source, 26

selecting data fields, 31

specifying criteria, 34

SQL syntax, 42

table joins, 32–33

updating, 68

verifying results, 42

Where clause, 34

datasets

renaming, 134

date format variables, 109–110, 652

add or subtract from date/time variables, 191

create date/time variable from set of variables, 191

create date/time variable from string, 191

extract part of date/time variable, 191

date formats

two-digit years, 652

date variables

defining for time series data, 207

dBASE files, 22, 24, 59–60

reading, 22, 24

saving, 59–60

debugging scripts, 708–709

break points, 708

debugging pane, 709

stepping through scripts, 708

Define Multiple Response Sets, 570

categories, 570

dichotomies, 570

set labels, 570

set names, 570

defining variables, 106, 109, 111, 113–115, 136

applying a data dictionary, 147

copying and pasting attributes, 114–115

data types, 109

missing values, 113

templates, 114–115

value labels, 111, 136

variable labels, 111

deleted residuals

in GLM, 433

in Linear Regression, 465

deleting multiple EXECUTES in syntax files, 352

deleting output, 268

dendrograms

in Hierarchical Cluster Analysis, 528

dependent *t* test

in Paired-Samples T Test, 405

descriptive statistics

in Descriptives, 359

in Explore, 366

in Frequencies, 355

in GLM Univariate, 435

in Ratio Statistics, 611

in Summarize, 384

in TwoStep Cluster Analysis, 521

Descriptives, 359

command additional features, 362

display order, 361

saving *z* scores, 359

statistics, 361

designated window, 6

detrended normal plots

in Explore, 367

deviation contrasts

in GLM, 427–428

DfBeta

in Linear Regression, 465

DfFit

in Linear Regression, 465

dialog boxes, 10, 633–634, 638, 703, 705

controls, 9

defining variable sets, 633

displaying variable labels, 8, 638

displaying variable names, 8, 638

optional specifications, 10

reordering target lists, 635

scripting, 703, 705

selecting variables, 10

subdialog boxes, 10

using variable sets, 634

variable display order, 638

variable icons, 10

variable information, 11

variables, 8

dictionary, 58

difference contrasts

    in GLM, 427–428

difference function, 210

differences between groups

    in OLAP Cubes, 399

differences between variables

    in OLAP Cubes, 399

Dimensions data, 53

    saving, 84

direct oblimin rotation

    in Factor Analysis, 507

Discriminant Analysis, 489

    analyzing held-out cases, 682

    command additional features, 498

    covariance matrix, 496

    criteria, 494

    defining ranges, 492

    descriptive statistics, 493

    discriminant methods, 494

    display options, 494, 496

    example, 489

    exporting model information, 497

    function coefficients, 493

    grouping variables, 489

    independent variables, 489

    Mahalanobis distance, 494

    matrices, 493

    missing values, 496

    plots, 496

    prior probabilities, 496

    Rao's V, 494

    saving classification variables, 497

    selecting cases, 492

    statistics, 489, 493

    stepwise methods, 489

    Wilks' lambda, 494

disk space, 86, 88

    temporary, 86, 88

display formats, 110

display order, 313

distance measures

    in Distances, 453

    in Hierarchical Cluster Analysis, 526

Distances, 451

    command additional features, 455

    computing distances between cases, 451

    computing distances between variables, 451

    dissimilarity measures, 453

    example, 451

    similarity measures, 454

    statistics, 451

    transforming measures, 453–454

    transforming values, 453–454

distributed mode, 91–93, 96–98, 100, 673

    available procedures, 100

    data file access, 96, 98

    Production Facility, 673

    saving data files, 97

    UNC paths, 100

division

    dividing across report columns, 590

DOLLAR format, 109–110

dollar sign, 335

   in pivot tables, 335

DOT format, 109–110

Draft Viewer, 301–303, 308–309, 638, 641

   box characters, 303

   cell borders, 303

   changing fonts, 308

   column borders, 303

   controlling default output display, 638

   display options, 641

   output format, 303

   printing, 308–309

   row borders, 303

   saving output, 309

   setting default viewer type, 302

Duncan's multiple range test

   in GLM, 430

   in One-Way ANOVA, 415

Dunnett's *C*

   in GLM, 430

Dunnett's *C*

   in One-Way ANOVA, 415

Dunnett's *t* test

   in GLM, 430

Dunnett's *t* test

   in One-Way ANOVA, 415

Dunnett's T3

   in GLM, 430

Dunnett's T3

   in One-Way ANOVA, 415

duplicate cases (records)

   finding and filtering, 157

Durbin-Watson statistic

   in Linear Regression, 468

editing data, 122–123

effect-size estimates

   in GLM Univariate, 435

eigenvalues

   in Factor Analysis, 504–505

   in Linear Regression, 468

embedding

   interactive charts, 274

   pivot tables, 274

entering data, 120–122

   non-numeric, 121

   numeric, 121

   using value labels, 122

environment variables, 638

   SPSSTMPDIR, 638

EPS files, 277, 286, 289, 668

   exporting charts, 277, 286, 289, 668

equamax rotation

   in Factor Analysis, 507

estimated marginal means

   in GLM Univariate, 435

eta

   in Crosstabs, 374

   in Means, 390

eta-squared

   in GLM Univariate, 435

   in Means, 390

Euclidean distance

   in Distances, 453

Excel files, 22–23, 59–60, 657

   adding menu item to send data to Excel, 657

   opening, 22–23

   saving, 59–60

Excel format

   exporting output, 277, 281

excluding output from Viewer with OMS, 727

EXECUTE (command)

   pasted from dialog boxes, 352

expected count
   in Crosstabs, 377
expected frequencies
   in Ordinal Regression, 476
Explore, 363
   command additional features, 369
   missing values, 369
   options, 369
   plots, 367
   power transformations, 368
   statistics, 366
exponential model
   in Curve Estimation, 487
export data, 59
exporting charts, 277, 286–290, 665, 668
   automated production, 665
   chart size, 286
exporting data, 657
   adding menu items to export data, 657
exporting output, 277, 281, 285, 668, 679
   Excel format, 277, 281
   HTML, 281
   HTML format, 277
   OMS, 713
   PDF format, 277, 282
   PowerPoint format, 277
   publishing to Web, 679
   Word format, 277, 281
extreme values
   in Explore, 366

Factor Analysis, 499
   analyzing held-out cases, 682
   coefficient display format, 509
   command additional features, 510
   convergence, 505, 507
   descriptives, 504

example, 499
extraction methods, 505
factor scores, 508
loading plots, 507
missing values, 509
overview, 499
rotation methods, 507
selecting cases, 504
statistics, 499, 504
factor scores, 508
file information, 58
file transformations, 241
   aggregating data, 229
   merging data files, 222, 226
   restructuring data, 241
   sorting cases, 219
   split-file processing, 233
   transposing variables and cases, 220
   weighting cases, 239
files, 272
   adding a text file to the Viewer, 272
   opening, 22
filtered cases, 126
   in Data Editor, 126
first
   in Means, 390
   in OLAP Cubes, 396
   in Summarize, 384
Fisher's exact test
   in Crosstabs, 374
Fisher's LSD
   in GLM, 430
fixed format, 43
fonts, 127, 272, 308, 331
   colors, 331
   in cells, 331
   in Data Editor, 127

in Draft Viewer, 308

in the outline pane, 272

footers, 294–295

footnotes, 326, 338–340

adding to a table, 340

in charts, 625

markers, 326, 338

renumbering, 339

forecast

in Curve Estimation, 488

formatting, 303

columns in reports, 582

draft output, 303

forward selection

in Linear Regression, 462

freefield format, 43

Frequencies, 353

charts, 357

display order, 358

formats, 358

statistics, 355

suppressing tables, 358

frequency tables

in Explore, 366

in Frequencies, 353

Friedman test

in Tests for Several Related Samples, 566

full factorial models

in GLM, 425

function procedures, 701

functions, 174

missing value treatment, 175

Gabriel's pairwise comparisons test

in GLM, 430

in One-Way ANOVA, 415

Games and Howell's pairwise comparisons test

in GLM, 430

in One-Way ANOVA, 415

gamma

in Crosstabs, 374

generalized least squares

in Factor Analysis, 505

geometric mean

in Means, 390

in OLAP Cubes, 396

in Summarize, 384

GET DATA, 86

versus DATA LIST command, 86

versus GET CAPTURE command, 86

GLM

model, 425

post hoc tests, 430

profile plots, 429

saving matrices, 433

saving variables, 433

sum of squares, 425

GLM Univariate, 421, 436

contrasts, 427–428

diagnostics, 435

display, 435

estimated marginal means, 435

options, 435

global procedures, 655, 702

global scripts, 702

Goodman and Kruskal's gamma

in Crosstabs, 374

Goodman and Kruskal's lambda

in Crosstabs, 374

Goodman and Kruskal's tau

in Crosstabs, 374

goodness of fit

in Ordinal Regression, 476

grand totals
    in column summary reports, 592
grid lines, 330
    pivot tables, 330
group labels, 314
group means, 387, 393
grouped median
    in Means, 390
    in OLAP Cubes, 396
    in Summarize, 384
grouping rows or columns, 314
grouping variables, 241
    creating, 241
growth model
    in Curve Estimation, 487
Guttman model
    in Reliability Analysis, 595, 597

Hampel's redescending M-estimator
    in Explore, 366
harmonic mean
    in Means, 390
    in OLAP Cubes, 396
    in Summarize, 384
headers, 294–295
held-out cases, 682
    analyzing, 682
Helmert contrasts
    in GLM, 427–428
Help button, 9
Help windows, 13
hiding, 267, 320–322, 658
    captions, 322
    dimension labels, 321
    footnotes, 322
    procedure results, 267
    rows and columns, 320

titles, 322
toolbars, 658
hiding (excluding) output from the Viewer with OMS, 727
Hierarchical Cluster Analysis, 523
    agglomeration schedules, 527
    cluster membership, 527, 529
    clustering cases, 523
    clustering methods, 526
    clustering variables, 523
    command additional features, 529
    dendrograms, 528
    distance matrices, 527
    distance measures, 526
    example, 523
    icicle plots, 528
    plot orientation, 528
    saving new variables, 529
    similarity measures, 526
    statistics, 523, 527
    transforming measures, 526
    transforming values, 526
hierarchical decomposition, 426
histograms
    in Explore, 367
    in Frequencies, 357
    in Linear Regression, 463
Hochberg's GT2
    in GLM, 430
    in One-Way ANOVA, 415
homogeneity-of-variance tests
    in GLM Univariate, 435
    in One-Way ANOVA, 417
Hotelling's T^2
    in Reliability Analysis, 595, 597
HTML, 277, 281, 668, 747
    adding customized code, 747

exporting output, 277, 281, 668
Huber's M-estimator
    in Explore, 366

ICC. *See* intraclass correlation coefficient, 597
icicle plots
    in Hierarchical Cluster Analysis, 528
icons, 662
    editing toolbar bitmap icons, 662
    in dialog boxes, 10
image factoring, 505
Immediate tab, 709
    script window, 709
import data, 22, 25
importance chart
    in TwoStep Cluster Analysis, 519
Independent-Samples T Test, 401
    confidence intervals, 405
    defining groups, 404
    grouping variables, 404
    missing values, 405
    options, 405
    string variables, 404
initial threshold
    in TwoStep Cluster Analysis, 517
input formats, 110
inserting group labels, 314
interaction terms, 426, 480
interactive charts, 273–274, 649
    copying into other applications, 273
    embedding as ActiveX objects, 274
    options, 649
    saving data with chart, 649
intraclass correlation coefficient (ICC)
    in Reliability Analysis, 597
inverse model
    in Curve Estimation, 487

iteration history
    in Ordinal Regression, 476
iterations
    in Factor Analysis, 505, 507
    in K-Means Cluster Analysis, 536

journal file, 638
JPEG files, 277, 286–287, 668
    exporting charts, 277, 286–287, 668
justification, 269, 640
    output, 269, 640

K-Means Cluster Analysis, 531
    cluster distances, 537
    cluster membership, 537
    command additional features, 538
    convergence criteria, 536
    efficiency, 535
    examples, 531
    iterations, 536
    methods, 531
    missing values, 537
    overview, 531
    saving cluster information, 537
    statistics, 531, 537
kappa
    in Crosstabs, 374
Kendall's tau-*b*
    in Bivariate Correlations, 439
    in Crosstabs, 374
Kendall's tau-*c*, 374
    in Crosstabs, 374
Kendall's *W*
    in Tests for Several Related Samples, 566
keyed table, 226

Kolmogorov-Smirnov *Z*
    in One-Sample Kolmogorov-Smirnov Test, 550
    in Two-Independent-Samples Tests, 555
KR20
    in Reliability Analysis, 597
Kruskal-Wallis *H*
    in Two-Independent-Samples Tests, 561
Kruskal's tau
    in Crosstabs, 374
Kuder-Richardson 20 (KR20)
    in Reliability Analysis, 597
kurtosis
    in Descriptives, 361
    in Explore, 366
    in Frequencies, 355
    in Means, 390
    in OLAP Cubes, 396
    in Report Summaries in Columns, 590
    in Report Summaries in Rows, 583
    in Summarize, 384

labels, 314
    deleting, 314
    inserting group labels, 314
    vs. subtype names in OMS, 720
LAG (function), 210
lambda
    in Crosstabs, 374
Lance and Williams dissimilarity measure, 453
    in Distances, 453
language
    changing output language, 638
last
    in Means, 390
    in OLAP Cubes, 396
    in Summarize, 384

layers, 291, 316–317, 325, 331
    changing, 317
    creating, 316
    displaying, 316–317
    in Crosstabs, 373
    in pivot tables, 316
    printing, 291, 325, 331
lead function, 210
least significant difference
    in GLM, 430
    in One-Way ANOVA, 415
level of measurement, 107, 140
    defining, 107
Levene test
    in Explore, 367
    in GLM Univariate, 435
    in One-Way ANOVA, 417
leverage values
    in GLM, 433
    in Linear Regression, 465
likelihood-ratio chi-square
    in Crosstabs, 374
    in Ordinal Regression, 476
Lilliefors test
    in Explore, 367
line breaks
    variable and value labels, 112
linear model
    in Curve Estimation, 487
Linear Regression, 457
    blocks, 457
    command additional features, 471
    exporting model information, 465
    missing values, 470
    plots, 463
    residuals, 465
    saving new variables, 465

selection variable, 463

statistics, 468

variable selection methods, 462, 470

weights, 457

linear-by-linear association

in Crosstabs, 374

link

in Ordinal Regression, 475

listing cases, 381

Loaded tab, 709

script window, 709

loading plots

in Factor Analysis, 507

location model

in Ordinal Regression, 478

logarithmic model

in Curve Estimation, 487

logging in to a server, 92

logistic model

in Curve Estimation, 487

Lotus 1-2-3 files, 22, 59–60, 657

adding menu item to send data to Lotus, 657

opening, 22

saving, 59–60

M-estimators

in Explore, 366

Mahalanobis distance

in Discriminant Analysis, 494

in Linear Regression, 465

Mann-Whitney U

in Two-Independent-Samples Tests, 555

Mantel-Haenszel statistic

in Crosstabs, 374

margins, 294, 337

in cells, 337

matched-pairs study

in Paired-Samples T Test, 405

maximum

comparing report columns, 590

in Descriptives, 361

in Explore, 366

in Frequencies, 355

in Means, 390

in OLAP Cubes, 396

in Ratio Statistics, 611

in Summarize, 384

maximum branches

in TwoStep Cluster Analysis, 517

maximum likelihood

in Factor Analysis, 505

McFadden R^2

in Ordinal Regression, 476

McNemar test

in Crosstabs, 374

in Two-Related-Samples Tests, 557

mean

in Descriptives, 361

in Explore, 366

in Frequencies, 355

in Means, 390

in OLAP Cubes, 396

in One-Way ANOVA, 417

in Ratio Statistics, 611

in Report Summaries in Columns, 590

in Report Summaries in Rows, 583

in Summarize, 384

of multiple report columns, 590

subgroup, 387, 393

Means, 387, 682

ASCII output from, 682

options, 390

statistics, 390

measurement level, 107, 140
  defining, 107
  icons in dialog boxes, 10
measurement system, 638
measures of central tendency
  in Explore, 366
  in Frequencies, 355
  in Ratio Statistics, 611
measures of dispersion
  in Descriptives, 361
  in Explore, 366
  in Frequencies, 355
  in Ratio Statistics, 611
measures of distribution
  in Descriptives, 361
  in Frequencies, 355
median
  in Explore, 366
  in Frequencies, 355
  in Means, 390
  in OLAP Cubes, 396
  in Ratio Statistics, 611
  in Summarize, 384
median test
  in Two-Independent-Samples Tests, 561
memory, 638
memory allocation
  in TwoStep Cluster Analysis, 517
menus, 7, 657
  customizing, 657
merging data files
  dictionary information, 226
  files with different cases, 222
  files with different variables, 226
  renaming variables, 225
metafiles, 277, 286, 668
  exporting charts, 277, 286, 668

methods, 699
  OLE automation objects, 699
Microsoft Access, 26
minimum
  comparing report columns, 590
  in Descriptives, 361
  in Explore, 366
  in Frequencies, 355
  in Means, 390
  in OLAP Cubes, 396
  in Ratio Statistics, 611
  in Summarize, 384
Minkowski distance
  in Distances, 453
missing values, 113
  defining, 113
  in Binomial Test, 546
  in Bivariate Correlations, 442
  in charts, 626
  in Chi-Square Test, 543
  in column summary reports, 592
  in Explore, 369
  in Factor Analysis, 509
  in functions, 175
  in Independent-Samples T Test, 405
  in Linear Regression, 470
  in Multiple Response Crosstabs, 576
  in Multiple Response Frequencies, 571
  in One-Sample Kolmogorov-Smirnov Test, 552
  in One-Sample T Test, 410
  in One-Way ANOVA, 417
  in Paired-Samples T Test, 407
  in Partial Correlations, 448
  in Report Summaries in Rows, 585
  in ROC Curve, 616
  in Runs Test, 549
  in Tests for Several Independent Samples, 564

in Two-Independent-Samples Tests, 556
in Two-Related-Samples Tests, 560
replacing in time series data, 211
string variables, 113
mode
    in Frequencies, 355
model file
    loading saved models to score data, 215
Moses extreme reaction test
    in Two-Independent-Samples Tests, 555
moving rows and columns, 313
mrInterview, 53
Multidimensional Scaling, 601
    command additional features, 606
    conditionality, 605
    creating distance matrices, 604
    criteria, 606
    defining data shape, 603
    dimensions, 605
    display options, 606
    distance measures, 604
    example, 601
    levels of measurement, 605
    scaling models, 605
    statistics, 601
    transforming values, 604
multiple comparisons
    in One-Way ANOVA, 415
multiple open data files, 131
multiple R
    in Linear Regression, 468
multiple regression
    in Linear Regression, 457
Multiple Response
    command additional features, 577
multiple response analysis
    crosstabulation, 574

frequency tables, 571
Multiple Response Crosstabs, 574
Multiple Response Frequencies, 571
Multiple Response Crosstabs, 574
    cell percentages, 576
    defining value ranges, 576
    matching variables across response sets, 576
    missing values, 576
    percentages based on cases, 576
    percentages based on responses, 576
Multiple Response Frequencies, 571
    missing values, 571
multiple response sets
    defining, 144
    multiple categories, 144
    multiple dichotomies, 144
multiple views/panes
    Data Editor, 127
multiplication
    multiplying across report columns, 590

Nagelkerke R^2
    in Ordinal Regression, 476
new features
    SPSS 15.0, 2
Newman-Keuls
    in GLM, 430
noise handling
    in TwoStep Cluster Analysis, 517
nominal, 107
    measurement level, 107, 140
nonparametric tests
    chi-square, 540
    One-Sample Kolmogorov-Smirnov Test, 550
    Runs Test, 547
    Tests for Several Independent Samples, 561
    Tests for Several Related Samples, 564

Two-Independent-Samples Tests, 553
Two-Related-Samples Tests, 557
normal probability plots
in Explore, 367
in Linear Regression, 463
normal scores
in Rank Cases, 186
normality tests
in Explore, 367
number of cases
in Means, 390
in OLAP Cubes, 396
in Summarize, 384
numeric format, 109–110

object browser, 700
objects, 693, 696–697, 700
overview, 693, 696
using in scripts, 693, 697, 700
variable-naming conventions, 696
observed count
in Crosstabs, 377
observed frequencies
in Ordinal Regression, 476
observed means
in GLM Univariate, 435
OK button, 9
OLAP Cubes, 393
statistics, 396
titles, 400
OLE automation, 681, 693, 696–697, 699–700
methods, 699
overview, 693, 696–697, 700
properties, 699
scripting with, 681
using objects, 693, 697, 700
variable-naming conventions, 696

OMS, 713, 741
command identifiers, 719
controlling table pivots, 721, 733
excluding output from the Viewer, 727
output object types, 717
SAV file format, 721, 727
SPSS data file format, 721, 727
table subtypes, 719
text format, 721
using XSLT with OXML, 742
variable names in SAV files, 735
XML, 721, 736
One-Sample Kolmogorov-Smirnov Test, 550
command additional features, 552
missing values, 552
options, 552
statistics, 552
test distribution, 550
One-Sample T Test, 408
command additional features, 410
confidence intervals, 410
missing values, 410
options, 410
One-Way ANOVA, 411
command additional features, 418
contrasts, 414
factor variables, 411
missing values, 417
multiple comparisons, 415
options, 417
polynomial contrasts, 414
post hoc tests, 415
statistics, 417
online Help, 13
Statistics Coach, 12
opening files, 22–26, 43
data files, 22

dBASE files, 22, 24

Excel files, 22–23

Lotus 1-2-3 files, 22

spreadsheet files, 22, 24

Stata files, 24

SYSTAT files, 22

tab-delimited files, 22

text data files, 43

options, 638, 640–641, 643, 645, 649–650, 652, 654–655

  charts, 645

  currency, 654

  data, 652

  Draft Viewer, 641

  general, 638

  interactive charts, 649

  output labels, 643

  pivot table look, 650

  scripts, 655

  temporary directory, 638

  two-digit years, 652

  Viewer, 640

ordinal, 107

  measurement level, 107, 140

Ordinal Regression , 473

  command additional features, 481

  link, 475

  location model, 478

  options, 475

  scale model, 480

  statistics, 473

outliers

  in Explore, 366

  in Linear Regression, 463

  in TwoStep Cluster Analysis, 517

outline, 269–271

  changing levels, 271

  collapsing, 270

  expanding, 270

  in Viewer, 269

output, 265, 267–269, 273, 276–277, 298, 301, 340, 640, 668

  alignment, 269, 640

  centering, 269, 640

  changing output language, 638

  copying, 268

  copying and pasting multiple items, 276

  copying into other applications, 273

  deleting, 268

  draft, 301

  exporting, 277, 668

  hiding, 267

  modifying, 340

  moving, 268

  pasting into other applications, 273

  saving, 298

  showing, 267

  Viewer, 265

Output Management System (OMS), 713, 741

output object types

  in OMS, 717

overview, 745

OXML, 742

page control

  in column summary reports, 592

  in row summary reports, 585

page margins, 294

page numbering, 294, 297

  in column summary reports, 592

  in row summary reports, 585

page setup, 294–295, 297

  chart size, 297

  headers and footers, 295

Paired-Samples T Test, 405
  missing values, 407
  options, 407
  selecting paired variables, 405
pane splitter
  Data Editor, 127
parallel model
  in Reliability Analysis, 595, 597
parameter estimates
  in GLM Univariate, 435
  in Ordinal Regression, 476
Partial Correlations, 445
  command additional features, 448
  in Linear Regression, 468
  missing values, 448
  options, 448
  statistics, 448
  zero-order correlations, 448
partial plots
  in Linear Regression, 463
password protection, 298
Paste button, 9
pasting, 275–276
  charts, 275
  pivot tables, 275
  pivot tables as tables, 275
  special objects, 276
pattern difference measure
  in Distances, 453
pattern matrix
  in Factor Analysis, 499
PDF
  exporting output, 277, 282
Pearson chi-square
  in Crosstabs, 374
  in Ordinal Regression, 476

Pearson correlation
  in Bivariate Correlations, 439
  in Crosstabs, 374
Pearson residuals
  in Ordinal Regression, 476
percentage sign, 335
  in pivot tables, 335
percentages
  in Crosstabs, 377
percentiles
  in Explore, 366
  in Frequencies, 355
performance, 88
  caching data, 88
permissions, 100
phi
  in Crosstabs, 374
phi-square distance measure
  in Distances, 453
PICT files, 277, 286, 288, 668
  exporting charts, 277, 286, 288, 668
pie charts
  in Frequencies, 357
  titles, 625
pivot tables, 267, 273–277, 291, 311–316, 320,
  322–323, 325–327, 329–332, 339–342, 650,
  668, 674
  adding captions, 340
  borders, 329
  cell formats, 327
  cell widths, 332
  changing appearance, 322
  changing display order, 313
  changing the look, 323
  continuation text, 331
  controlling table breaks, 341
  copying, 273

copying and pasting multiple tables, 276

copying into other applications, 273

creating charts from tables, 342

default column width adjustment, 650

default look for new tables, 650

deleting group labels, 314

displaying hidden borders, 330

editing, 311

editing two or more, 311

embedding as ActiveX objects, 274

exporting as HTML, 277, 668

finding label definitions, 315

footnote properties, 326

format control for production jobs, 674

general properties, 325

grid lines, 330

grouping rows or columns, 314

hiding, 267

identifying dimensions, 312

inserting group labels, 314

layers, 316

manipulating, 311

moving rows and columns, 313

pasting as metafiles, 275

pasting as tables, 273, 275

pasting as text, 275

pasting into other applications, 273

pivoting, 311–312

printing large tables, 341

printing layers, 291

properties, 325

resetting defaults, 315

rotating labels, 314

scaling to fit page, 325, 331

selecting rows and columns, 339

showing and hiding cells, 320

transposing rows and columns, 313

ungrouping rows or columns, 314

using icons, 312

pivoting

controlling with OMS for exported output, 733

PLUM

in Ordinal Regression, 473

PNG files, 277, 288

exporting charts, 277, 288

polynomial contrasts

in GLM, 427–428

in One-Way ANOVA, 414

port numbers, 93

portable files

variable names, 60

post hoc multiple comparisons, 415

PostScript files (encapsulated), 277, 289, 668

exporting charts, 277, 289, 668

power estimates

in GLM Univariate, 435

power model

in Curve Estimation, 487

PowerPoint, 281

exporting output as PowerPoint, 281

PowerPoint format

exporting output, 277

predicted values

saving in Curve Estimation, 488

saving in Linear Regression, 465

prediction intervals

saving in Curve Estimation, 488

saving in Linear Regression, 465

price-related differential (PRD)

in Ratio Statistics, 611

principal axis factoring, 505

principal components analysis, 499, 505

printing, 128, 291–292, 294–295, 297, 308–309, 325, 331, 341
  chart size, 297
  charts, 291
  controlling table breaks, 341
  data, 128
  draft output, 308–309
  headers and footers, 294–295
  layers, 291, 325, 331
  page numbers, 297
  page setup, 294
  pivot tables, 291
  print preview, 292
  scaling tables, 325, 331
  space between output items, 297
  text output, 291
prior moving average function, 210
procedures, 701
  scripts, 701
Production Facility, 638, 640–641, 665, 668, 670, 672–674, 676, 678
  command line switches, 676
  exporting charts, 665, 668
  exporting output, 668
  format control for pivot tables, 674
  format control with command syntax, 675
  macro prompting, 672
  options, 673
  output files, 665
  publishing output, 678
  publishing to Web, 678
  running multiple production jobs, 676
  scheduling production jobs, 676
  specifying a remote server, 673
  substituting values in syntax files, 670
  syntax rules, 665
  using command syntax from journal file, 638

  using command syntax from log, 640–641
profile plots
  in GLM, 429
programming with command language, 345
properties, 325, 699
  OLE automation objects, 699
  pivot tables, 325
  tables, 325
proportion estimates
  in Rank Cases, 186
Proximities
  in Hierarchical Cluster Analysis, 523
publishing output, 679
  with Production Facility, 678

quadratic model
  in Curve Estimation, 487
Quancept, 53
Quanvert, 53
quartiles
  in Frequencies, 355
quartimax rotation
  in Factor Analysis, 507

r correlation coefficient
  in Bivariate Correlations, 439
  in Crosstabs, 374
R statistic
  in Linear Regression, 468
  in Means, 390
R-E-G-W F
  in GLM, 430
  in One-Way ANOVA, 415
R-E-G-W Q
  in GLM, 430
  in One-Way ANOVA, 415

R^2
  in Linear Regression, 468
  in Means, 390
  R^2 change, 468
random number seed, 175
random sample, 34
  databases, 34
  random number seed, 175
  selecting, 237
range
  in Descriptives, 361
  in Frequencies, 355
  in Means, 390
  in OLAP Cubes, 396
  in Ratio Statistics, 611
  in Summarize, 384
rank correlation coefficient
  in Bivariate Correlations, 439
ranking cases, 184
  fractional ranks, 186
  percentiles, 186
  Savage scores, 186
  tied values, 187
Rankit estimates, 186
Rao's V
  in Discriminant Analysis, 494
Ratio Statistics, 609
  statistics, 611
recoding values, 160, 179–182, 188
reference category
  in GLM, 427–428
regression
  Linear Regression, 457
  multiple regression, 457
  plots, 463
regression coefficients
  in Linear Regression, 468

related samples, 557, 564
relative risk
  in Crosstabs, 374
Reliability Analysis, 595
  ANOVA table, 597
  command additional features, 599
  descriptives, 597
  example, 595
  Hotelling's T^2, 597
  inter-item correlations and covariances, 597
  intraclass correlation coefficient, 597
  Kuder-Richardson 20, 597
  statistics, 595, 597
  Tukey's test of additivity, 597
remote servers, 91–93, 96–98, 100, 673
  adding, 93
  available procedures, 100
  data file access, 96, 98
  editing, 93
  logging in, 92
  Production Facility, 673
  saving data files, 97
  UNC paths, 100
removing group labels, 314
renaming datasets, 134
reordering rows and columns, 313
repeated contrasts
  in GLM, 427–428
replacing missing values
  linear interpolation, 213
  linear trend, 213
  mean of nearby points, 213
  median of nearby points, 213
  series mean, 213
Report Summaries in Columns, 587
  column format, 582
  command additional features, 593

grand total, 592
missing values, 592
page control, 592
page layout, 585
page numbering, 592
subtotals, 592
total columns, 590
Report Summaries in Rows, 579
break columns, 579
break spacing, 584
column format, 582
command additional features, 593
data columns, 579
footers, 587
missing values, 585
page control, 584
page layout, 585
page numbering, 585
sorting sequences, 579
titles, 587
variables in titles, 587
reports
column summary reports, 587
comparing columns, 590
composite totals, 590
dividing column values, 590
multiplying column values, 590
row summary reports, 579
total columns, 590
Reset button, 9
residual plots
in GLM Univariate, 435
residuals
in Crosstabs, 377
saving in Curve Estimation, 488
saving in Linear Regression, 465

restructuring data, 241, 244–247, 250–253, 255, 257–258, 260, 262
and weighted data, 262
creating a single index variable for variables to cases, 252
creating index variables for variables to cases, 250
creating multiple index variables for variables to cases, 253
example of cases to variables, 245
example of one index for variables to cases, 251
example of two indices for variables to cases, 251
example of variables to cases, 244
options for cases to variables, 260
options for variables to cases, 255
overview, 241
selecting data for cases to variables, 257
selecting data for variables to cases, 247
sorting data for cases to variables, 258
types of restructuring, 241
variable groups for variables to cases, 246
rho
in Bivariate Correlations, 439
in Crosstabs, 374
risk
in Crosstabs, 374
ROC Curve, 613
statistics and plots, 616
rotating labels, 314
row percentages
in Crosstabs, 377
rows, 339
selecting in pivot tables, 339
running median function, 210
Runs Test, 547
command additional features, 549
cut points, 547–548

missing values, 549
options, 549
statistics, 549
Ryan-Einot-Gabriel-Welsch multiple *F*
  in GLM, 430
  in One-Way ANOVA, 415
Ryan-Einot-Gabriel-Welsch multiple range
  in GLM, 430
  in One-Way ANOVA, 415

S model
  in Curve Estimation, 487
S-stress
  in Multidimensional Scaling, 601
sampling
  random sample, 237
SAS files
  opening, 22
  saving, 59
SAV file format
  routing output to an SPSS data file, 721, 727
Savage scores, 186
saving charts, 277, 286–290, 649, 668
  BMP files, 277, 286, 288, 668
  EPS files, 277, 286, 289, 668
  JPEG files, 277, 286–287, 668
  metafiles, 277, 286, 668
  PICT files, 277, 286, 288, 668
  PNG files, 288
  PostScript files, 289
  saving interactive charts with data, 649
  TIFF files, 288
  WMF files, 277, 286, 290
  WMF format, 668
saving files, 59–60, 97
  data files, 59–60, 97
  database file queries, 42

SPSS data files, 59
saving output, 277, 281, 285, 298, 309–310, 668, 679
  draft output, 309
  Excel format, 277, 281
  HTML, 277, 281, 668
  HTML format, 277
  password protection, 298
  PDF format, 277, 282
  PowerPoint format, 277, 281
  publishing to Web, 679
  saving draft output as text, 310
  text format, 277, 285, 668
  Word format, 277, 281
saving pivot table views, 319
scale, 107
  in Multidimensional Scaling, 601
  in Reliability Analysis, 595
  measurement level, 107, 140
scale model
  in Ordinal Regression, 480
scale variables
  binning to create categorical variables, 160
scaling
  exported charts, 286
  pivot tables, 325, 331
scatterplots
  in Linear Regression, 463
Scheffé test
  in GLM, 430
  in One-Way ANOVA, 415
scientific notation, 109, 335, 638
  in pivot tables, 335
  suppressing in output, 638
scoring
  displaying loaded models, 218
  loading saved models, 215

models supported for export and scoring, 213
script window, 686, 688, 700
    Debug menu, 708
    Immediate tab, 709
    Loaded tab, 709
    object browser, 700
    properties, 688
    Stack tab, 709
    Watch tab, 709
scripting tips, 681, 686, 688, 693, 695, 697,
  699–701, 703, 708
    adding a description, 703
    custom dialog boxes, 703
    debugging, 708
    getting automation objects, 697
    how scripts work, 693
    object browser, 700
    procedures, 701
    properties and methods, 699
    script window, 686
    starter scripts, 688
    variable declarations, 695
scripts, 655, 657, 661, 681–683, 685, 688, 690,
  703, 708
    adding a description, 703
    adding to menus, 657
    autoscript file, 655, 693
    autoscripts, 683, 690, 693
    creating, 685, 690
    debugging, 708–709
    declaring variables, 695–696
    dialog boxes, 703, 705
    global procedures file, 655, 702
    overview, 681
    running, 681
    running with toolbar buttons, 661
    script window, 686, 688

    starter scripts, 688
    using automation objects, 693, 696–697, 700
    with command syntax, 710–712
seasonal difference function, 210
select cases, 235
selecting cases, 235
    based on selection criteria, 237
    date range, 238
    random sample, 237
    range of cases, 238
    time range, 238
selection methods, 339
    selecting rows and columns in pivot tables, 339
selection variable
    in Linear Regression, 463
servers, 92–93
    adding, 93
    editing, 93
    logging in, 92
    names, 93
    port numbers, 93
session journal, 638
shading, 337
    in cells, 337
Shapiro-Wilk's test
    in Explore, 367
shared drives, 100
showing, 267, 321–322, 658
    captions, 322
    dimension labels, 321
    footnotes, 322
    results, 267
    rows or columns, 321
    titles, 322
    toolbars, 658
Sidak's *t* test
    in GLM, 430

in One-Way ANOVA, 415
sign test
    in Two-Related-Samples Tests, 557
similarity measures
    in Distances, 454
    in Hierarchical Cluster Analysis, 526
simple contrasts
    in GLM, 427–428
size difference measure
    in Distances, 453
sizes, 271
    in outline, 271
sizing exported charts, 286
skewness
    in Descriptives, 361
    in Explore, 366
    in Frequencies, 355
    in Means, 390
    in OLAP Cubes, 396
    in Report Summaries in Columns, 590
    in Report Summaries in Rows, 583
    in Summarize, 384
smoothing function, 210
Somers' *d*
    in Crosstabs, 374
sorting cases, 219
space-delimited data, 43
Spearman correlation coefficient
    in Bivariate Correlations, 439
    in Crosstabs, 374
Spearman-Brown reliability
    in Reliability Analysis, 597
speed, 88
    caching data, 88
split-file processing, 233
split-half reliability
    in Reliability Analysis, 595, 597

splitting tables, 341
    controlling table breaks, 341
spread-versus-level plots
    in Explore, 367
    in GLM Univariate, 435
spreadsheet files, 22–24, 62
    opening, 24
    reading ranges, 23
    reading variable names, 23
    writing variable names, 62
SPSS
    basic steps, 11
SPSS data file format
    routing output to a data file, 721, 727
SPSSTMPDIR environment variable, 638
squared Euclidean distance
    in Distances, 453
Stack tab, 709
    script window, 709
standard deviation
    in Descriptives, 361
    in Explore, 366
    in Frequencies, 355
    in GLM Univariate, 435
    in Means, 390
    in OLAP Cubes, 396
    in Ratio Statistics, 611
    in Report Summaries in Columns, 590
    in Report Summaries in Rows, 583
    in Summarize, 384
standard error
    in Descriptives, 361
    in Explore, 366
    in Frequencies, 355
    in GLM, 433, 435
    in ROC Curve, 616

standard error of kurtosis
   in Means, 390
   in OLAP Cubes, 396
   in Summarize, 384
standard error of skewness
   in Means, 390
   in OLAP Cubes, 396
   in Summarize, 384
standard error of the mean
   in Means, 390
   in OLAP Cubes, 396
   in Summarize, 384
standardization
   in TwoStep Cluster Analysis, 517
standardized residuals
   in GLM, 433
   in Linear Regression, 465
standardized values
   in Descriptives, 359
starter scripts, 688
Stata files, 24
   opening, 24
   reading, 22
   saving, 59
Statistics Coach, 12
status bar, 7
   hiding, 8
   showing, 8
stem-and-leaf plots
   in Explore, 367
stepwise selection
   in Linear Regression, 462
stress
   in Multidimensional Scaling, 601
strictly parallel model
   in Reliability Analysis, 595, 597
string format, 109

string variables, 113, 121
   breaking up long strings in earlier releases, 60
   computing new string variables, 174
   entering data, 121
   in dialog boxes, 8
   missing values, 113
   recoding into consecutive integers, 188
Student-Newman-Keuls
   in GLM, 430
   in One-Way ANOVA, 415
Studentized residuals
   in Linear Regression, 465
Student's *t* test, 401
subgroup means, 387, 393
subroutine procedures, 701
subsets of cases
   random sample, 237
   selecting, 235, 237–238
subtitles
   in charts, 625
subtotals
   in column summary reports, 592
subtypes, 719
   vs. labels, 720
sum
   in Descriptives, 361
   in Frequencies, 355
   in Means, 390
   in OLAP Cubes, 396
   in Summarize, 384
sum of squares, 426
   in GLM, 425
Summarize, 381
   options, 383
   statistics, 384

syntax, 345, 351, 638, 640–641, 661, 665, 710, 712
   accessing *SPSS Command Syntax Reference*, 13
   journal file, 350, 352, 638
   log, 640–641
   output log, 348
   pasting, 347
   pasting into scripts, 712
   Production Facility rules, 665
   running, 351
   running command syntax with toolbar buttons, 661
   syntax rules, 345
   with scripts, 710–712
SYSTAT files, 22
   opening, 22

t test
   in GLM Univariate, 435
   in Independent-Samples T Test, 401
   in One-Sample T Test, 408
   in Paired-Samples T Test, 405
T4253H smoothing, 210
tab-delimited files, 22–23, 43, 59–60, 62
   opening, 22
   reading variable names, 23
   saving, 59–60
   writing variable names, 62
table breaks, 341
table chart, 342
table subtypes, 719
   vs. labels, 720
TableLooks, 323–324
   applying, 323
   creating, 324
tables, 341
   controlling table breaks, 341

Tamhane's T2
   in GLM, 430
   in One-Way ANOVA, 415
target lists, 635
tau-*b*
   in Crosstabs, 374
tau-*c*
   in Crosstabs, 374
templates, 114–115, 645
   in charts, 628, 645
   using an external data file as a template, 147
   variable definition, 114–115
temporary active file, 88
temporary directory, 638
   setting location in local mode, 638
   SPSSTMPDIR environment variable, 638
temporary disk space, 86, 88
test of parallel lines
   in Ordinal Regression, 476
tests for independence
   chi-square, 374
tests for linearity
   in Means, 390
Tests for Several Independent Samples, 561
   command additional features, 564
   defining range, 563
   grouping variables, 563
   missing values, 564
   options, 564
   statistics, 564
   test types, 563
Tests for Several Related Samples, 564
   command additional features, 567
   statistics, 567
   test types, 566

text, 43, 272, 277, 285, 301, 310, 340, 668
adding a text file to the Viewer, 272
adding to Viewer, 272
creating text output, 301
data files, 43
exporting draft output as text, 310
exporting output as text, 277, 285, 668
in cells, 340
TIFF files, 288
exporting charts, 277, 286, 288, 668
time series analysis
forecast, 488
predicting cases, 488
time series data
creating new time series variables, 209
data transformations, 206
defining date variables, 207
replacing missing values, 211
transformation functions, 210
titles, 272
adding to Viewer, 272
in charts, 625
in OLAP Cubes, 400
tolerance
in Linear Regression, 468
toolbars, 658, 660–662
creating, 658, 661
creating new tools, 661
customizing, 658, 661
displaying in different windows, 660
editing bitmap icons, 662
showing and hiding, 658
total column
in reports, 590
total percentages
in Crosstabs, 377

totals, 682
automatically bolding in output, 682
transformation matrix
in Factor Analysis, 499
transposing rows and columns, 313
transposing variables and cases, 220
tree depth
in TwoStep Cluster Analysis, 517
trigger events, 692
autoscripts, 692
trimmed mean
in Explore, 366
Tukey estimates, 186
Tukey's *b* test
in GLM, 430
in One-Way ANOVA, 415
Tukey's biweight estimator
in Explore, 366
Tukey's honestly significant difference
in GLM, 430
in One-Way ANOVA, 415
Tukey's test of additivity
in Reliability Analysis, 595, 597
Two-Independent-Samples Tests, 553
command additional features, 557
defining groups, 556
grouping variables, 556
missing values, 556
options, 556
statistics, 556
test types, 555
Two-Related-Samples Tests, 557
command additional features, 561
missing values, 560
options, 560
statistics, 560
test types, 559

two-sample *t* test
  in Independent-Samples T Test, 401
TwoStep Cluster Analysis, 513
  options, 517
  plots, 519
  save to external file, 521
  save to working file, 521
  statistics, 521

uncertainty coefficient
  in Crosstabs, 374
unstandardized residuals
  in GLM, 433
unweighted least squares
  in Factor Analysis, 505
user-missing values, 113

V
  in Crosstabs, 374
value labels, 111, 122, 127, 136, 643
  applying to multiple variables, 142
  copying, 142
  in Data Editor, 127
  in merged data files, 226
  in outline pane, 643
  in pivot tables, 643
  inserting line breaks, 112
  using for data entry, 122
values, 335
  pivot table display format, 335
Van der Waerden estimates, 186
variable attributes, 114–115
  copying and pasting, 114–115
  custom, 116
variable declarations, 695–696
  in scripts, 695–696

naming conventions, 696
variable importance plots
  in TwoStep Cluster Analysis, 519
variable information, 631
variable labels, 111, 638, 643
  in dialog boxes, 8, 638
  in merged data files, 226
  in outline pane, 643
  in pivot tables, 643
  inserting line breaks, 112
variable lists, 635
  reordering target lists, 635
variable names, 106, 638
  generated by OMS, 735
  in dialog boxes, 8, 638
  mixed case variable names, 106
  portable files, 60
  rules, 106
  truncating long variable names in earlier releases,
    60
  wrapping long variable names in output, 106
variable pairs, 241
  creating, 241
variable sets, 633–634
  defining, 633
  using, 634
Variable View, 105
variables, 10, 106, 124–125, 241, 631, 633, 638
  defining, 106
  defining variable sets, 633
  definition information, 631
  display order in dialog boxes, 638
  in dialog boxes, 8
  inserting new variables, 124
  moving, 125
  recoding, 179–182, 188
  renaming for merged data files, 225

restructuring into cases, 241

selecting in dialog boxes, 10

variable information in dialog boxes, 11

variance

in Descriptives, 361

in Explore, 366

in Frequencies, 355

in Means, 390

in OLAP Cubes, 396

in Report Summaries in Columns, 590

in Report Summaries in Rows, 583

in Summarize, 384

variance inflation factor

in Linear Regression, 468

varimax rotation

in Factor Analysis, 507

vertical label text, 314

Viewer, 265, 267–272, 276, 297–298, 640, 643

changing outline font, 272

changing outline levels, 271

changing outline sizes, 271

collapsing outline, 270

copying output, 268

deleting output, 268

display options, 640

displaying data values, 643

displaying value labels, 643

displaying variable labels, 643

displaying variable names, 643

excluding output types with OMS, 727

expanding outline, 270

hiding results, 267

moving output, 268

outline, 269

outline pane, 265

pasting special objects, 276

results pane, 265

saving document, 298

space between output items, 297

virtual active file, 86

Visual Bander, 160

Wald-Wolfowitz runs

in Two-Independent-Samples Tests, 555

Waller-Duncan *t* test

in GLM, 430

in One-Way ANOVA, 415

Watch tab, 709

script window, 709

Web, 679

publishing output to, 679

weighted data, 262

and restructured data files, 262

weighted least squares

in Linear Regression, 457

weighted mean

in Ratio Statistics, 611

weighted predicted values

in GLM, 433

weighting cases, 239

fractional weights in Crosstabs, 239

Welch statistic

in One-Way ANOVA, 417

wide tables

pasting into Microsoft Word, 273

Wilcoxon signed-rank test

in Two-Related-Samples Tests, 557

Wilks' lambda

in Discriminant Analysis, 494

window splitter

Data Editor, 127

windows, 5

active window, 6

designated window, 6

WMF files, 277, 286, 290, 668
    exporting charts, 277, 286, 290, 668
Word format
    exporting output, 277, 281
    wide tables, 277
wrapping, 325
    controlling column width for wrapped text, 325
    variable and value labels, 112

XML
    OXML output from OMS, 742
    routing output to XML, 721
    saving output as XML, 713
    table structure in OXML, 736
XSLT
    using with OXML, 742

Yates' correction for continuity
    in Crosstabs, 374
years, 652
    two-digit values, 652

z scores
    in Descriptives, 359
    in Rank Cases, 186
    saving as variables, 359
zero-order correlations
    in Partial Correlations, 448